T0329640

Modern Electricity Systems

Modern Electricity Systems

Engineering, Operations, and Policy to address Human
and Environmental Needs

Vivek Bhandari
Rao Konidena
William (Bill) Poppert

Registered Offices
John Wiley & Sons, Inc., 111 River Street, Hoboken, NJ 07030, USA
John Wiley & Sons Ltd, The Atrium, Southern Gate, Chichester, West Sussex, PO19 8SQ, UK

Editorial Office
The Atrium, Southern Gate, Chichester, West Sussex, PO19 8SQ, UK

For details of our global editorial offices, customer services, and more information about Wiley products visit us at www.wiley.com.

Wiley also publishes its books in a variety of electronic formats and by print-on-demand. Some content that appears in standard print versions of this book may not be available in other formats.

Library of Congress Cataloging-in-Publication Data applied for
ISBN: 9781119793496 [Hardback]

Cover Design: Wiley
Cover Image: © Monty Rakusen/Getty Images

Set in 9.5/12.5pt STIXTwoText by Straive, Chennai, India
Printed and bound by CPI Group (UK) Ltd, Croydon, CR0 4YY

C9781119793496_190722

Endorsements of "Modern Electricity Systems: Engineering, Operations, and Policy to address Human and Environmental Needs"

There has never been a greater need for serious energy policy makers and strategists to understand the principles of our electricity grid. We can't transform our system by blagging our way through the details, but only by understanding the fundamentals concepts. This book explains those concepts clearly and understandably and has come just in time.

Dr Jemma Green
Co-founder and Executive Chairman
Powerledger.

I have thoroughly enjoyed reading the two chapters of your newly released Textbook "Modern Electricity Systems:" I wish I had access to such a fine introduction and deep dive into most of these complex issues when I first became a PUC Commissioner. Every Commissioner (and staff) should have this textbook to help them understand energy markets and the various agencies and organizations that regulate and impact those markets. I really like the international emphasis of the textbook. I also like the challenging exercises that you have provided for students. There is no better way to learn than to have students tackle some real-life situations. Bravo!!

Phyllis A. Reha, Principal Consultant
PAR Energy Solutions LLC
St. Paul, Minnesota.

This looks like a valuable contribution to today's global conversation about energy. As a microgrid operator, we benefit from having the ability to dispatch power generation, energy storage, and demand assets in real-time. This requires understanding of our relationships with our thermal and electric customers, our local utility, and the regional ISO. The concepts addressed in this book should be studied by anyone in an energy leadership role including: utility regulators, policy makers, utility managers, power grid operators, energy project developers, through and including electrical power engineers and microgrid owner/operators.

Ted Borer, PE, CEM, LEED AP
Princeton University; Director, Energy Plant
Princeton, NJ.

The book provides an accessible overview of key engineering and economics topics that are an important foundation for someone advocating for sustainable market changes. The book's attention to energy poverty highlights the crucial links between social justice, climate change, and energy security, and provides helpful framing for those thinking critically about policies to promote equitable access to cleaner energy.

Jessica Bell, Deputy Director
State Energy & Environmental Impact Center
NYU School of Law.

Certainly, the information in its format and layout is an excellent resource for those seek the knowledge in the energy sector as well as learners. The work provides referenced historical perspective and several study cases that imbue intensive critical thinking and problem-solving experience.

Khaled Nigim, Ph.D., PEng, Professor
ICET program, Lambton College
Sarnia, Ontario.

This book takes a timely look at the modern issues of electric systems. As a Development Officer at a large humanitarian organization, of note are the authors' comments in chapter 10 on "Energy policy should include considerations of Energy Poverty." Given that almost half the world's population still lives in poverty on under $3 a day, the book is quick to identify that up to a quarter of the world's population doesn't have a reliable electric supply resulting in energy poverty. The book helps us to understand that weaving in an element of energy in design could ultimately help address the United Nation's Sustainable Goals of providing affordable and clean energy around the world

Nashmeen Moslehuddin
Senior Director of Development
CARE USA.

The authors have leveraged their combined 70 years of experience in the energy sector to develop a book that is unique among treatments of modern electric energy systems. It integrates engineering methods and concepts, financial requirements, business perspectives, policy drivers, and social implications to provide readers with a deep and broad understanding of the electric power industry in the US and around the world. The book appropriately concludes by defining energy poverty and describing the need for nations worldwide to direct policy formation towards addressing this complex subject. This book is a must-read for all energy sector engineers and policy makers. It is also an exceptionally useful book for university coursework at both the senior undergraduate and graduate levels.

James McCalley, Distinguished Professor
London Chaired Professor of Power Systems Engineering
Department of Electrical and Computer Engineering
Iowa State University, Iowa.

It is a unique book that holistically brings together the pressing issues in modern electricity systems. The authors begin with fundamentals of electrical engineering, economics, and tradeoffs made in real-life planning and operations. The subsequent chapters explain the role of people and institutions, energy poverty, the actual workings of power systems, wholesale, retail, and local energy markets, their relevance now, and their future direction. I especially liked the diversity of global examples in this book.

The contribution of Kathmandu University in Electricity system is also reflected in this book through the research and education engagement of the author at Kathmandu University. Their combined international, diverse experience is the secret sauce for making this book readable and timely. It is a must-read for engineers, policymakers, managers, and other professionals working in the electrical energy sector.

Prof Dr. Bhola Thapa
Vice Chancellor, Kathmandu University
Nepal.

Rao is one of the most knowledgeable energy industry experts I've ever encountered in all my years covering the energy transition. Whenever I have a question about a new topic or when I'm doing research for a new article, he patiently explains complex engineering concepts in a way that is easy to understand. I have no doubt that this book will serve as a resource for anyone entering the energy industry for the first time and will probably help some seasoned experts, too.

Jennifer McKelvey Runyon
Senior Content Director
POWERGRID International

Contents

Foreword

I am an internationally recognized expert and thought leader in energy policy, electric grid operations and markets, and the interface of disruptive energy systems with traditional utility structures. My firm, GridPolicy, Inc., is dedicated to furthering and investing in clean, sustainable distributed energy resources (DERs) and efficient grid enhancing technologies (GET). I am also the Chief Regulatory Officer of Voltus, Inc., the world's leading platform for DERs. In 2006, I received my first of two Presidential nominations from President Bush to serve as a Commissioner on the Federal Energy Regulatory Commission (FERC) and was confirmed by the Senate. I served as a Commissioner from 2006 to 2013. In 2006, I was designated by President Obama to be Chairman of FERC. I was FERC's longest-serving chairman (2009–2013). In 2016, *Public Utilities Fortnightly* named me one of the 10 individuals since 1990 who have had the greatest impact on the electric utility industry.

I met Rao Konidena while working on developing testimony for the Cardinal Hickory Creek transmission line on behalf of the Environmental Law & Policy Center. As a Midcontinent Independent System Operator (MISO) transmission expert, Rao testified on the reliability aspects of that MISO multivalue project. On a recent client engagement, Rao and I worked together on PJM Interconnection's generator interconnection reform process.

Modern Electricity Systems: Engineering, Operations, and Policy to Address Human and Environmental Needs is a necessary and valuable contribution to the understanding of our evolving electric grid and its operational needs. Given the growing complexity of wholesale electric market products and structures, the increasingly technical nature of grid engineering and power flow modeling issues, and the need for power system operations to keep pace with an ever-more renewables-based grid, electric system planning can be complex and mind-numbing for outsiders.

The authors begin with the fundamentals of electrical engineering and economics in Chapter 1 and 2 followed by real-life cases in Chapter 3. Those cases lay the foundation for a thorough discussion of wholesale markets in Chapter 4, followed by details in Chapters 5–9 of the evolution of FERC-administered energy, capacity, and ancillary services markets.

Those chapters weave into the discussion notes on US institutions such as FERC and European grid operators under ENTSO-E. This in-depth review of regulatory policy informs the reader of the key roles these institutions play in setting energy policy. Chapter 5–8 discuss the myriad stakeholders involved in policy and technical decisions while describing in detail how power systems work.

Finally, Chapter 10 touches on energy poverty by addressing the issues of power quality and accessibility that are at the heart of today's modern electrical engineering systems.

I am honored to write the foreword for this book on modern electricity systems. It stands as a unique text, enabling the reader to understand how energy market structures and the modern energy grid possess the potential in the US and internationally to strive to meet human objectives, such as addressing climate change, and to solve human needs, such as energy poverty.

Jon Wellinghoff

Preface

A well-informed group of individuals in charge of energy production and use is essential to create a sustainable, economic, and greener tomorrow.

Technologies and costs are rapidly changing, and this book widely debates environmental goals. The future of energy is at a crossroads. Therefore, there is a great need for a professional text in the energy industry to provide an overview of all three major areas in sufficient depth: engineering, policy, and operations. This book stresses the increasing importance of these areas to our societal and climate needs. Our careers and professional insights have evolved over our working lives, allowing readers to benefit from our six plus decades of combined experience. This includes the rapidly changing trends of distributed generation, storage, and local energy markets and topics important to humanity – such as energy poverty.

This book's perspectives do not come from the authors' work in one place alone. The key messages have been framed from international endeavors on various continents and following interaction with energy advocates, engineers, environmentalists, humanitarians, regulators, businesses, and policy professionals from diverse countries and continents. Additionally, the authors' interaction with international grid operators and nongovernmental organizations (NGOs) enables a broad viewpoint valuable for a reader starting out in the industry or only moderately versed in the industry.

Finally, writing this textbook is a way for the authors to give back to the industry and those served by it. We hope the reader appreciates our collective efforts.

The Need for a Textbook

There is a need for a textbook/professional text in the energy markets industry to offer an overview of engineering, policy, and operational issues. Some textbooks in the market go in-depth on a topic, but we feel none covers all three in sufficient depth for the reader who wants and needs certain information.

There is another apparent reason why this textbook is needed. The electric utility industry is changing in front of our eyes from a centralized generation model to a decentralized generation model. Fossil-based fuel types are being replaced by fossil-free (i.e. renewable) generation. As utility professionals retire, a new human generation faces new technological and social challenges. Environmentalists, policymakers, investors, and NGOs need the tools to evaluate and address these changes. Hence, this textbook aims to help readers understand these changes, to provide context and a path forward for energy professionals.

Why now? Because this energy transformation is happening much faster now than 20 years ago. We need context now more than ever. This book provides the context to understand how this transformation is happening, the tradeoffs, and where we can help each other. We offer a toolbox for the reader to learn essential things to make informed decisions. The growing global impacts and social issues can only be adequately discussed and advocated with this basic level of comprehension. However, fundamental change in this sector can only be achieved through understanding the systems. It is now up to you!

Vivek's Professional Journey

Vivek reflects on his 15 years of professional global working experience; he started as a schoolteacher, then became a junior engineer doing house/industrial wiring and building microgrids in South Asia, working his way up to leading the business and technology sectors for world-class giants like Siemens and budding startups like mPrest Systems and Powerledger that are placed to becoming global leaders in technologies like artificial intelligence/machine learning and crypto/blockchain, respectively.

Vivek wants to share his technical and business experience working primarily with for-profit organizations in this sector. He shares what others have shared with him over the years in his professional careers, such as making informed decisions, writing summaries/briefs, negotiating contacts to advance solutions, and the business development and technical insights of the generation, transmission, distribution, and end-use of energy.

Rao's Professional Journey

Rao reflects on his 20 years of professional working experience in the US. As one might expect, his thoughts have evolved from a junior engineer's mindset of solving a load flow in the PowerWorld tool to an adviser's mindset of advising a client on when to send a letter to a regional transmission organization's (RTO) independent board.

Rao wishes to share what others have shared with him over the years in his professional career, such as speaking up in meetings to flag a concern, writing blogs to market an idea, making contacts to advance a solution, and asserting a client's viewpoint at RTO stakeholder committee meetings.

Bill's Professional Journey

Bill's experiences over his 30+ years started with the first hotel active solar system in the US plus instructing energy engineering in Tunisia 30 years ago. Since then, he has led award-winning corporate energy engineering teams working with utilities and energy users. He has financed renewable energy projects, built companies, and carried out mergers and acquisitions. His passion is for reducing world energy poverty. We must remember that up to a quarter of the world's population does not have a reliable electricity supply. This has led to his work with several NGOs and United Nations agencies, including on-site work in Africa.

Why We Wrote This Book

Here are some of the salient ideas that have led us to write this book.

- There is a need for graduate-level coursework on distributed generation and to plant the seed for a lifelong appreciation for new and emerging technologies.
- Working with international grid operators, utilities, multinational companies, and NGOs allowed us to include our decades' worth of experience and professional interactions in writing.
- The lack of a holistic book led to the creation of this one, which the authors envision to be used by engineers, advocates, lawyers, financial personnel, and anyone interested in the energy industry.
- We have a passion for giving back and training future energy leaders.

We wanted to paint a holistic picture based on our combined global industrial and academic experiences. Hence, we set out to write this book. Should you decide to set out to read it, we are sure you will find it fulfilling. Finally, if you have suggestions, please feel free to reach out. We will try our best to address them in the next revision/edition.

Who is Jay?

As one reads this book you may notice that many of our Exercises involve helping "Jay" address the world's energy issues as they progresses in their career. Jay is Everyperson. A stand-in for those learning from and contributing to the solutions for the issues presented in this book. Good Luck!

Acknowledgments

The authors would like to acknowledge the continual help and support from several professionals, university professors, and corporate leaders during our book-writing journey.

We would also like to thank Chris Poppert (Digital Marketing Specialist, Tolomatic, USA), Kaiyang Sun (Senior Electrical Engineer, Diality, USA), César Ulate (Economist and Professor, Banco Central de Costa Rica and Universidad de Costa Rica, Costa Rica), Daniel Wilson (Research Fellow at the University of Minnesota, USA), Yugal Kishore (Power System Consulting Manager, Siemens, Australia), Swapna Konidena (Project Delivery Manager, Deloitte, USA), Christophe Druet (Electricity Grid Consultant, Belgium), Ming Ni (Managing Director, Leidos, China/USA), Daniel Møller Sneum (Postdoctoral Researcher in the Energy Economics and System Analysis Group, Technical University of Denmark, Denmark), and Nashmeen Moslehuddin (Director, CARE, Bangladesh/USA) for proofreading and providing valuable and critical feedback for several chapters.

Finally, we would like to acknowledge the continuous support from our families and would like to dedicate this book to them.

About the Authors

The authors jointly bring seven decades of global experience from over two dozen countries.

Vivek Bhandari is a visionary and strategic business leader with more than 15 years of global experience in engineering, information technology, operations technology, sales, general management, consultation, and teaching. He has secured, led, planned, delivered, and supported numerous digitalization, energy, and sustainability mega projects in Asia, North America, Europe, and the Asia Pacific region. He has worked and led teams at Fortune 200 companies like Siemens (by leading grid software portfolio for Digital Grid business in ANZ), and budding startups like mPrest (as a global Technical Director) and Powerledger (as a Chief Technology Officer).

He has a strong work ethic, leadership skills, a talent for building things, the capacity to create/maintain/improve relations, and a history of taking, executing, and leading the most complex orders in the electric power industry. He has unbeaten global experience in lifecycle projects (R&D/development, pre-sales, sales, delivery, and postdelivery services).

Bhandari holds a doctorate and master's degree from the University of Minnesota in the US and a bachelor's degree from Kathmandu University in Nepal. He is a registered professional engineer in Nepal and Australia and a senior member of the IEEE. He likes writing about engineering, policy, management, and leadership topics.

Rao Konidena of Rakon Energy LLC (https:// rakonenergy.com/) is an independent consultant with more than 25 years of experience focused on providing policy and testimony support, business development, and training in wholesale energy markets. Rao likes helping consumers and environmental advocate clients. Most recently, Rao was with Midcontinent Independent System Operator ISO (MISO) as Principal Advisor for Policy Studies, working on energy storage and distributed energy resources. At MISO, Rao worked in management and nonmanagement roles around resource adequacy, economic planning, business management, and policy functions.

Rao is Co-President of the Finnish American Chamber of Commerce – Minnesota (FACC-MN) and the Board of Ever Green Energy and Minnesota Solar Energy Industries Association (MnSEIA).

He holds an MBA and MSEE degree from the University of Minnesota and the University of Texas, respectively, in the US and a bachelor's degree from Bangalore University in India. He also frequently contributes to the Renewable Energy World blog (https://www.renewableenergyworld.com/author/rao-konidena/).

William (Bill) Poppert is a veteran with over three decades of experience in the energy sector. He has worked in the US and internationally in the operation, design, financing, and development of energy projects, programs, and businesses. Bill is enthusiastic and energetic. A lifelong entrepreneur, he is a practical and pragmatic problem-solver and leader.

Bill's energy experience started with design engineering and construction supervision, including the oversight and operation of the first large-scale hotel solar system in the US. He also instructed on professional energy efficiency engineering in Tunisia 30 years ago for USAID. Bill built and headed a three-time national award-winning energy management program overseeing $50 million annual energy use in over 500 Carlson Companies Inc buildings. He has also financed renewable energy projects, built companies, and carried out mergers and acquisitions for a Fortune 500 company. His passion is reducing world energy poverty, which has led to his work over the last decade with several NGOs and UN agencies, including on-site work in Africa. Bill's current ventures, such as www.worldenergyutility .com, reflect his view that a complete energy plan approach is critical for long-term success.

Bill has a background in Design Engineering and an MBA from the University of Minnesota encompassing coursework from the London Business School. He has served on the University of Minnesota Alumni Council and founded the International Forum business lecture series. Bill has served on the Minnesota Chamber of Commerce's Electric Rate Task Force, was past chair of Environmental Initiative's Board of Directors, and served as CARE Minnesota's Fourth District Chair. He also serves with UN energy development task forces.

About the Companion Website

This book is accompanied by a companion website:

https://www.wiley.com/go/bhandari/modernelectricitysystems

The website includes:

- Case studies

1

Essentials of Power and Control

1.1 Introduction

The electrical grid is one of the most amazing hidden parts of our modern technological society. Is it not amazing that the small wires can move enough electrons to power massive machines and light millions of square feet in a fraction of a second? Further, it is done with some semblance of control without worrying about its underlying factors. So, what are the physical, mathematical, and sociopolitical dimensions of electrical energy? Traditionally, we have been unaware of these dimensions because we were just the end-user, aka an "electron-taker" (like "price-taker" in economics). At most, we would worry about getting reliable and affordable power. For example, how many of us know that when a branch of a tree (literally) falls in a distribution line the distribution utility's (including in the most advanced countries like the United States) major source of information is the phone call that you make to the utility? Yes, the distribution utilities might still not have much automation or many sensors and so must rely on phone calls. Even in a place where there are wholesale markets, many of us are not aware that electricity is traded like shares on the stock exchange. We would not want to be involved in the power system operations. We would not care about power system operations or attributes such as environmental values in the past.

These days, consumers are becoming sophisticated consumers or even "prosumers" (a consumer who also produces energy). The prosumer (we) wants reliable, affordable, cleaner, and resilient power. We want to improve the efficiency of our use, and some even want to generate local energy and contribute to the grid. Our decisions, actions, and involvement affect the physics, mathematics, and sociopolitical aspects of traditional power systems. Therefore, we must have a basic understanding of electrical systems and electricity flow primarily because this prosumer "we" is much intertwined with the system compared to our historical counterpart.

Modern Electricity Systems: Engineering, Operations, and Policy to address Human and Environmental Needs,
First Edition. Vivek Bhandari, Rao Konidena and William Poppert.
© 2022 John Wiley & Sons Ltd. Published 2022 by John Wiley & Sons Ltd.
Companion website: www.wiley.com/go/bhandari/modernelectricitysystems

This chapter is meant for today's "us." After reading this chapter, you (e.g. graduate student, engineer, non-engineer, decision-maker, policymaker, and members of the informed public) will be:

- Acquainted with the electric power system, including generation, transmission, distribution system operations, and control.
- Acquainted with the general principles that guide this operation.
- Use this knowledge to make informed decisions in their day-to-day lives and work.

1.2 Basic Principles of Power and Control

An electrical power system is a network of components that supplies, transfers, and uses electrical energy. The electrical grid of Texas in the USA, or microgrid at a supermarket in Finland, or an interconnected network of micro hydro units in Nepal are all examples of electrical power systems. The power system generally consists of generators that generate the electricity and the transmission and distribution system (lines, towers, insulators, transformers, etc.) that carry and feed the power to homes, industries, and other end-users. In most places worldwide, the generators generate electricity that fluctuates between highs and lows over time, called the alternating current (AC). In exceptional cases, the electricity generated is constant over time (e.g. electricity used by trains, ocean liners, and most electric automobiles) called direct current (DC). We look at some examples later in this chapter. Note that electricity generated by centralized power plants can be used directly in trains and other vehicles. Sometimes it can also be stored in batteries, or by pumped hydropower and even geothermally. Pumped hydro is an arrangement of multiple water reservoirs that are located at different heights. When water flows from top to bottom, it generates electricity. It also includes a mechanism that pumps water back to the upper reservoir, especially when the electricity prices are low, to store it and use it when the electricity generation prices are high. Similarly, geothermal storage uses earth's underground heat for energy generation and storage. Electricity can also be co-generated on board and used in trains, boats, buses, and automobiles. These are often referred to as diesel-electric or hybrid systems (see Chapter 7 on different forms of generation).

1.2.1 Energy and Power

Energy is the capacity to do work. Work is the act of applying force over a distance, e.g. pushing a desk over a distance or moving electrons over a distance. Watt-hours or British thermal units (BTUs) are the most used units of measurement in electrical engineering.

Exercise 1.1   We need to measure force and distance before calculating energy. Let us take an example. If Sandra lifts 1 kg of weight for a distance of 1 m, how much energy does she spend? (Hint: $E = Fd = (ma)d = (mg)d$, where E = energy, F = force, d = distance, m = mass, a = acceleration, and g = gravity.)

Several other units measure energy, including N-m (Newton meter), joules, and calories. Watt is the instantaneous unit of power, also called electrical demand. Electrical energy is generally measured in watt-hours and is also called electrical consumption. Power is defined as the rate of producing or consuming energy and the time factor is important. For example, Sandra must move from point A to B. She can use a tiny scooter engine or she could use a jet engine. Which one do you think will move her faster? Her weight is the same, the distance is the same, and everything else is the same. Though both will require the same amount of energy, the jet engine will move faster because it has more power to do the work quicker. Figure 1.1 compares different units of energy and power.

Let us take the example of electrical power. If Sandra's light emitting diode (LED) light consumes 1 W of electrical power for one hour, this equals one watt-hour of energy. A thousand of these units are called kilowatt-hours (kWh). So, if Sandra's fan consumes one thousand watts for one hour, her fan consumes 1 kWh or one unit of electrical energy consumption. Power can also be calculated from voltage and current. Electrical power is the product of current and voltage. For example, $1\,W = 1\,V \times 1\,A$.

Section 1.2.2 considers voltages and currents.

Exercise 1.2   Jay wants to understand his electricity bill. Can you help him? To do so, you need to understand it yourself first. Look at your electricity bill. We will do some calculations from the bill.

1) How many units of electricity did you consume last month? Also, from the same bill, can you find how much you are paying for every unit of electricity you consume?
2) Now, identify the watt ratings of two electrical devices you use every day. How often do you use them? Use this information to calculate the energy usage per month, i.e. how many units per month are these two devices consuming?
3) Now, these devices from question (b) are damaged and need to be replaced. How much would you like to pay for this? How much would the newer and more-efficient versions cost?
4) Let's say you had chosen a 15 W CFL (compact fluorescent lamp) bulb for question (b). A newer version of these bulbs are LED bulbs that give you the same amount of light but only consume 11.25 W. Imagine these newer versions are 25% more efficient. How much money would you save per year if you replaced the old ones with the newer, more-efficient version?

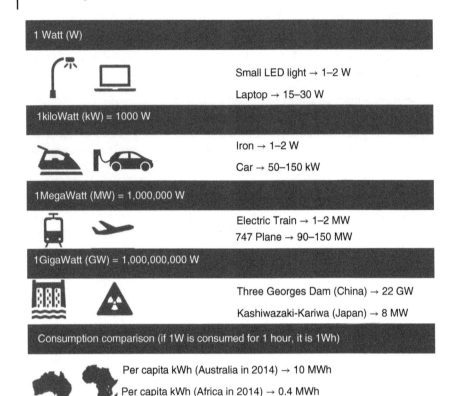

Figure 1.1 Different units of energy and power comparison [1].

5) Why should you and why should you not replace it? Justify your answer.

6) As an example of information given to customers about their energy use and its cost, one of the authors' utility bills says, "For an average residential customer, 49% of your bill refers to power plant costs, 13% to high voltage line costs, and 38% to the cost of local wires connected to your home." If you have this data, what percentage of delivered energy cost is from generation/transmission and distribution?

1.2.2 Voltage, Current, and Impedance

Voltage is the amount of potential energy (a form of energy) difference between two points. Such a potential difference is due to the difference in charge between

Figure 1.2 The beer mugs.

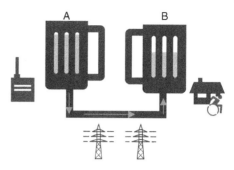

these two points. Voltage is measured in volts. Voltage is the difference in electric potential between two points in a circuit. Current is the flow of charge and is measured in amperes. Impedance, also sometimes called resistance (e.g. in DC circuits), is the property of any material that impedes current flow. It is measured in ohms.

Let us take an example of two interconnected beer mugs, as shown in Figure 1.2. Since there is a difference in beer level between the two mugs, the beer will flow from mug A to mug B. You will have a generator in an electrical system instead of mug A and loads (i.e. equipment that use electricity like a light or a fan) instead of mug B. The pipe connecting mugs A and B is the transmission and distribution system consisting of wires and transformers. The potential energy difference between these two mugs can be compared to voltage. The flow of actual beer can be compared to the current.

Another widely used way to explain voltage, current, and impedance is the fluids analogy. In this analogy, voltage is like pressure, and current is like flow.

The relationship between voltage, current, and impedance was described by Georg Ohm in the late 1700s. He stated that the current flow through a conductor is directly proportional to the voltage and inversely proportional to the impedance. If we represent voltage as V, current as I, and impedance as Z, according to Ohm's law

$$V (\text{volts}) = I (\text{amps}) \times Z (\text{ohms})$$

Exercise 1.3 You and Jay live in Nepal. Jay has learned the concepts of voltage, current, and impedance and installed a fan. You need to do the same. The fan has a motor. You bought a 24 V, 0.5 kW motor. You are an expert in wiring and can connect the wires, switches, and motors. Remember $P = V \times I$.

1) What should the current rating of the wire be?
2) What happens if you choose a lower rating (thinner wire) or a higher rating (thicker) for the wire?

1.2.3 Alternating Current vs. Direct Current

Most of us know about the band AC/DC and their rock and roll. However, we are on a slightly different topic (equally interesting, like the band). Energy.gov describes this as a "war of currents." Thomas Edison generated electricity as DC (electrical current that does not change over time). It quickly gained popularity in the United States. However, it was hard to transmit it over long distances at that time. Nicola Tesla worked on another version: AC (electrical current that fluctuates and reverses over time). It was easy to convert it into higher or lower voltages and could be transmitted over long distances. General Electric vouched for Edison's DC and Westinghouse for Tesla's AC. Over time, both methods gained popularity. Fast forward to today. We generally use AC at our homes and DC in small electronics or while transmitting power at high voltages over long distances, etc. AC still has the advantage of being easy to convert into higher or lower voltages and is easy to transmit. DC also has its advantages and is still used in many places. The choice of AC and DC is primarily a tradeoff between their usages. For example, at low voltage and extreme high voltages, DC is better for general applications than AC. We do not explicitly discuss the tradeoff between AC and DC in Chapter 3. However, the reader is encouraged to read that chapter to understand the tradeoffs.

AC can be produced by an alternator (an electricity generator). A wire loop is rotated (using wind, hydro, or steam, etc.) inside a magnet creates voltage and currents that alternate over time. The output waveform can be sinusoidal, square, triangular, etc. The most common is sinusoidal. Nonsinusoidal waves are used only for special purposes, like the internal drives of a motor or switching signals internal to an inverter, and sinusoids are used for everything else, including for our purposes here. This can be represented mathematically as

$$V(t) = Vp \sin (2\pi \times f \times t \pm \varphi)$$

where $V(t)$ is the voltage as a function of time, Vp is the amplitude (the maximum that voltage reaches in either direction, i.e. $+Vp$ to $-Vp$; sin() is describing the sinusoidal function; 2π is the conversion factor for frequency f in Hertz into radians; t is the time, and φ is the phase. The phase is a relative measurement of time-shift from another wave (we will understand this more with the power factor). Note that the voltage (and current) generated from a polyphase machine will have multiple $V(t)$s.

Exercise 1.4 In the United States, you might have heard that the AC voltage is 120 V and the frequency is 60 Hz. In other parts of the world, the power used in homes and commercial buildings is often 220 V, and the frequency is 50 Hz.

Assuming the phase is zero:

1) How does the voltage equation above change? Use your country's voltage. (Hint: in a 120 V system, 120 V is the root mean square [RMS] value.) Explore what RMS value is and note that 120 V RMS ~ =170 V peak (peak value = $\sqrt{2}$ RMS value).

Unlike AC, DC does not oscillate back and forth. The voltages and currents are constant over time. DC is generated using batteries and rectifiers or commutators that convert AC into DC. This is represented mathematically as

$$V(t) = Vp$$

where Vp is the steady value of DC over time. For a simple AA battery, Vp is ~1.5 V.

DC is used by most electronics, e.g. cell phones, flat-screen TVs, computers, electric vehicles (EVs), etc. This is primarily because the batteries generally work with DC, logic gates generally require DC (zero or one, on or off), no need to worry about pulsation, no need to worry about frequencies, etc. AC is sometimes loosely referred to as line voltage. However, it is used in larger items like AC motors, pumps, refrigerators, dishwashers, etc. DC is used in small electronics because these devices require a nonfluctuating voltage/current to perform the logical operation and have specific voltage/current levels that define their memories, logics, timers, etc. DC is also used because it can easily be stored in small batteries.

Unlike DC, AC can easily be stepped up (e.g. increase the voltage from 11 to 132 KV) or stepped down using a transformer. Hence it is used extensively. Once it comes to our houses, AC is typically used through small plug-in transformers and power electronic devices to power small electronics and to charge batteries for those electronics. Keeping the power constant (remember the power is the product of voltage and current), increasing voltage means decreasing current, which means less heat due to current flow (less loss).

Exercise 1.5   Now that you know AC and DC, it's time to ask whether you know about high-voltage DC or HVDC? What is HVDC? Why is it used in some places? For example, there is a 370 km long HVDC line between Tasmania and Melbourne in Australia. It is called the Basslink. Why do you think this was used for transmitting power?

1.2.4   Single Phase vs. Multiphase

In our homes, two wires come from the electricity retailer/distributor. In the case of industries and commercial venues, multiple wires are used. If there is one live (or hot) wire, it is a single-phase supply, and if there are multiple live (or hot)

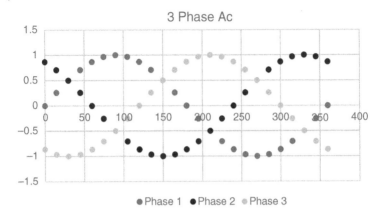

Figure 1.3 Three-phase AC.

wires, it is a multiphase supply. For example, when there are three live wires, it is a three-phase supply, and there are three ACs through the same circuit. A three-phase supply is uniformly separated in phase-angle and provides more reliable and consistent power for a powerful industrial appliance. The main electrical grid uses three-phase because it is more efficient when transmitting energy through smaller wires at a lower cost. See Figure 1.3.

Various parts of the world provide different voltage levels for single- and three-phase supplies. For example, in Australia, single-phase electricity is typically connected at 230 V (Western Australia and Queensland use 240 V). Similarly, in Nepal, it is 220 V, and in the USA, it is 120 V. A similar case is with the supply frequency. Some countries have 50 Hz; others have 60 Hz supply.

Remember, one form is not better than the other. Such supply voltages or frequencies were a result of a legacy system. For example, engineers at the German company AEG decided to fix the frequency at 50 Hz when they built the first of their European generating facilities. It is said that they did this because the number 60 did not fit the metric standard unit sequence (1, 2, 5). Since AEG was a pioneer, it enjoyed a monopoly. Hence, most of the rest of the Continent adapted to a 50 Hz system. For our readers, this information should not matter much other than to help you decide what equipment to buy in which part of the world. For example, if you are in Nepal, you will buy a light bulb that operates in 220 V, 50 Hz. If you are in the US, you will buy a light bulb that operates in 120 V, 60 Hz. Unless there is an internal electronics converter circuit, the 50 Hz system will not work in a 60 Hz system, and vice versa.

Exercise 1.6 We looked at single- and three-phase supplies. Is there a two-phase power? Why? Why not? Why do you think multiphase generators are manufactured instead of many single-phase generators? Are the voltages and currents generated from the multiphase machines perfectly synchronized? Or do they lead or lag one another?

Some renewables, like rooftop solar, can be integrated into a single-phase supply, whereas others like larger wind farms are integrated into a three-phase system? Why?

1.2.5 Active, Reactive, Apparent Power, and Power Factor

We have already briefly discussed the concept of phase. When there are multiphase generators, the voltages and currents may or may not be perfectly synchronized. The time difference between any two waveforms is called the phase-angle. For current and voltage waveforms, when the phase-angle difference is zero, the zero, maximum, and minimum of both waveforms occur simultaneously. The phase-angle difference is introduced due to nonlinear components like inductors and capacitors. In DC and pure AC circuits, the voltage and currents are in phase. Figure 1.4 compares AC and DC waveforms.

Exercise 1.7 Phase-angle is an interesting topic. One of the universally used devices is the phasor measurement unit (PMU). It measures the phase angle. Explore more about PMUs. And find out about the benefits they bring and at what cost. Do not forget to revisit this while reading Chapter 10's discussion of energy poverty. Do you think we should focus our policies on adding more PMUs or on improving energy access?

Power is the product of voltage and current. If voltage $V(t) = Vp\sin(2\pi \times f \times t)$ and current $I(t) = Ip\sin(2\pi \times f \times t \pm \varphi)$.

(Note: The phase of voltage is shown as zero here. It can be represented by φ_v, and the current can be φ_i. The result of the discussions below will remain the same.)

Now while converting peak values into RMS

$$V(t) = \sqrt{2}\,V\sin(2\pi \times f \times t) \text{ and } I(t) = \sqrt{2}\,I\sin(2\pi \times f \times t \pm \varphi).$$

Figure 1.4 AC vs. DC waveforms: wind generators generate voltages and currents that alternate over time, aka AC, whereas the photovoltaic generators generate voltages and currents that do not change over time, aka DC. The oscilloscopes are showing graphs of voltages and currents (y-axis) over time (x-axis).

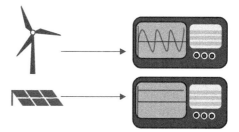

Therefore, instantaneous power, $p = V*| = \sqrt{2}\,V\,\sin(2\pi \times f \times t) * \sqrt{2}\,|\,\sin(2\pi \times f \times t \pm \varphi)$

$$= 2\,VI\,\sin(2\pi \times f \times t)\,\sin(2\pi \times f \times t \pm \varphi)$$

$$= VI\,(\cos\varphi - \cos(2\omega t \pm \varphi))$$

where $\omega = 2\pi \times f \times t$.

The term $VI\,(\cos\varphi)$ equals the actual amount of power that is being dissipated and measured in watts. It is a constant value that depends on V, I, and the phase angle between them. Over several cycles, this quantity is also the average instantaneous power value.

The second term is a time-varying function whose frequency equals twice the angular frequency ω. Since the time-varying part is extremely hard to measure, only the first component, $VI\,(\cos\varphi)$, is generally measured and referred to as the active power.

Exercise 1.8   Your AC supply is rated as $V(t) = 120\,\sin(\omega t)$ volts, and the current is given as $I(t) = 5\,\sin(\omega t - 10°)$ amps. You have connected a 0.5 kW fan to this supply.

1) What is the average power that this circuit will absorb?
2) If $P = VI\,\cos\varphi$ and from Ohm's law, $V = IZ$, where Z is the impedance of your fan. What is the value of Z?
3) Suppose you unplug the fan and plug a 2 kW refrigerator into this power supply. What will happen? Why?

The instantaneous power equation from above can further be broken into $VI\,\cos\varphi\,(1 - \cos 2\omega t) \pm VI\,\sin\varphi\,\sin 2\omega t$. The first term is proportional to $VI\,\cos t\varphi$ producing an average value of $VI\,\cos\varphi$. And the other term is proportional to $VI\,\sin\varphi$, which is pulsating at twice the frequency, producing an average of zero over a cycle. Q represents the value that is proportional to $VI\,\sin\varphi$. It is called reactive power and is measured in volt-amperes-reactive (VAR).

The concept of reactive power is interesting. This power travels from source to reactive elements during the positive cycle and flows back during the negative cycle. Therefore, the energy is lost and never consumed. Reactive power exists only in an AC system (in DC, voltage and current phases are the same and hence $VI\,\sin\varphi = 0$) and helps maintain the voltage within limits (e.g. within $\pm5\%$ of its value).

Imagine a pint of beer (see Figure 1.5). The actual drink can represent active power. It quenches the thirst (does actually work). However, there is also foam on the top, which can represent reactive power. In this example, the actual content

Figure 1.5 Beer mugs return: active power (kW); beer, reactive (kVAR); beer foam and apparent power (KVA).

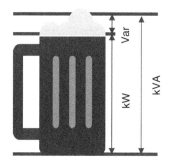

(actual beer and foam) equals apparent power. In other words, apparent power is the summation of real and reactive power.

S in complex representation represents it. Apparent power can be represented as

$$S = \text{Real power} + j\text{Reactive power} = VI\cos\varphi + jVI\sin\varphi = P + jQ$$

$$\text{Magnitude of } S = \sqrt{((Q^2 + P^2))}.$$

With all of this in the background, what is the power factor? The power factor is the cosine of the angle between voltage and current. It is the ratio of active (or real) and apparent power.

Exercise 1.9 At Jay's school in Nepal, the power outlet has 220 V at 50 Hz. He likes to experiment with things. He saw a motor with a rating of $20 + j50$ ohms. Can you help him answer the following questions?

1) Can you find the current, then find the active/real power, apparent power, and power factor?
2) What will happen to the quantities in (a) if he uses a 10 ohms electrical heater?
3) What are the leading and lagging power factors?
4) If more motors are added at Jay's school, what will happen to the power factor of the circuit? What are the disadvantages?
5) In many areas, there are rate penalties for uncompensated low-power factors. Power factor correction techniques can be applied. What are such techniques? Please investigate them.

1.3 Control Overview

Electrical power systems need to be controlled at various levels in order to securely and reliably meet electrical demand. There are multiple layers of control. For example, a local controller in a power plant will maintain the immediate voltage swings. Otherwise, it may damage the plant or even impact the grid. In another layer of control, an operator sitting thousands of miles away might want to control a circuit breaker, or a generator dispatcher's system might need to be fixed if there are automatic frequency changes. Yet, the trader might want to find the optimal bids and commitment for their units in another control layer.

Such hierarchical control can be seen in micro, mini, or larger power grids. Typically, this can be divided into three layers. Layer 1 is the generation plant's (like solar, diesel, nuclear, etc.) specific control. This is also called primary control. Such controllers are responsible for maintaining stability at a subsecond timescale (when we look at the reactions from a controller in milliseconds and microseconds). These controllers often come built into the generation plant. Layer 2 is the secondary control that consists of the automation (e.g. automatic generation control and supervisory control and data acquisition [SCADA] type operations). Layer 3 is the tertiary control that consists of look-ahead features and advanced visualization/control capabilities. Figure 1.6 shows this control hierarchy – as it is used in a typical microgrids and virtual power plant scenario (see Section 1.9).

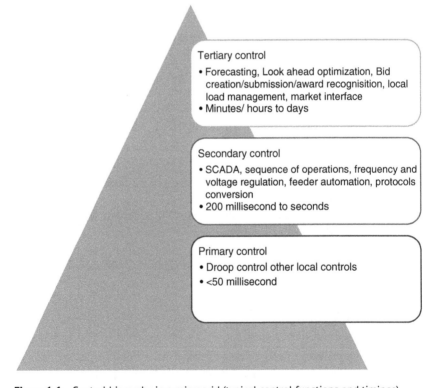

Figure 1.6 Control hierarchy in a microgrid (typical control functions and timings).

1.4 Power Generation and Grid: Operation and Control

Grid operations are complex and critical for any country's operation. The control centers are staffed 24/7, 365 days a year, to ensure safety, reliability, and availability. The operators must be licensed, and they must undergo several hundred hours of training, including rigorous simulation exercises. For example, some operators in the US must be trained for four to six years before they can independently work on controlling a generation desk. In some ways, this is analogous to the rigor of an air traffic controller. The air traffic controller should help operate the flights without causing havoc. Similarly, the operators must ensure that the generated power is sufficient to meet the load without making the grid unstable and without resulting in a blackout. Let us now understand some of the concepts related to these operators' work.

We have learned the fundamentals of power engineering in Sections 1.2 and 1.3. Now, let us look at Figure 1.7. This should help us understand how electricity is generated and where and how it is used. The figure shows power generation by the power plants (AC or DC). Such powerplants can be centralized or decentralized (as discussed in Section 1.9). In the case of the central powerplant, the electricity is transmitted over long distances by using transmission lines at large voltage levels (hundreds of kilovolts). The voltage levels are increased to decrease the currents (and reduce the transmission losses, remember $V = IZ$, and active losses $P = I^2R$). Once this electricity is near the end-use, it is stepped down to a low voltage and fed through the distribution lines to industries, commercial complexes, houses, etc. The reason for stepping the voltage down is that not all appliances run on high voltage. The end-use appliances are a mixed bag of low/medium/high voltage appliances. The end-use appliance may be AC or DC, single-phase or three-phase

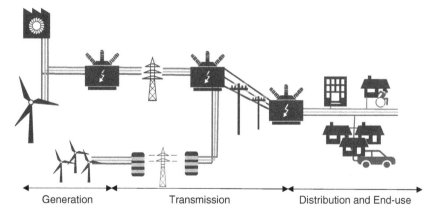

Figure 1.7 Simplified representation of a power system.

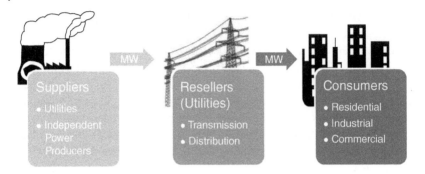

Figure 1.8 Flow of electricity in the wholesale and retail electricity market [2].

(see Section 1.2.4). Depending on this need, the stepped-down voltage is further transformed at the device level. Traditionally, the flow of electricity is unidirectional and is decoupled from other systems (see Figure 1.8).

Further, it has normally not been cost-effective to store electricity. Exceptions in the past include water stored above hydroelectric dams. It appears we may be entering an age where battery storage on some scale (both utility and end-used scale) is becoming more cost-effective.

In a simplified representation, the electricity is generated with a three-phase AC, transmitted over long distances (either using HVDC or AC), and generally distributed using single-phase (see Section 1.2.4 on single vs. three phases) AC lines. In Figure 1.7, Jay would get a single-phase (voltage between one of the colored wires to neutral) AC supply.

Let us see an example. Jay returns from work and turns on the light. Since electricity was not stored, some generators were cranked up to generate electrons to power this light. After being generated, these electrons are pushed through the transmission and distribution infrastructure to his home, through the switch, and into the light bulb. All of this happens before Jay notices any delay. Is this not amazing?

Exercise 1.10 Why does increasing voltage decrease the current? And why does decreasing current mean lower line losses? Find/estimate the line losses as a percentage of your total power generation. (Hint: in many parts of the world estimates are readily available.) What would happen if you generated onsite? Would you be able to avoid such losses? Why? Why not?

Figure 1.7 is a simplified representation. The power does not flow just from a set of traditional large generators, through a few sets of transmission and distribution lines, and to the end-users. These days the power is also flowing from decentralized generators and microgrids distributed throughout the network. Some of these distributed generators (see Section 1.9) are coupled with the heating infrastructure. Transmission rules are different from distribution because the workings, goals and mandates of these systems are different. Entities that own the transmission may not necessarily own the distribution, and they may not necessarily own the generators. There could be wholesale traders and retailers, market and reliability operators, and network infrastructure providers.

1.5 Generation Dispatch and Balancing the System

If you listen to the radio (or are used to listening to it), you will understand that it is important to tune it to the right frequency. The sound quality suffers if the frequency is slightly up or down. The same principle applies to the system frequency in a power system. The consequence, however, is more severe. Depending on where you live, the system frequency of the power system is either 50 or 60 Hz. Operating the power system within a few millihertz around the system frequency is OK. But getting too far off, especially for longer periods, can be catastrophic. The system operator (SO) or the balancing authority has the authority to maintain a stable system frequency. Figure 1.9 shows a system with a balanced frequency of 50 Hz.

If the load is larger than the generation, system frequency decreases. The SO could take remedial action, like deploying the reserve generators or curtailing the load by using load-shedding or demand response (DR) techniques. If the load is smaller than the generation, system frequency increases. The SO could take remedial action like curtailing the reserve generators or restoring the shed loads. All this balancing act requires the SO to take control (see Section 1.3) and to consider the present state of the transmission and distribution network.

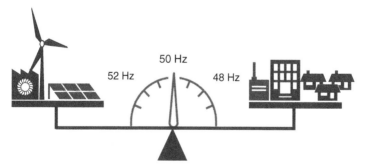

Figure 1.9 Supply (generation – left side of the balance) and demand (load – right side of the balance) need to match.

1.6 Transmission and Distribution Network

Transmission and distribution comprise the network of wires, transformers, switch gears, substations, and other types of equipment that carry the electricity from powerplant to consumer.

1.6.1 Transmission Network

A transmission system moves the power over long distances with lower energy losses. The transmission lines are huge rocket-like skeletal towers near the highway and far away from large urban areas. The voltages are stepped up to lower losses during long-distance transmission. (Remember: $V = IR$ and $I2R$ losses.) Transmission systems are also highly interconnected to enhance the reliability of the overall system (see Chapter 9 for additional details on reliability in terms of interconnection support). They are the backbone of the electrical power system. Transmission lines that connect different regions or different utilities are called tie-lines. The transmission lines allow the power system to operate as a large interconnected system. For example, most of North America, east of the Rockies (exception Quebec and Texas), is the Eastern Interconnect; west of the Rockies, including British Columbia, Alberta, and part of Baja Mexico is the Western Interconnect; the electric grids of Nepal and Texas have their own interconnect. The definition of interconnection allows the utilities and operators within an interconnect operate their portion of the system, referred to as the balancing area.[1] The transmission operators in this interconnect are called transmission system operators. Transmission is regulated by a mix of national and local rules (depending on the part of the world you are in, this might change slightly).

The transmission operators conduct power system and other planning studies to determine where to build the transmission lines ($T +$ years) and conduct power flow/contingency analysis for forward-looking operational planning for determining how the power can be securely flown and what to do when there is a power system event (e.g. $T + 24$ hours or $T + 48$ hours).

1.6.2 Distribution System

The distribution system is a local and lower voltage infrastructure. Historically, it distributed the power to the end-use (household or industrial/commercial). The voltages that were stepped up (in transmission) are stepped down because the

---

1 In many parts of the world, interconnection balancing authorities (BAs) are responsible for balancing system frequency. They maintain the frequency at 50 or 60 Hz.

end-user cannot consume electricity in several hundred kilovolts.[2] Distribution systems are radial or meshed. Urban distribution systems are generally underground and sometimes in common utility ducts. They could also be above the ground with utility poles, wires, and distribution transformers. The power comes from these distribution transformers, through the wires and the meters to the end-customer. Usually, a 120 or 240 V supply is used at residential places.

In a radial distribution system, each customer is fed from one feeder (one branch in a substation that feeds the loads, like with a tree branch). In a mesh network, there are loops and interconnections. Traditionally, distribution systems would only have simple lines and components that take power from the transmission and give it to the end customer. However, the distribution system is changing. Today's distribution system is heavily integrated with distributed energy resources (DERs) like solar, microturbines, home, commercial battery storage, and wind. Like transmission, the distribution system is also regulated by a mix of national and local rules (depending on the part of the world that you are in, this might change slightly).

Distribution operators conduct power systems and other planning studies for determining where to build the new lines and conduct power flow and state estimator plus advanced applications for better management of their distribution system. Advanced functions include, but are not limited to, (i) Fault Location Isolation and Service Restoration (FLISR) for locating fault(s), conducting switching to isolate the faulty network, finding the remedy automatically, and quickly restoring it; (ii) optimal feeder reconfiguration (OFR) for optimally reconfiguring the networks to switch the load from an overloaded part of the network to another part; and (iii) distributed energy resource management system (DERMS) for conducting DR, operating virtual power plants (VPPs) and offering flexibility services.

1.6.3 One-line or Single-line Diagram

Power engineers use single-line diagrams (SLDs) or one-line diagrams to represent the power system (see Figure 1.10). SLDs are simplified notations where schematic symbols represent electrical elements. Only one conductor is shown instead of representing three phases with three lines. It is extremely easy and comprehensive because it can show a DC or an AC system. Elements on the diagram do not represent the physical size, shape, or location. Rather, it is a common convention to

---

2 Note: in some jurisdictions, like in the United States, the transmission system fall under federal jurisdiction and the distribution system fall under state/local jurisdictions. The incentives are different at federal and state/local levels. Therefore, to get a federal rate of return, transmission owners can play games in classifying their T&D facilities as "T."

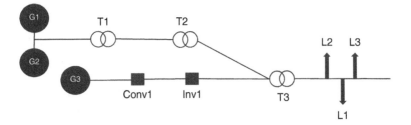

Figure 1.10 SLD for power system network from above. Such diagrams are typically used for power flow studies or supervisory control and data acquisition (SCADA) operations.

organize elements that are read from top to bottom or left to right. For example, Figure 1.7 can be represented in SLD in Figure 1.10.

In Figure 1.10, G represents the generators, T represents the transformers, Conv and Inv represent the converter and the inverter (aka switching devices), and L represents the loads.

1.7 Wholesale and Retail Markets

Electricity can be bought and sold at wholesale and retail. This can be done over vastly different periods, from minutes to years. The resellers buy electricity in wholesale markets and resell it in retail markets. Utilities can act as the resellers, e.g. Xcel Energy is a utility in the US. Or dedicated resellers do it, e.g. Origin Energy is a retailer in Australia. The end-users are residential, industrial, and commercial customers who consume electricity. They generally purchase electricity through retail markets. Some larger ones can also buy directly from the suppliers (through wholesale markets) or generate locally.

1.7.1 Wholesale Market

In the past (and at present in some parts of the world), transmission, distribution, and generation assets were owned by a single entity. This entity is often a national utility, especially in fragile economies. For example, in Nepal a single utility called the Nepal Electricity Authority owns most of these assets. In such places, real-time dynamic buying and selling power is rarely performed. Elsewhere, especially when these assets are separately owned, competition exists. In these regions, the assumption that electrical energy can be traded as if all generators and loads were connected to the same busbar is not tenable. Therefore, wholesale markets exist. Many of them are centralized, and many are decentralized.

1.7.1.1 Centralized Market

Centralized wholesale markets exist when competing suppliers offer their electricity through a market mechanism to resellers who resell the electricity to end-users. These markets are active in several countries, like the United States, Canada, Australia, the European Union, etc. In such a market, there is a central market operator. Some examples of such markets can be found in Chapters 4 and 9.

Figure 1.11 illustrates the wholesale markets in the European Union.

A grasp of timescales is important for understanding market transactions. Trades happen at different timescales, and planning happens at different timescales. Timescales, among other things, impact reliability decisions while taking uncertainties and perception of the future into account. Markets that operate at timescales of years (long-term bilateral trades) need to focus on the future and make predictions with lots of uncertainty. In contrast, markets that operate at timescales of minutes (e.g. spot-type markets) focus more on reliability and deal with more certainties.

For example, California ISO in the US, among other things, runs both forward (referred to as day-ahead) and real-time (referred to as spot market) markets. In a spot-type market, the trade happens in near-real-time. In the forward market, the buyer and seller decide on the amount to trade and the price in advance. This is done to avoid the variability of trading only in the "spot price."

Depending on the timescales, these markets can offer assorted products and services, e.g. energy, capacity, or ancillary services. Ancillary services are system support services like frequency response, voltage regulation, system strength, restart services, and black-start capabilities, etc. The type of ancillary service and its compensation mechanism also depends on the market that one is operating and its timescale.

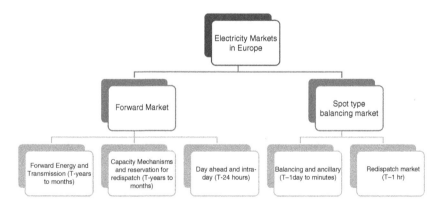

Figure 1.11 Schematic representation of wholesale markets in the European Union. Source: Modified from [3].

Regarding modes of trades in these markets, trade can happen in several ways, such as bilateral contracts, contracts through power exchanges, forward and spot through markets operators, etc. Let us take a fictitious example of EnergyZ (a supplier) participating in a centralized wholesale market. This market determines its price and megawatt (MW) clearance (how much MW, when, and at what price will EnergyZ supply them). The cleared price is called the spot price. Spot prices are determined by matching supply and demand and subjecting them to physical constraints (e.g. matching generation with load, not overloading the line, etc.) and commercial constraints (e.g. a reserve of 10 MW is required, the cleared price cannot be larger than Y$/MWh, etc.). Such supply and demand matching requirements are taken care of by system operators (SOs) or market operators (MOs). See Figure 1.9. These operators can determine operating reserves, network constraint management, and a balancing mechanism. They collect the supply bids from the generators and match them with demand. These bids are in the form of price/MW pairs, and these bids are used to construct a supply curve for each generator and their maximum and minimum power values. The MO then runs some form of security constrained economic dispatch (or unit commitment) algorithms to dispatch the least-cost generation to meet demand and determine prices. The cleared prices could be nodal (referring to power system buses or nodes) or regional (referring to a region). If the former, they are called locational marginal price (e.g. spot market prices in Pennsylvania, New Jersey, and Maryland, collectively known as PJM). If the latter, they are called regional prices (e.g. regional spot prices in Australian Energy Market Operator, or AEMO).

Another example involves a bilateral contract in a centralized market. EnergyZ, a fictitious company in Australia (a seller), signs long-term contracts with a mining company (a buyer).

In the above example of a contract, EnergyZ and the mine have agreed to set a price of 40$/MWh. Figure 1.12 shows this price vs. the spot price over a hypothetical day. The mine (buyer) pays EnergyZ the agreed and spot price delta when the spot price is less than the agreed price. Similarly, when the spot price is higher, EnergyZ pays the mine the delta of the agreed price and the spot price for the mine to purchase the electricity from the pool of suppliers. In another example of a bilateral contract, the mine and EnergyZ could enter into a bilateral agreement with a fixed price (irrespective of the spot price). In either case, the bilateral prices are private between the mine and EnergyZ. Therefore, the details do not need to be disclosed. However, a transmission line is a shared resource in a centralized market. Hence, the buyer and the seller must report the amount of power transmitted. This allows them to make adequate payments. The SO mandates this information to ensure the power flow's security and uses it to determine the system security along with the other bids/offers that it will be managing.

Figure 1.12   Hedging contract and spot price example.

Exercise 1.11   In a centralized wholesale market from above, what happens if EnergyZ owns most of the generators?

1.7.1.2   Decentralized Market

In a decentralized market, there is no central MO. Hence, a bilateral trade-based market is an example of a decentralized market. Another example is peer-to-peer (P2P) trading in a distribution system. In such markets, with the computational ability of today, it is easy to identify transactions that will make the power system insecure. Deciding which transactions are to be curtailed requires some predetermined rules, e.g. nature of transaction firm vs. flexibility to change the load and generation, etc. Such decentralized (bilateral-only) trading cannot factor in the economic benefits. One way to overcome this is to price power system security. The buyer and seller are the best places to identify and make the most economical purchase. If they also can buy transmission rights (e.g. through public auctions), they can make a rational decision as whether to have this additional cost or be ready for curtailment.

Unlike bilateral trading, a decentralized P2P setting could include multilateral trading, i.e. multiple buyers and sellers can come together, use their rulesets to settle on the market price, and conduct the trade without a central entity.

Exercise 1.12   In the decentralized bilateral market from Section 1.7.1.2, the owner of the transmission rights has the right to transmit a certain amount of power through a given network at a certain time. This owner can use them or sell them. They can also decide not to sell them (especially in a less-than-competitive market). What happens if the transmission right is improperly used? (Hint: one way to solve this is to introduce an approach "use them or lose them." Another way to solve this is centralized trading.)

1.7.2 Retail Market

Retail electricity markets are more common than wholesale electricity markets. The basic principle of a retail market is to buy wholesale and sell the electricity to the end customers [4]. Typically, in a retail sale (irrespective of the model), the profit margins are narrow, but the sales volume is large. In the retail market, the end-user:

- Can choose their supplier from competing retailers (e.g. retail consumers in Melbourne can buy electricity from one of more than five retailers).
- May not choose their suppliers but can still have a government regulation on their tariff (e.g. consumers in urban Minnesota must buy from Xcel Energy and the Minnesota Public Utilities Commission sets the Minnesota tariff for Xcel). Similarly, the Australian energy regulator sets the default market offer (which is an electricity tariff that is likely to serve as a price cap for energy retailers).
- Buy directly from the wholesale market (e.g. Power Club in Australia, offers its customers, after they have paid a membership fee, direct access to wholesale electricity prices). The retail supply can use various intermediaries like retailers, utilities, traders, and brokers. It should also be noted that retail electricity sales may be just electricity or bundled a multi-utility sale.

Metering is an important function in electricity retail (for end-users, retailers, and distribution companies). Smart meters are used for measuring kWh, doing settlements, measuring power quality, and performing DR activities (please refer to Section 1.8 on smart meters). The retail supplier must perform customer analysis to estimate the demand by looking at the sales contract, smart meter readings, or the demand. It should then estimate the market price, considering the purchase cost. In doing so, the retail supplier hedges the price risk (the difference between the estimated and actual price), quantity risk (the difference between the estimated and actual demand), collection risk (the risk of protecting the revenue), and regulatory risk (arbitrariness and assumptions made by the regulator). After accounting for these factors, the retail supplier provides electricity to the end customer.

1.7.2.1 Competitive Retailing

In a competitive retail environment, the retail suppliers buy wholesale and resell the electricity (often bundled with other services) to the end customer. The retail suppliers may or may not own the transmission and distribution infrastructure required to provide this service. The prices in the competitive retail environment might be static or dynamic. Static prices change infrequently, and static prices can be:

- **Flat rates**: In this structure, the customers pay a flat fee per kWh consumption, e.g. for small residential customers in Nepal. In this model, the retail supplier absorbs the full risk of market price uncertainty. Flat rates could also be tiered. In a tiered flat-rate structure, the customers pay a slightly different rate based on the amount of use. For example, Alectra Utilities in Canada use such rates in two tiers. A lower rate is charged for the electricity usage up to a certain threshold, and a second higher rate is used for additional use.
- **Time-of-use (ToU) rates**: In this structure, customers pay a different rate depending on the day. The retail supplier sets the ToU prices considering long-time trends in its purchase prices. ToU pricing generally offers customers a more economical alternative to flat rates. ToU can also be seasonal, especially for customers who only operate their facility in certain seasons.
- **Demand rates**: This is a published fixed-rate structure, normally for commercial customers, that bills separate charges for (i) instantaneous peak demand (kW) and (ii) consumption charges (kWh). The dynamic prices frequently change on short notice.

Dynamic prices could be:

- **Peak rates**: In this structure, the retailer can occasionally declare an unusually high retail price for a limited number of hours, reflecting peak demand on the grid system. Demand rates can also be coupled with interruptible rates. In this case, the retail supplier may interrupt the service to consumers at short notice and up to a maximum number of times per year. These price signals represent actual instances of scarcity in the production of power.
- **Demand response rates** [3]: In this structure, the retail supplier may contact the consumer to offer a payment to reduce consumption below a baseline. For example, Great Britain and Australia are actively conducting DR trials.

1.7.2.2 Regulated Retailing

The regulator generally sets (or guides) the retail rates in a regulated tariff environment. This is a tricky position because if the rates are set too high, the retail resellers will drain the consumers' money. Similarly, if the rates are set too low, the retailer supplier might be bankrupt. There is not yet a foolproof method to set it accurately. The regulators, however, follow some form of the options described below.

---

3 Please note that for structures like demand pricing, ToU, and DR rates the use of a smart meter system is necessary (please refer to Section 1.8).

- **Spot market**: In this option, the retail prices reflect the hourly spot or day-ahead wholesale market.
- **Competitive procurement**: In this option, tenders (quotations) are put out to secure competitive supply. This option is widely used in the United States.
- **Default tariffs**: This option is often used when there are not enough bidders for competitive procurement or when the regulator wants to put a price cap on the retail sale of electricity. Such prices are used as indicative prices by the regulator. The regulators sometimes also use a "last resort tariff" to protect a consumer whose retail supplier disappears (e.g. because of bankruptcy), and this consumer must be transferred to another retail supplier.

1.7.2.3 Retailing at Wholesale Prices

Is a retailer (or a retail supplier) needed in the electricity market? Or is it only needed when it adds value like providing metering services, implementing control technologies, bundling other products, etc.? If there is no additional value, retail suppliers are just middle people and add extra charges to cover their services. Giving consumers direct access to wholesale prices mitigates this problem in principle. Therefore, retailing at wholesale prices is an option. For example, a handful of retailers in Australia give their customers access to wholesale energy prices. However, they do charge a membership fee. These retailers can introduce mechanisms to share the risk and reward of exposing their consumers to wholesale prices.

1.7.2.4 Consumers as Producers

Most of the pricing structures above indicate that the reseller is selling and the consumer is buying. However, with distributed generation and grid edge devices emerging, the consumer becomes a prosumer (a proactive consumer). Therefore, there are tariff structures like feed-in-tariff (FIT), net-metering (NM), and local P2P trading mechanisms in places.

- **FIT**: In this mechanism, eligible distributed generators can have guaranteed payments ($/kWh) for their electricity generation. For example, German FITs that guaranteed a fixed price for generating electricity [5].
- **NM**: NM is another mechanism to facilitate distributed generation like the FIT. In this mechanism, the distributed generator can contribute to the electricity that it adds to the grid. For example, a household with rooftop solar panels will typically generate more than it consumes, especially during daylight hours. The electricity meter will run backward and provide credit to the owner during this time. Therefore, at the end of the month, the household will only pay the net amount to the resell supplier. In addition to the US and other markets, Israel and the UAE have had NM mechanisms since 2013 and 2015, respectively. Similarly,

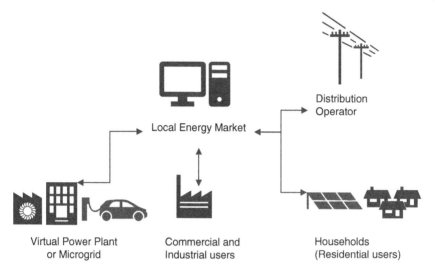

Figure 1.13   Example P2P trading in a local energy market.

Morocco, Nigeria, and Ghana have also introduced such mechanisms [6]. Nepal also has NM mechanisms.

- **P2P trading and local energy market (LEM)**: P2P trading mechanisms are emerging globally as a potential solution to address the issues of DERs. Peers can directly buy and sell among them or through an LEM. LEMs can be used to activate flexibilities from the DGs and prevent grid congestion at a local scale. LEMs can be operated by a distribution operator or by an independent entity. They can be run at a community level or as a system of such markets in a layered approach. Figure 1.13 shows a local community level market. The VPPs and the households will submit offers to buy and sell electricity in this market. The distribution operator will also send signals indicating forecasted grid congestions or send requests to procure flexibilities. The LME will then run a multicriteria optimization to maximize local exchanges and harvest the flexibilities. For example, Siemens and its partners ran a project called Pebbles in Germany to realize this concept [7].

Exercise 1.13   What kind of transmission, distribution network, and market structure exist in your country, your state, and your community? Can you draw a high-level conceptual drawing, like shown in Figure 1.7 and SLD as shown in Figure 1.10? Can you show what voltage levels exist? Show the MO and the retailer (if any). Show the utility responsible for transmission and distribution. The diagram should give somebody who does not know about your region a clear idea of the electricity infrastructure, operations, and governance.

1.8 Smart Meters

Smart meters are key to smart pricing. A smart meter is a standard utility meter with varying degrees of built-in intelligence. Before looking at the suitability of smart meters for a municipal utility, it is necessary to have a solid background understanding of smart meters and their operations. This section aims to provide this understanding through a comparison of smart meters with traditional meters, a description of their capabilities, an outline of the development of the utility meter, and an explanation of their functioning.

Figure 1.14 shows a non-time-specific evolutionary timeline of the utility meter. With the introduction of the electromechanical meter in the late nineteenth century, the metering of electricity began. The functioning of these meters was based on the rotation of a metal disk due to magnetic forces. It had a mechanical counter, and the meter reader had to read the meter and compare it with the previous month's reading to determine the consumption. With advancements in solid-state electronics, electromechanical meters were replaced by newer electronic meters in the mid-1970s. Electronic meters opened a new horizon in metering because of the digitization of meter data. As a result, electrical utilities developed an automatic meter reading (AMR) system that allowed a meter reader to use a car fitted with a receiver to wirelessly transfer the data from individual households when they were within range. Currently, smart grid technology is also gaining popularity. Thus, the so-called advanced metering infrastructure (AMI) was introduced, and the new era for modern-day smart meters began. This technique has added higher values to the AMR technique. In this way, the smart meter was developed from its early predecessor, the electromechanical meter.

The built-in intelligence that distinguishes smart meters from solid-state/electromechanical electric utility meters also allows for a range of different services, dependent on the smart meter model. Various capabilities that the smart meter can provide include bidirectional communication among other meters and a central server, the capability to show tariffs for a particular period, the capability

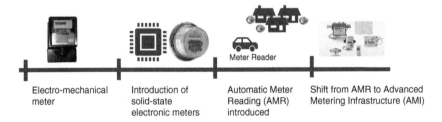

Figure 1.14 Evolution of smart meters [8].

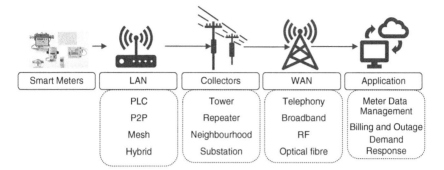

Figure 1.15 Indicative high-level architecture for smart meter communication [8].

to show the projected monthly bill, the storage of electricity consumption data for future use, and the control of home appliances remotely.

Figure 1.15 shows the basic smart meter system architecture for data transfer. To understand smart meters operating more fully, let us consider an example where smart meters collect and transmit data to data collectors via local area networks (LANs). Data collectors could be substations, communication towers, or other head meters. These collectors respond to the end application, like meter data management, billing, or outage management through wide area networks (WANs).

Smart meters are the key to smart grids. They provide cost savings and save time for onsite personnel to visit for meter reads by detecting problems and reducing disruptions, allowing usage of dynamic tariffs, facilitating DR and virtual power plants, facilitating settlement processes, and protecting revenues. Though they have these benefits, even in the twenty-first century, they are often blamed for creating privacy and radiation issues; the latter claim is baseless and the former can easily be overcome by following standard cybersecurity and quality engineering practices.

1.9 Distributed Generation and Grid Edge

The power system started as a decentralized generator near the load centers. At those times, these generators were mostly nonrenewable. Over time, their size grew, and therefore they were taken away from the load centers like cities. Power was then generated far away, carried to the load centers using transmission and distribution infrastructure, and distributed. However, things are changing now. Solar, batteries, and other DERs are becoming increasingly popular and affordable. These technologies are being installed by both customers and energy businesses worldwide. Communities are increasingly seeking to install microgrids, minigrids, and other forms of distributed generation. Where there is a reliable grid, these microgrids rely on the main grid most of the time. But in several cases, they are self-sufficient and even exchange services with the grid. These technologies at the nontraditional edge of the power system (aka consumer edge) are called grid edge. If used carefully, they can easily improve the grid's reliability, improve power quality, and help defer costly network upgrades. In Section 1.9.1, we look at some of the grid edge examples.

1.9.1 Microgrids in Kenya and Other Locations

Microgrids in fragile economies (e.g. Nepal, Bangladesh, and Kenya) have a separate appeal. Most of the population in fragile economies does not have access to the national grid. Therefore, the communities start to build micro- and minigrids. Kenya already has over 25 microgrids and plans to deploy another 23 of them. For Kenya, every 1 MW of power from such a grid translates to 180 new jobs and substantial energy savings in rural household energy costs [9]. The public sector leads such development in Kenya. They have instituted clear policies and guidelines to help mitigate investment risks and thus attract investment. The dispersed settlements, rural landscape, and the terrain with marginal counties make such solutions the only viable solution for access to reliable electricity in Kenya.

Another trend in fragile economies and rural locations off the grid are the various the solar home systems (SHSs). These are self-contained solar collectors with a small battery sized for a small household. The positives are their low cost, portability, and being easily installable first electrification. The negatives include insufficient size or storage for growth, limited life, and often not being robust enough to join a future mini- or national grid.

1.9.2 Microgrids in the United States

The United States has a reliable power system. Nevertheless, there are over 160 operational microgrids. The drivers for microgrids are mostly resiliency, going green, and proactiveness for handling DERs in the United States. Microgrids currently only provide a tiny fraction of US electricity, but their use will grow significantly in the future. This is not only because traditional microgrid counts are increasing but also because newer generations of them are also emerging. The newer generations of microgrids can communicate with one another or look ahead and automatically plan their operation. For example, Commonwealth Edison uses a microgrid controller to integrate multiple microgrids to create a cutting-edge microgrid cluster or a microgrid-of-microgrids. Similarly, Blue-lake Rancheria in California can automatically plan for a day-ahead future to balance its load, generation, and import/export from the grid and automatically implement the plan in real time.

1.9.3   Flexibility Services in Europe

A core element of Europe's energy transformation is increased generation from renewables. In Europe, renewable generation comes from intermittent sources like wind and solar – most of which is behind the meter (imagine yourself standing outside your house looking at your electric meter; you are sitting in its front, and behind the meter is everything that is behind it, like your house, or inside it, such as your solar panels, battery energy storage systems, etc.). In the distribution system, such large penetration of renewables could easily create uncertainty in overgeneration (especially in places with lower demand) or uncertainty in load management (especially in places that see a sharp rise in demand). Such uncertainty could lead to inefficiencies, curtailments, and even rolling outages. The distribution operator needs to make considerable investments to upgrade its distribution infrastructure and distribution management system to resolve this. Another option is to do little to moderate investment in flexibility services. Flexibility, i.e. "change in generation or consumption pattern as a reaction to an external signal," can help minimize or postpone such investments. Flexibility services, voltage control, congestion management, optimal capacity investments, reduction in technical losses, and reduced curtailment of renewables can be achieved. Flexible services are quickly gaining popularity in Europe. Such services can be obtained using P2P trading techniques and LMEs. Please refer to Section 1.7.2 on an example implementation in Germany.

1.9.4   Virtual Power Plants in Australia

Australia's energy network is evolving as residential, industrial, and commercial generation and storage assets slowly eclipse the centralized generation model and introduce a new one. These assets at the residential, industrial, and commercial levels can be grouped as a pseudo generation called virtual power plant (VPP). VPPs can be an ecosystem of residential solar and battery infrastructures. In a community in southern Australia that is operated by the Australian Gas Light Company (AGL), VPPs are like traditional forms of generation, except (i) generation and load assets in a VPP mechanism are owned by consumers, (ii) these assets could be physically at separate places, but they work together for one objective, and (iii) the fuel that powers a VPP generally comes from cleaner sources like solar, wind, hydrogen, etc.

1.10   Changes to the Grid

Not more than a decade ago, few people would have thought that buildings would participate to the utility grid to the extent that they do now. Still, in many parts

Figure 1.16   Distribution infrastructure in Jogbani (Bihar, India).

of the world, such a concept is radical. For example, in the state of Bihar in India, getting unidirectional reliable electricity is the key (see Figure 1.16), whereas the adjacent state of Uttar Pradesh the power grid is looking to allow P2P trading.

Overall, things are changing. Together with the onsite generation (and load management facilities), buildings are emerging as grid-edge devices – many form VPPs. For example, the Sello shopping mall in Finland looks like an ordinary mall from the outside. However, electrically speaking, it has distributed generators (solar, diesel, and batteries) and building loads. Therefore, it participates in the Fingrid and Nord Pool markets as a VPP [10]. With more nonconventional generation and loads – e.g. EVs, fuel cells, microturbines, and small-scale hydrogen production – this landscape is more dynamic than ever. When this is coupled with technologies like trading in LEMs (e.g. buying and selling power among geographically distant prosumers using a distribution grid and its market mechanisms), P2P trading (e.g. buying and selling electricity locally among two households), VPPs, wholesale markets, the boundaries between generation, transmission, and distribution are more dynamic than ever. The opportunities to evolve into a more reliable, affordable, and cleaner grid are limitless.

1.11   Visioning

In this chapter, you learned enough engineering concepts to advocate for an energy policy or put together a business case. Foundational engineering concepts like active and reactive power, inertia, frequency, voltage, current, markets, wholesale, retail, and local markets, along with foundational economic concepts like supply, demand, and equilibrium, were discussed. This chapter laid the foundation and set the same background for management, engineering, and policy students.

In this chapter and this book, we are helping the student and practitioner to create a toolkit of knowledge to address the new challenges in modern electricity. As you have seen, there are complex technological, environmental, and humanitarian issues that need to be faced. In the next 20–30 years, the electrical grid will become cleaner, more technologically advanced, and more available to all of humankind.

Case 1.1 *Microgrids and Transmission and Distribution (T&D)* The current discourse on microgrids is mainly limited to enhancing resiliency or going green. It misses the importance of the existing T&D infrastructure required for microgrid operations. Does the microgrid rely on T&D? What about islanded microgrids? You need to support a business case for your company planning to invest in microgrid technology in sub-Saharan Africa. Create a slide deck (with a maximum of 10 slides) illustrating if it would be wise for your company to invest in microgrid technologies in sub-Saharan Africa – given the status of their T&D infrastructure.

Case 1.2 *Distributed Generation (Renewables, Microgrids, and Hybrids)* What kind of policies currently exist in your country for supporting distributed generation? Create a chart to summarize the past, present, and future policies. What is currently lacking? Please add your recommendations. We learned about power systems, policy, and economics in this chapter. Do you know about power electronics? Power electronics are the electronic devices used in a power system, for example inverters and converters. What is the difference between an inverter and a converter? Please explore. If you need a little in-depth understanding, please refer to https://cusp.umn.edu/courses and refer to the power electronics, power systems, and renewables courses.

Case 1.3 *Market Power in a Centralized Wholesale Market* In 2000 to 2001, an electricity crisis emerged in California. California had 45 GW of installed capacity. With a demand of 28 GW, it experienced blackouts, and the wholesale prices soared up to over 10× their average value. Have you learned about the centralized wholesale market? How could this happen? What could happen if the wholesale prices roared but the retail prices were capped? Imagine living through this crisis and your company filed for bankruptcy. After the crisis, you were assigned a particular investigative assignment. Investigate the California electricity crisis and write a management summary with a proposed solution that you will recommend to the regulators. If your institution does not recommend a format, please use the following format. Keep it short (2–3 pages maximum) and avoid using complicated language.

- Introduction
- Issue in question
- Findings

- Potential solutions
- Recommendation.

Case 1.4 *Pricing Carbon*   One of the methods to encourage renewable penetration is to put a price on carbon. You, and your team of five people, are tasked to identify different methods to put a price on carbon. You need to prepare a policy brief. Policy briefs are more common in political science, sociology, public health, and public policy. It is different than an essay or a report. Policy briefs need to be more concise, clear, and evidence-based and should persuade the target audience of your key message. Could you prepare a policy brief by taking examples of pricing carbon from your neighboring countries? If your institution does not recommend a format, please follow the following structure

- Executive summary
- Purpose
- Critical analysis
- Recommendations.

References

1 The World Bank (2014). Electric power consumption (kWh per capita): sub-Saharan Africa (excluding high income), Australia. https://data.worldbank .org/indicator/EG.USE.ELEC.KH.PC?locations=ZF-AU (accessed 11 December 2021).

2 Bhandari, V. (2019). Analysis of engineering, socio-political and market aspects of energy policies using examples from carbon tax, market diffusion of combined heat and power and vehicle-to-grid services. https://conservancy .umn.edu/bitstream/handle/11299/206330/Bhandari_umn_0130E_20166.pdf? sequence=1&isAllowed=y (accessed 11 December 2021).

3 Florence School of Regulation (2021). Electricity markets in the EU. https://fsr.eui.eu/electricity-markets-in-the-eu.

4 Pérez-Arriaga, I.J. (2010). Electricity retail. https://ocw.mit.edu/courses/ institute-for-data-systems-and-society/ids-505j-engineering-economics-and- regulation-of-the-electric-power-sector-spring-2010/lecture-notes/MITESD_ 934S10_lec_18.pdf (accessed 11 December 2021).

5 Sutton, I. (2021). Germany: will the end of feed-in tariffs mean the end of citizens-as-energy-producers? https://energypost.eu/germany-will-the- end-of-feed-in-tariffs-mean-the-end-of-citizens-as-energy-producers/#:~: text=1%20January%202021%20marked%20the,via%20wind%2C%20solar%20or %20biomass (accessed 11 December 2021).

6 Smith, T. (2020). Net-metering gaining favour throughout Middle East and Africa. https://www.esi-africa.com/industry-sectors/generation/solar/net-metering-gaining-favour-throughout-middle-east-and-africa (accessed 11 December 2021).

7 Gebhardt, S. (2021). Pebbles. https://pebbles-projekt.de/en (accessed 11 December 2021).

8 Siemens (2021). Meters: meters for exact consumption data acquisition. https://new.siemens.com/global/en/products/buildings/hvac/meters.html (accessed 11 December 2021).

9 Burger, A. (2019). Kenya continues rollout of off-grid minigrids. https://microgridknowledge.com/kenya-off-grid-minigrids (accessed 11 December 2021).

10 Bujnoch-Gross, C. (2019). Siemens expands green energy potential with virtual power plants. https://press.siemens.com/global/en/pressrelease/siemens-expands-green-energy-potential-virtual-power-plants (accessed 11 December 2021).

2

Basic Discounting and Levelized Costs Concepts

2.1 Introduction

Electrical energy demand is changing around the world. Countries around the globe are producing electricity from fossil fuels, nuclear, and renewables and conducting load management activities to meet this changing demand. Depending on the source of generation and end-use, these generation and load management activities are directly related to CO_{2e} (carbon dioxide equivalent) emissions and to human welfare. To understand these technologies and their impact, we look at economics even for the analysis of environmental attributes and effects. Economics plays a vital role – by assuming monetary value as the universal medium of exchange. Hence, this chapter provides foundations for some of the most widely used economic principles and their application in power systems.

As a new reader of the topics related to economics, it might sometimes feel like the ideas and explanations are connected. However, the intention here is to introduce the concepts first and let you use them in the exercises and other chapters. For example, in Chapter 3, we investigate the tradeoffs made in US electricity markets using the law of supply and demand mentioned here.

In this chapter, as a concept-building exercise, we will learn about cashflows and discount rates and investigate some of the applications while making business decisions, taxes, and examples of environmental taxes like a carbon tax, subsidies in and around electrical energy, and the questions around reforming them, present, annual, and future values, financing, risks and uncertainties around projects, benefits-to-cost ratios, and return on investments (ROIs) for making investment and financial decisions. We will also discuss additional economic topics like power plant economics, market economics, and renewable energy economics. These concepts are equally important in stronger and fragile economies.

In fragile economies, they are at a budding stage in developing utility-scale infrastructure. For example, the country of Papua New Guinea is currently

Modern Electricity Systems: Engineering, Operations, and Policy to address Human and Environmental Needs, First Edition. Vivek Bhandari, Rao Konidena and William Poppert.
© 2022 John Wiley & Sons Ltd. Published 2022 by John Wiley & Sons Ltd.
Companion website: www.wiley.com/go/bhandari/modernelectricitysystems

building infrastructures to electrify 70% of its population by 2030, and current electrification is less than 15%. In doing so, they need to decide whether to invest in smart meter infrastructure and microgrids or build utility-scale transmission or distribution. Detailed analysis using the tools and principles in this chapter could help such countries in their investment decisions. Here we introduce the concept, and these concepts also iteratively appear in other chapters. For example, would a carbon tax or a price or a constraint alleviate energy poverty? Would it increase the price? What could the government do to reduce the increase in price? Should energy be subsidized? What is energy subsidy reform? These topics are discussed in Chapter 10.

Similarly, robust and stronger economies are going into decentralization, digitalization, and decarbonization (commonly known as the 3Ds of energy). For example, Australia is looking at procuring flexibility services at the distribution level. The utilities there need to decide whether to upgrade existing distribution infrastructure or allow virtual power plants and microgrids to provide flexibility locally, i.e. they need to compare and consider investment alternatives. Detailed analysis using the tools and principles in this chapter could help such countries in their investment decisions. These tools not only help to make decisions on a large scale. They also help individuals in their day-to-day lives. You might be thinking about installing a solar water heater or solar panels for electricity or buying an electric vehicle. The tools presented here can help in making such decisions. The concepts are important, and the applications are limitless.

In Chapter 1, we introduced fundamental concepts around electrical engineering. Here, we introduce the fundaments of electrical engineering economics. After reading this chapter, the reader (e.g. graduate student, engineer, nonengineer, decision-maker, and members of the informed public) will be

- Acquainted with the basic economic principles.
- Acquainted with the application of such principles to power and energy systems.
- Able to see the cost/benefit analysis as it applies to carbon reduction and other environmental benefits.
- Use this knowledge to understand and make decisions related to the economics of powerplants, renewables, and electricity markets.

2.2 Fundamentals

2.2.1 Cashflow and Discount Rate

Discount rates and cashflows are critical factors in making decisions related to energy systems. Why is that? This is because, generally, energy systems require relatively high upfront costs, and the returns need to be analyzed over time. Arguably, the capital cost has become even more important with the advent of significant renewable energy. Since ongoing fuel costs are not an issue, capital cost plays an even greater role.

Discount rates are needed to analyze the time value of money, and cash flows are used to analyze the net amount of cash (or cash equivalents) that are being transferred into and out of this business. Discount rates and cash flows are relevant not only for making investment decisions but also for the effective management of ongoing projects. The concepts of cash flow and discount rates assume that the investment or ongoing project will be implemented according to the pre-analyzed scenario of events (done during the moment of assessment) according to the pre-defined inputs and expected statistics.

One of the factors that affect the scenario is called the discount rate. For our purposes, we will consider this to be the percentage rate of return expected on the weighted average cost of capital, which is debt borrowed and/or equity invested in a project. In other words, a discount rate is like an interest rate on the money used to build or by a project. It attributes a value to future cash flows. The higher these rates are, the lower the value we assign to future savings, or vice versa. Discount rates or cash flows are not the only key indicators used in decision-making. Discount rates are typically used to calculate indicators, like net present value (NPV), internal rate of return (IRR), and payback period (PP), which are discussed in this chapter. For now, if the discount rate and cash flows are so important, how do you calculate them? How do you use them and what are the impacts? Let's take an example using a concept called a discount factor, which is defined as

$$\frac{1}{(1 + discount\_rate)^{years}}$$

For payment in 10 years at a discount rate of 1%, the discount factor is 0.905. For the same at a discount rate of 10%, the discount factor is 0.386. It means that cash worth $100 10 years from now is equal to $90.5 (at 1% discount rate) and $38.6 (at 10% discount rate) today. In other words, to have the same relevance as that of a $100 investment today, a payment 10 years from now must yield $110.46 (at 1% discount rate) and $259.37 (at a 10% discount rate).

Some energy projects are financed by the government, others are financed by the private sector, and yet others are financed by individuals. Normally, the investments made by the government are not as risky as those made by private investors. Therefore, the discount rates are different. Both discount rates are influenced by the availability of capital and the cost of borrowing. However, for the private sector, including individual investments, there is usually an additional risk of returns from the financial market.

Let's take a fictitious example of building and operating a power plant in a data center. To understand this example, a few basic topics need to be understood. Fixed costs or capital costs are the initial investment costs required for building the power plant, e.g. the costs to buy and install the generator. It takes several years to build a power plant. After that, the system goes live and starts operating. Operations and maintenance (O&M) costs are the costs for running the power plant, e.g. fuel costs. Revenue is obtained by selling power, e.g. selling power in the wholesale market (energy and ancillary services – see Chapter 1) or using bilateral contract (see Chapter 1). The power plant does not always operate to its full capacity. The average operational capacity is called the capacity factor.

In this example, it takes three years to build this power plant (permits, planning, actual build, etc.). After the plant becomes operational in year 3, it can be operated for 50 years. The annual revenues are obtained by electricity sales. There are also annual O&M costs. It is a 1 MW plant. The construction costs are $3 million. The investment costs are linearly spread over three years. The O&M costs per year are 3% of the investment costs. O&M costs are assumed to increase by 2% over time steadily. The average revenue from electricity sales is $30/MWh at 70% capacity factor, and an additional 5% revenue is obtained from providing the ancillary services (refer to Chapters 1, 4, and 9 for discussions on ancillary services). Figure 2.1 shows the accumulated cash flow at a 0% discount rate. It shows that the investment becomes cash positive at around year eight.

Exercise 2.1   How will the graph in Figure 2.1 look like for higher discount rates? Will the investment become cash positive before or after year eight? Assuming the revenue remains the same, what happens if O&M costs increase, e.g. because of a government tax?

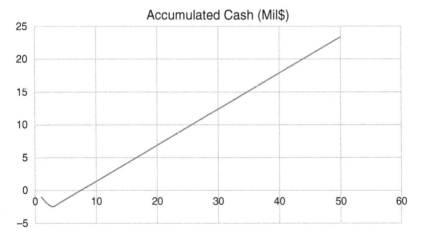

Figure 2.1 Accumulated cash flow example at a discount rate of 0% (if no expected interest or investor cost for the capital to construct the project).

2.2.2   Market Failures and Externalities

Generating electricity has environmental and societal impacts. For example, building dams for hydropower, preventing passage of fish, burning fossil fuel emitting CO_{2e}, or mining for gold polluting water resources are examples of negative externalities. Negative externalities are costs of doing business that are not borne by the company but by society and the environment. Electricity production is filled with such externalities. If such externalities are left unchecked, they will damage our world. Such externalities are there because there are market failures. For example, in region Y, coal is abundant and cheap. Therefore, the CEO of GenCo (generating company) establishes coal power plants. Burning coal would pollute the air, water, and soil, and it sickens the people. The harm done by burning coal is unaccounted for. The price of this generation does not reflect the real cost to society. Therefore, the CEO benefits at the cost of unaccounted damage to society and the environment. So, what can be done?

Exercise 2.2

1) Summarize the negative externalities from generating electricity from coal.
2) How can the negative externalities be accounted for?
3) What would such accounting do to the price of electricity from coal?
4) Are there negative externalities of renewables? Why (not)? If there are, what could they be? Should we also account for those?

   The negative externalities can be accounted for. The dire air emissions from a coal fire plant take the productive life of people and damage the environment. So, if the coal generator was asked to control the emissions like particulate matter, sulfur dioxide, nitrogen oxide, mercury, etc., by installing filtering devices, the coal-fired electricity would no longer be cheap. Another method would be to add the value of pollution abatement as a tax to this coal electricity. There are several other methods. Whatever the means, accounting for negative externalities makes the coal more expensive whilst helping to protect society and the environment. Both long- and short-term impacts need to be understood to price the negative externalities adequately. The short-term impacts, local externalities, are relatively easier to price than pricing a longer-term impact, like climate change. It should be noted that accounting for all the externalities is not trivial. There are several methods and models available, for example, the addition of devices and abatement measures or the imposition of tax. Some economists put a value on the negative externalities by using a variable called statistical life. It is the amount of money that our society is willing to spend in life. This could be a good means of calculation. However, it is very difficult to calculate this value, and it is based on fallacies.

For example, (i) every society would have subjective views on the values of life, (ii) ethical issues of putting a value needs to be addressed, e.g. "What is the value of not killing?", and (iii) the principle rests on a notion that human life is more expensive than other life (which may [not] be the case). Like statistical life, several other methods are available. All these methods do a job, but there is no silver bullet. Nevertheless, if we want to envision a future where electricity production is not only reliable and secure but also cleaner, we need to put a value on the negative externalities. Only then can we bring all the technologies to a level playing field.

2.2.3 Tax and Subsidy

As illustrated in Figure 2.2, the electricity (and heat) sector is by far the largest contributor of CO_{2e} emissions. Economic instruments could act as carrots or sticks and change the dynamics of electricity generation, transmission, distribution, and end use. Therefore, it could significantly alter the emission profile of this sector. Let us look at some examples.

2.2.3.1 Carbon Tax

Carbon pricing is an economic instrument that captures the externalities which the external costs (e.g. damage to health, crops, property, etc.) and ties them to the sources that create such damages. Such a price could shift the burden of damages caused by emissions from the end-users to the emitters.

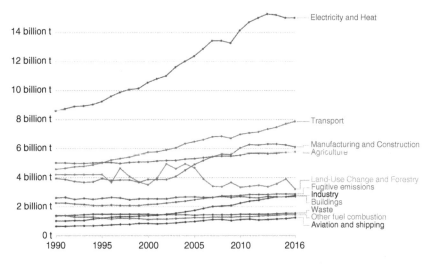

Figure 2.2 Greenhouse gas emissions by sectors. They are measured in tonnes of CO_{2e} [1]/Our World in Data /CC BY 4.0.

- **Carbon tax**: Pigouvian tax, which is a tax on activities that generate negative externalities, is the basis of creating a carbon tax.

In the electricity sector, most of the emissions from the electricity sector come from fossil fuel combustion at the power plant. The location of the power plants is known, and the emissions can be accurately recorded. Therefore, an instrument like a carbon tax (i.e. a levy on its emissions) could increase the electricity price, and in the longer term could discourage the use of fossil fuel (see the law of Supply and Demand in Section 2.3.8). According to the World Bank, currently 25 national jurisdictions have implemented a carbon tax [2].

- **Emissions trading scheme or system** (ETS): This is another example of carbon pricing. In ETS schemes like cap-and-trade, a yearly cap is imposed for emissions. These are referred to as carbon credits. The supply and demand for such credits create a market for trading them. Thus, these credits can be sold and bought between low and high emitters. Due to the market mechanisms, like in the previous case, cap and trade could reduce emissions. Currently, seven national jurisdictions have some form of emission trading schemes implemented.
- **Other schemes or systems**: A few other carbon pricing mechanisms are also popular, for example: (i) offset mechanisms whereby the emission reductions from a project or program can be sold domestically or internationally according to predefined accounting principles and (ii) results-based climate finance, where payments are made if predefined emission reductions are delivered and verified.

Finally, the reader should be aware that, in many cases, countries could commit to such taxes and trading or other schemes. However, there is generally no "stick" (repercussion) if they don't implement these.

Exercise 2.3   Jay has a short-term consulting assignment as a policy adviser to the Minister of Energy in Malawi. Malawi is considering putting a carbon tax. The Minister is knowledgeable about supply-and-demand curves. Use this concept to create a two-page policy brief explaining:

1) Why do we need to internalize the effect of externalities?
2) How does this internalization relate to a carbon tax? (Hint: research Pigouvian tax and look carefully at the graph shown in Figure 2.3.)
3) What are the pros and cons of a carbon tax?
4) Should the minister consider alternative policies? Why (not)?

2.2.3.2 Subsidy
Like putting a price on carbon, subsidizing generation from noncarbon sources is another way to curb emissions. For example, the Victorian government in

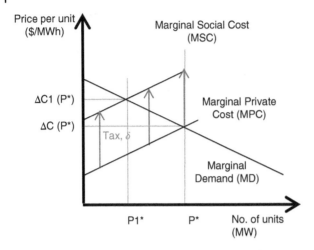

Figure 2.3 Supply-and-demand curve with illustrative tax.

Australia gives a rebate of ~$2200 (AUD) for installing a solar home system. Additionally, the consumers are also eligible for an interest-free loan for up to four years to pay the costs that are above the rebate amount. Similarly, the US government provides $0.01–$0.02/kWh as production tax credits for supporting electricity generation from eligible generation sources like wind. Another example is the feed-in-tariff in Japan imposed to support renewable generation, primarily from solar. Subsidies generally lower production costs (capital costs or O&M costs), lower the price for the consumer, or raise the price for a disfavored producer. Based on this classification, renewable subsidies (and in general energy subsidies) can be the following kinds.

- **Direct financial transfer**: Examples of this type of subsidy include grants or low-interest or preferential loans to producers or consumers. These subsidies lower the cost of production or lower the price paid by the consumer. Victorian government's subsidy for solar from above is an example of the direct financial transfer.
- **Preferential tax treatment**: Examples of this type of subsidy include rebates or exemptions on royalties, sales taxes, producer levies and tariffs, investment tax credits, production tax credits, state-sponsored loan guarantees, etc. Like the previous type, these subsidies also lower the cost of production or lower the price paid by the consumer. Production tax credits in the US are an example of preferential tax treatment.
- **Trade restrictions**: Examples include fixing quotas, technical restrictions, trade embargoes, import duties, and tariffs. These measures raise the price for disfavored producers or ban the product (as imports of diesel cars in

France). For example, to promote local manufacturing in 2012, the United States imposed tariffs (antidumping tariffs) on Chinese solar manufacturers. An antidumping tariff is a tariff placed by a domestic government on imports that are believed to be priced lower than their fair market value. Over the years, with similar motivations, several administrations have changed tariff schemes. One recent move was the repealing of a tariff exemption for imported bifacial solar modules – meaning these modules became subject to the same tariff as other silicon solar panels. These tariffs are currently 20%.

- **Energy-related services to lower the full cost**: Examples include direct investment in energy infrastructure, research and development (R&D) sponsored by public funds, liability insurances, free storage of waste or fuel, and free transport. These measures lower the cost of production. Government investment in hydrogen projects (please refer to Chapters 3 and 7 on different power generation techniques), especially for R&D and infrastructure development, is an example of energy-related services provided by the government at less than full cost.

- **Regulatory approaches**: Examples include demand guarantees, mandated deployment rates, price controls/caps, and restrictions to market access. These measures not only lower the cost of production and lower the price paid by the consumer but also raise the price for the disfavored producer. Renewable energy targets set by governments around the world are examples of these measures.

Exercise 2.4   Jay's consulting assignment as a policy adviser to the Minister of Energy of Malawi has been extended. Now the Minister also wants to explore renewable subsidies. Create a two-page brief explaining:

1) Potential renewable subsidies that can be implemented.
2) What happens if carbon tax and renewable subsidies are implemented together?
3) Generally, fossil fuel subsidies are existent in almost all countries. Can the fossil fuel subsidy and renewable subsidy interact? Why (not)?

2.2.4   Present Value and Future Value

Typical energy projects (especially the larger generation, transmission, distribution, and end-use projects) take a long time to build (several years), and they operate for decades. Things change over time, and there is a time value of money. Therefore, it is important to compare various alternatives over time. Hence, economic concepts like the calculation of present, annual, and future value are important for comparing alternatives and evaluating projects. To understand this, we need to understand the interest rate (for the purposes here, it can be considered like discount rates) and the period of compounding. For example, Jay and his second cousin Adaliya both decide to invest $1000. Jay's bank gives an annual interest rate of 4.5%. Aaliyah's bank gives an interest rate of 4.45%, but it compounds monthly. Therefore, at the end of year 1, Jay will receive $1000 \times (1 + 0.045)^1 = \$1045$. Adaliya, at the end of year 1, will receive $1000 \times (1 + 4.45\%/12)^{12} = \$1046$. Adaliya earns one extra dollar. Though the amount looked small, we hope the message is clear: *The more the compounding periods per year, the greater the total amount of interest paid.* In this example, $r = 4.45\%$ is the nominal[1] interest rate, and $m = 12$ is the number of compounding periods. Therefore, period interest rate $i = r/m$. If $r/m = 1$ and the compounding happens every year, the future value F after n years of investment is $P \times (1 + \text{interest rate})^n$.

Let us take another example. Adaliya has financed a small solar startup in India. Her returns are not tied to the startup performance. She gets a fixed return. The investment pays her $5000 at the end of the first, second, and third year for an annual rate of 12%, which compounds quarterly. How much did she invest in the beginning?

Quarterly period interest rate $= 12/4 = 3\%$.

$$P = 5000 * ((1/(1 + 0.03)^4) + (1/(1 + 0.03)^8) + (1/(1 + 0.03)^{12})) = \$11\,896$$

(Hint: since the interest is compounded quarterly, we must construct the calculations every quarter.)

Exercise 2.5 If the interest continuously compounds, m in the above equation becomes very large, and, in the limit, as m goes to infinity, period interest i approaches zero. In this case, the formula above becomes $F = P\,(e^{rn})$, where n is the number of discrete valuation periods and r is the nominal interest rate, as above. Adaliya is at a stage to renegotiate the contract at the end of year 3. She wants to invest the same principal as before, $11\,896$, but she wants the interest to continuously compound at 12%.

1) What will her payment be at the end of the first, second, and third year. Will it be higher or lower than the previous case? Why? Hint:

$$11896 = F \times (1/e^{0.12} \times 1 + 1/e^{0.12} \times 2 + 1/e^{0.12} \times 3)$$

2) If the interest rate is the same 12% and it compounds annually, how long will it take for her to double her money? Hint:

$$F = 2P = P\,(1 + \text{interest\_rate})^n \rightarrow n = \ln(2)/\ln(1 + \text{interest\_rate})$$

3) Like present and future values, there is also annual value. What is an annual value? Please explore.

---

[1] The nominal interest rate is the interest rate before taking inflation (or fees or compounding interest) into account.

4) Interest rate is a very useful topic in financing.
   a) Typically, two types of financing are available: debt and equity. From a business perspective, debt means the issuance of bonds to finance the business. Equity means issuing stock to do the same. What is the role of the interest rate here?
   b) Government or the private sector could finance a project or a business. How could interest rates play in government vs. private financing?

2.2.5   Risk and Risk Management

Risk management is an important part of electrical engineering because it may increase the project's cost. For example, risk in a project could be mitigated, accepted, or avoided. If the former two strategies are taken, the costs are generally transferred to the end customer. This may even shut the project down, for example if the risk exposure is so high that the policy or the investment becomes a no-go. Therefore, risk identification, assessment, and mitigation techniques are important.

2.2.5.1   Identification

In this foundational stage of risk management, risks are identified by identifying sources of events that cause a threat to the project. Events are possible occurrences that might impact the project, sources are the elements that cause them, and threats are the results of events. For example, in a microgrid delivery project, let us look at the following technical risks. First, one of the sources of risk could be overcommitted human resources. It could lead to an event – lower-quality project delivery and, therefore, a missed delivery milestone with a threat – liquidated damages that must be paid for as a missed milestone. Second, other sources of risk could be the quality of the product that could lead to similar events and threats.

Generally, in electrical engineering projects, identified risks can be classified as technical risks, financial risks, legal risks, environmental risks, natural disasters, country/political risks, and others. There is no single bulletproof risk identification and classification method. However, if one learns from past projects, thinks about the objective of the current project, and then thinks about the risks or events that could impede achieving these objectives, a list of suitable and probable risks can be created.

2.2.5.2   Assessment

Not all risks are equal. Therefore, once the risks are identified, further assessment is required. To do so, one needs to assign monetary values and probabilities. This is combinedly referred to as risk exposure. In the example in Section 2.2.5.1, if the liquidated damages are $5000 per day and the schedule would stretch by 24 days. The cost impact would be $120 000. For example, in a microgrid delivery project, one of the sources of risk could be overcommitted human resource. It could lead to an event – lower quality project delivery and, therefore, a missed delivery milestone with a threat – liquidated damages that must be paid for a missed milestone. Let us call this risk 1. Second, other sources of risk could be the quality of the product that could lead to similar events and threats. Let us call it risk 2.

The probability of risk 1 could be 60%, but the probability of risk 2 could be 2% because the product is matured. Therefore, without taking any mitigating measures, the impact to the project (risk-exposure) from risk 1 (overcommitted human resource) would be $72,000, and from risk 2 (lower quality of product) would be $2,400. Hence, the project may decide to swallow risk 2 but manage risk 1 and thus create a risk mitigation plan for it.

2.2.5.3 Mitigation

Risk mitigation is the most dynamic step in managing risk. The risk manager now has a list of risks with their anticipated impacts and priorities. Large projects generally have a risk manager or an entire risk department. Several multinational companies have risk management departments in which the resources are pooled for large, risky projects. However, in smaller projects, they don't have the luxury of having a risk manager. The small project developer or their project manager would act as a risk manager.

Irrespective of this, the risk manager (or a proxy for them) must investigate mitigation strategies. Possible strategies that are used in mitigation are avoidance (to take action to remove the probability of occurrence), reduction (to take action to reduce the impact), and acceptance (to take the risk as it is). For example, for risk 1 from Section 2.2.5.2, to avoid resource overload, the risk manager might work with the resource manager to preassign project priorities to the resources working in multiple projects (reduce the probability[2] to 10% hence reduce the risk exposure to $12 000). They may also work with the resource to only assign the resource to their project (avoid), or given the situation they may just accept to carry the risk (acceptance).

The purchasing of insurance is one risk mitigation strategy used with power projects. This can cover natural disasters and other things that the underwriter is familiar with in order to be able to provide a premium. One issue between project developers and insurers is new technologies. Things such as the life of new technologies (e.g. energy storage projects) are more difficult to estimate since these technologies are not yet time-tested.

Another form of risk mitigation is dealing with country risk in fragile, politically unstable, or developing economies. For humanitarian purposes, international

---

2 Probability is the branch of mathematics that deals with the likelihood of an event or proposition to occur. It is a number between 0 and 1 (can also be expressed as a percentage) and is expressed as $P(A)$ or $p(A)$ or $Pr(A)$.

development banks such as the World Bank or the African Development Bank (ADB) can provide partial financing for projects in situations where private capital may not be comfortable with the political or economic risk of a particular region.

There is no ultimate answer to dealing with risks. Awareness of the factors and ultimately keeping project goals and objectives in mind throughout are the keys to managing project risks. There are tradeoffs that project managers need to make between benefits, costs, and risks. Once all of this is done, they need to propose a plan and document it.

We are risk takers, and we manage risks every day. A graduate student skipping lessons, a stockbroker investing in a hot commodity, or a child trying to bike for the first time are all examples of risks that those individuals will have to manage. Similarly, in electrical engineering projects, products, and policies, there are risks to be managed, i.e. we need to identify, assess, and mitigate them.

2.3 Simple Applications

Now that we understand risks, discounting, taxes, subsidies, and basic economic principles, let's apply them. Applications are extremely important in our day-to-day personal or work lives.

Whether you're thinking about diversifying your home or farm energy portfolio or making informed policies or business decisions, you need to answer questions like, "When does the investment make sense?" This is not easy to answer, but several economic and financial tools can help. These tools use the concepts explained elsewhere in this chapter. Note that, sometimes, the decisions could also be based on other factors. For example, because you valued energy independence more than anything else, you might want to install the microgrid even if the economics didn't pencil out as expected. Nevertheless, these tools are important so that you can cut out the guesswork and make informed decisions. Additionally, these tools and techniques are also sometimes important for experienced (but not familiar with economic techniques) professionals in the electrical engineering field to be able to say whether a project is viable financially or not by doing simple calculations like the ones mentioned in Section 2.3.

2.3.1 Simple Payback

A simple payback period is the number of years it takes for the savings to offset the initial investments. It assumes that the shorter the payback period the more economically attractive the investment option.

Exercise 2.6 Jay's employer is looking into making investment decisions.

1) In option 1, they could invest $1 million in a solar project. The project will generate $200 000 annually. What is the payback period?

2) There is another project to upskill their employees. It requires a $100 000 investment and will increase the productivity of the employees that translates to $50 000 productivity savings each year for 10 years (total: $500 000). What is the payback period?

3) If the investment decision is merely based on the payback period, which option would the employer choose? Is it a wise decision? Why (not)?

2.3.2 Return on Investment

ROI is the ratio between net profit and costs of investment over a period. High ROI means the investment does much better on the costs. Unlike with payback period, ROI is generally expressed as a percentage. For example, if an investment of $1 000 000 has an ROI of 25%, one should expect that the gains will be 25% more than the initial investment. In other words

$$ROI = Revenues - Costs/Costs =$$
$$(125\,000\,000 - 1\,000\,000)/1\,000\,000$$

Exercise 2.7

1) If another investment option has an ROI of 30%, which one would you choose?
2) For 25% ROI, how long does it take to realize that?
3) How about the time value of money? Do you need to consider it? Why (not)?

2.3.3 Gross Margin

Like ROI, for businesses gross margin (GM) is another important measure. GM is generally comprised of Overhead costs and profit. It is the gross profit as a percent of the sales price. Businesses generally have GM targets that they must meet for any project or investment they do. If the profit on a $1 000 000 job is $250 000, the GM is 25%. A GM of 25% simply means that the total overhead (e.g. salaries of CEOs, business development, marketing, rent, etc.) and profit equals 25% of the total sales. Therefore

$$Gross\ Margin = Overhead + Profit$$

or

$$Gross\ Margin = (Price - Total\ Cost)/Price$$

Exercise 2.8 For the solar project described in Exercise 2.6, with $1 000 000 as the price, Jay's company also builds the secondary control system (see Chapter 1 on control hierarchy). The total cost (including licenses, hardware, software, and engineering) is $150 000. The control business has an overhead of 15% and a profit target of 7%.

1) How much is his company likely to price this controller at?
2) If the market price is $165 000, and to be in the business they have to sell at the market price, what will the GM for the secondary control job be?

2.3.4 Net Present Value

Net present value (NPV) provides today's dollar value of an investment by calculating the costs and benefits by considering the time value of money. It is the sum of discounted present values throughout the project's lifetime.

Exercise 2.9 In the example described in Exercise 2.8, assume a discount rate of 8% and a period of 10 years for both options.

1) Calculate NPV for both options. (Hint: NPV is the sum of present values of benefits and costs for each year. The present value for each year $= F/(1 + \text{interest\_rate})^{\text{years}}$.) Did the investment decision change? Why (not)?
2) How big should NPV be?

2.3.5 Levelized Costs

Levelized costs are another good tool.[3] Not everyone agrees with the values, but there is, in general, an agreement on the levelized concept. Levelized costs bring the comparison to a similar level playing field for comparing options that produce similar outputs. For example, while deciding to buy a car $/mile-driven could be a good measure to compare petrol and diesel options. Similarly, while deciding whether to install renewables or buy from the grid, $/kWh could be a good measure.

Exercise 2.10 Jay's company now wants to further investigate the solar project. The O&M costs of the project are $50 000 per year, which increases by 1.5% every year.

1) What is the NPV? Are your net costs larger than the revenues? If the costs are higher, what measures can be taken to make this a profitable business? (Hint:

---

3 Note that not everyone agrees with the concept of levelized cost values. But there is, in general, an agreement on the levelized concept. Exercise 2.10 further expands on differences in values of levelized costs.

consider government subsidies or market mechanisms like flexibility services or putting a tax on nonrenewable generation. Be creative.)
2) What is the levelized cost of energy (LCOE) ($/kWh) for this project? Assume that the plant produces 1.6 million kWh on average every year.

Exercise 2.11   After understanding LCOE, Jay stumbles upon a rate case and sees the concept of "average costs." What are the similarities and differences with LCOE?

2.3.6 Benefits and Cost Analysis

Benefits and cost analysis (BCA or CBA) is an approach to comparing alternatives for completed or potential courses of action or to estimate or evaluate the value against the costs for a project, a product, or a policy. Using discounting principles as described in Section 2.2.1, all the costs and benefits are expressed in terms of their present value. After that, a ratio of costs and benefits is evaluated. This ratio is used to determine if the benefits of a policy outweigh its costs (and by how much compared to other alternatives). Other similar techniques are cost–utility analysis, risk–benefit analysis, social ROI analysis, etc.

Exercise 2.12   For the solar project example from Exercise 2.6, calculate:

1) The benefit–cost ratio.
2) How does this ratio look like for the option for productivity increase from (a)?
3) What are the possible limitations of this method?

2.3.7 Lifecycle Cost

Lifecycle costs are an approach to assess the total costs of an asset (could also be a project) over its lifecycle. This includes capital costs, O&M costs, and any residual value or costs of cleanup and disposal at the end of a project's life. These are sometimes also called whole-life costs. For example, you buy 30 A4 batteries for $20. However, the price that you are paying includes the capital and O&M costs incurred by the producer and the seller and their GMs. Suppose it requires $30 extra to dispose of them and another $15 for accounting for the pollution caused during extraction and transportation. Adding all these costs would make the battery pack cost $65. How would it impact your buying decision? Similarly, if you were investing in a battery production plant and, owing to these costs (e.g. environmental costs), your estimated investment rose from $20 million to $65 million (hypothetically). Would you still invest in this technology? Would your decision change? Why (not)? This is an example of lifecycle costing considering the environment. Lifecycle costing is also used in economic appraisals associated with evaluating asset acquisition proposals, e.g. in the IT industry, this concept is loosely called the total costs of ownership (TCO) and is used in evaluating IT hardware and software acquisition.

2.3.8 Supply and Demand

Like any other commodity, electricity is also traded in various markets. Sometimes in the wholesale market (e.g. Meridian Energy in Australia and New Zealand trades energy and ancillary services in the wholesale markets of Australia and New Zealand, respectively) and at other times in the retail (e.g. Energy Australia trades the electricity in the retail market in Australia). The nature of the trade is sometimes through a monopoly (e.g. Xcel energy resells electricity in urban Minnesota as a regulated monopoly; similarly, Nepal Electricity Authority resells electricity throughout Nepal as a vertically integrated monopoly). At other times it is not (e.g. tens of electricity resellers resell electricity in Victoria in Australia). Since trade is involved, like any other commodity, the basic principles of economics also apply to electricity. Let us understand some of these principles in this section. First, we need to understand the functional relationship between price, demand, and supply.

In microeconomics, the supply curve is the relationship between the price of a good and the quantity the suppliers are willing to supply. According to the law of supply, a supply curve is upward sloping, meaning as price increases suppliers are willing to sell more. Similarly, a demand curve is a relationship between the price of a good and the quantity that buyers are willing to buy. According to the law of demand, a demand curve is normally downward sloping. This means that as the price decreases consumers will buy more goods. These curves are useful in understanding market equilibrium. Market equilibrium is a situation in the market when both the suppliers/sellers and the consumers/buyers are willing to sell/buy the given quantity at the same price. At this price, the market is cleared (see Chapters 4 and 9 for discussion on the practicalities associated with the markets).

Economists use P or p for price and q or Q for quantity. Since we are already using P for active power and Q for reactive power, going forward, we will use π for representing price, x for quantity, and π^* for the equilibrium price (aka market-clearing price). When the look-ahead or spot market settles using optimization techniques, the bids submitted by the buyer and sellers are used to form cost curves. These curves are matched to find the equilibrium point when subjected to the power system and other regulatory constraints.

Let's summarize and apply the concepts from this section in solving Jay's newer dilemmas in Exercise 2.13.

Exercise 2.13   Jay is planning to buy a refrigerator. He currently has a Model M. It is rated at 150 W and requires a minor repair of $100. After the repair, the life of his Model M is one additional year, i.e. at the end of year two, he must replace the refrigerator. He looked at purchasing a Model X. He can buy it immediately. A Model X is priced at $1000 and is rated at 100 W. His friend recommended a Model Z, which comes at a slightly higher price of $1100, has a one-time government rebate of $50, and is rated at 90 W. However, it will only be in the market next year. The electricity costs are 10C/kWh, and the fridge runs 24/7 (remember discussion of power and energy in Chapter 1).

1) Considering Jay does not know economics and discount rates, which fridge is he likely to buy?
2) For the cases below, calculate initial investment and operations costs up to year two if:
   a) He wants to buy Model X immediately.
   b) He wants to buy Model X in two years (Hint: remember, he will still have to invest in Model M's repair till he buys Model X.)
   c) He wants to buy Model Z in two years.

If he buys Model Z, the government announced that he would get a one cent per kWh rebate on annual consumption. What will be his initial investment and the annual cost for up to two years?

Exercise 2.14   Now that you have solved Jay's problem, he has another problem at work. His employer's local organization in Europe is investigating the option to invest in solar. Now, the problem has slightly changed. They want to compare the costs of solar against other gas peakers. (Gas peakers are power plants designed to balance fluctuating power requirements, especially by operating during periods of high demand or supply shortfalls.) They asked for him to give advice. He stumbled upon the following chart from Lazard.

1) Why is there is a band of prices (instead of just one price for one region) in the LCOE calculations below?
2) For Europe, how is LCOE changing over time? Please explore.
3) Other organizations also calculate and publish LCOE. Are the results the same? Different? Please explore.
4) What could impact the LCOE calculations?
5) Should they make decisions just by looking at the graph in Figure 2.4? Why (not)?
6) What else should they consider?
7) What should Jay's advice be?

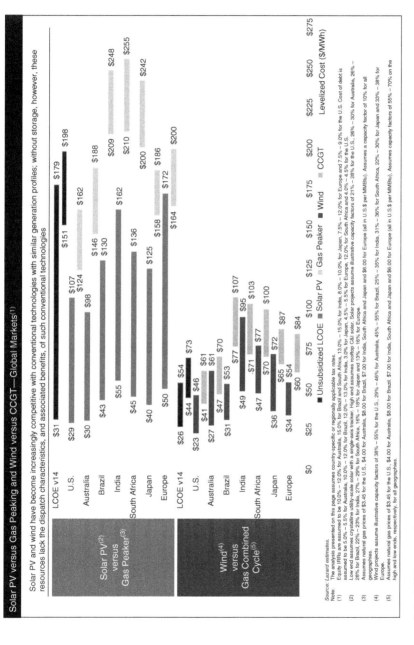

Figure 2.4 LCOE comparison [3]/Lazard Ltd.

2.4 Extended Applications

2.4.1 Wholesale Market

Until now, we have learned about electrical energy systems and their underlying economics. We have also looked at the wholesale markets and other markets in several sections in this book. Here we will learn about the advanced techniques that are used in these markets.

The first concept is power flow. It is the numerical analysis of the flow of electric power in an interconnected system. Sometimes it is also referred to as load flow. Power flow is generally used for two purposes: planning and operations. For example, for operations, when the operator is dispatching the generators, they need to know if such dispatch would overload (or burn) any lines. The most important information that comes as an output of a power flow is voltage magnitude and phase (see Chapter 1 for the concept of phases) and the active and reactive power (see Chapter 1 for the concept of active and reactive power) at each node in the power system. The answer that the operator or engineer is trying to answer from power flow is if the power system can reliably and adequately supply the connected demand. There is no economics involved in power flow. So why are we talking power flow in this chapter? Power flow is often coupled with additional analyses like unit commitment (UC) and economic dispatch (ED), which use economic principles described in this chapter. Additionally, along with the power flow, ED, and UC, additional engineering analyses are conducted, for example short-circuit calculations and fault analysis, stability analysis, contingency analysis, etc. Some of these are mentioned in Chapter 7.

When generators are dispatched in a power system, as explained in Chapter 1, the least-cost generators need to be dispatched. (Generators bid in the market. Their bids reflect their costs. So the least-cost generator is the cheapest generator.) To achieve least-cost dispatch, either an optimal power flow (optimal, for example, to make sure power flows correctly assuming least-cost generators are dispatched) or an iterative power flow and UC/ED techniques are used. The UC/ED problem tries to achieve the least-cost dispatch. Least-cost dispatch can be achieved using various techniques like creation of a merit order list using business and other rules, or a mathematical cost minimization. In an example merit order list-based dispatch, the generators are arranged in a least-cost to highest-cost ranking, and depending on the load and their flexibilities they are chosen from the pool. In a mathematical cost minimization, the total system costs are minimized so that the projected demand is met at minimum total system costs. This optimization is based on the law of supply and demand, as described in Chapter 1.

The aim is to dispatch the least-cost portfolio of generators when subjected to power-system constraints. If the costs of individual i^{th} generators are listed as

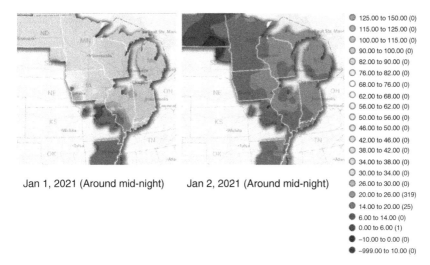

Jan 1, 2021 (Around mid-night)   Jan 2, 2021 (Around mid-night)

125.00 to 150.00 (0)
115.00 to 125.00 (0)
100.00 to 115.00 (0)
90.00 to 100.00 (0)
82.00 to 90.00 (0)
76.00 to 82.00 (0)
68.00 to 76.00 (0)
62.00 to 68.00 (0)
56.00 to 62.00 (0)
50.00 to 56.00 (0)
46.00 to 50.00 (0)
42.00 to 46.00 (0)
38.00 to 42.00 (0)
34.00 to 38.00 (0)
30.00 to 34.00 (0)
26.00 to 30.00 (0)
20.00 to 26.00 (319)
14.00 to 20.00 (25)
6.00 to 14.00 (0)
0.00 to 6.00 (1)
−10.00 to 0.00 (0)
−999.00 to 10.00 (0)

Figure 2.5 LMPs in Mid-Continent Independent System Operator of the United States. The legend on the right shows the prices in $/MWh.

equations relating dollars and electric power, i.e. $C_i(P_i)$, this problem can be solved using standard cost minimization techniques, and the results give the dispatch order and prices. In a nodal system, like that used in the United States, they are called locational marginal prices (LMPs). These prices represent the marginal cost of delivering (or consuming) an additional megawatt-hour at a particular location in the system. LMPs vary in space (e.g. the prices in Minneapolis vs. Saint Louis were different on 1 January 2021; see Figure 2.5) in time (e.g. the prices in Omaha on 1 January 2021, were hundreds of dollars vs. tens of dollars on 2 January 2021; see Figure 2.5). Both the locational aspect and the time aspect are very important. Up to a certain point, the more discrete one becomes, the more accurate results will be obtained. However, every market has a slightly different philosophy on this. For example, the Australian Energy Market Operator (AEMO) for its National Electricity Market (NEM) currently solves for regional prices, i.e. Victoria has one price, New South Wales has another price, and these prices change every five minutes. California ISO for the Western Interconnection of the United States, solves nodal (not regional) prices at similar discrete intervals.

Here is an example method that could be used, for example, in the United States that solves nodal wholesale prices. The explanation assumes that the reader is versed in the basic optimization techniques. Details of these techniques and their applications in power systems would require another book. Therefore, we don't explain them here. Rather, for unfamiliar readers, first, please do the following exercise.

Exercise 2.15 Several cost minimization techniques are used in power systems. Please:

1) Explore the ED concept.
2) Explore the UC concept.
3) What is the difference between ED and UC?
4) How is this related to power flow analysis?
5) Explore high-level details of optimization techniques. What is a Lagrangian?

Now that you have a high-level understanding, let us look at an example. For a given time of day, if the costs of all the generators are collected, a cost-minimization algorithm can be run as

$$\text{Minimize:} \qquad f(\underline{x}) = \sum_{i=1}^{N} C_i(P_i)$$

which is subject to

$$h(\underline{x}) = 0 \text{ and, } g(\underline{x}) \leq 0$$

where

1) $C_i(P_i)$ is the cost of the market participant in \$/h. Each $C(P)$ can be represented as
$$C(P) = cP^2 + bP + a$$

where a, b, and c are the coefficients of the curve and the unit of a is \$/h, the unit of b is \$/MWh, and the unit of c is \$/MW^{2}h.

2) $g(\underline{x})$ is the set of inequality constraints.
3) $h(\underline{x})$ is the set of equality constraints.

The Lagrangian for this optimization problem is

$$L(x, \lambda, \mu) = f(x) + \lambda^T h_i(x) + \mu^T g_i(x)$$

An optimality condition requires that $\Delta L = 0$, i.e.

1) $\Delta L_x(x, \lambda, \mu) = 0 \rightarrow \Delta f_x(x) + \Delta h_x(x)\lambda + \Delta g_x(x)\mu = 0$
2) $\Delta L_\lambda(x, \lambda, \mu) = 0 \rightarrow h(x) = 0$
3) $\Delta L_\mu(x, \lambda, \mu) = 0 \rightarrow g(x) \leq 0$
4) $\mu^T g_i(x) = 0$ and $\mu^T \geq 0$

The equality constraints come from balancing the load/demand, i.e. $P_{total\ load} + P_{total\ loss} - \sum_{i=1}^{N} P_{geni} = 0$, and the inequality constraints come from the minimum and maximum limits of the entities in the power system like:

1) Active power limit of the market entities, i.e. $P_{min_i} \leq P_i \leq P_{max_i} \forall i = 1$ to N
2) Reactive power limit of the market entities, i.e. $Q_{min_i} \leq Q_i \leq Q_{max_i} \forall i = 1$ to N

3) Voltage limits of the buses, i.e. $V_{min_i} \leq V_i \leq V_{max_i} \forall i = 1$ to M
4) Flow limits in each transmission line i and j, i.e. $MW\,flow_{ij} \leq MW\,flow_{ji}^{max} \forall i, j = 1\ to\ P$ and $i \neq j$ where N represents the total number of generators and loads, M represents the total number of buses, and P presents the total number of transmission lines.

The solution to this optimization results in the dispatch of the least-cost generators and determines the incremental prices, called LMP, at different nodes. LMP can be formulated as the marginal change in the cost to the overall system to supply an additional megawatt of load at bus k. $LMP_k = \Delta T_{cost}/\Delta P_k$ where ΔT_{cost} is the marginal change in cost and ΔP_k is the change in load. Using this definition LMP of bus k is given by

$$LMP_k = \frac{\partial C_{ref}(P_{ref})}{\partial P_{ref}} - \frac{\partial P_{loss}}{\partial P_k}\frac{\partial C_{ref}(P_{ref})}{\partial P_{ref}} - \mu a_{lk}$$

$$LMP_k = LMP_{ref} - \left(\frac{\partial P_{loss}}{\partial P_k} LMP_{ref} + \mu a_{lk} \right)$$

where LMP_{ref} is the LMP of the reference bus, $\frac{\partial P_{loss}}{\partial P_k} LMP_{ref}$ is the loss component/sensitivity of an LMP, and μa_{lk} is the congestion component. μ is the Lagrangian multiplier for the inequality in the optimization equation and $a_{lk} = (\partial Pflow\_l/\partial Pk)$ is the sensitivity of power flowing in line l to a 1 MW change at bus k.

Exercise 2.16

1) If you are living in a country with a centralized wholesale market, for your region, do you have a nodal market? Do you have a regional market? What are the advantages and disadvantages of these?
2) If you are living in a country/region without a wholesale electricity market, what are the advantages and disadvantages of a wholesale market? What are the advantages and disadvantages of day-ahead and spot markets?
3) In Figure 2.5, the legend shows that the prices can be positive or negative. What do negative prices represent? (Hint: use the equation for LMP.)
4) Explore another pricing mechanism that exists in your country for the buying and selling of wholesale electricity.

2.4.2 Retail Market

Retail markets are more common than wholesale markets. Setting prices in a competitive retail market, setting prices in a government-regulated retail monopoly, and the direct transfer of price from a wholesale to a retail market (with a top-up membership fee) all require economics. The economic principles and methods described in previous sections can be used. Let us see some examples and do some exercises. UtilityM is also a retailer. It is regulated by a utility commission. It has 1.5 million retail customers. It is a regulated monopoly. It has a steady growth of ~1% customers year on year. Its current retail rate is $0.15/kWh. It is planning to invest $50 million, $55 million, and $65 million in distribution system upgrades over the next three years. Therefore, it has put together a rate case. In a very simplistic case, it could propose to recover this investment by proportionally increasing the rates of its 1.5 million + new customers. It can use the discount rates (see Section 2.2.1), and NPV principles (see Section 2.3.4). If the utility commission decides that the investment is necessary, can it recover this initial outlay with a rate increase?

Exercise 2.17 In Exercise 2.15:

1) Suppose the utility commission does not approve the rate increase. The investment is necessary. What could happen to the utility and its customers?
2) If the utility commission does not investigate/challenge the rate case and approves the rate increase as is, what could happen to the utility and the consumers?
3) How can the utility commission protect vulnerable customers from this rate increase?
4) If the utility is not regulated and the retail rates are set by the free market, how could the utility make its investment? Why (not)?

Australia, Germany, Great Britain, New Zealand, and Texas are some of the regions where the prices are set by the market. Among them, Texas is the one with little to no price regulation. The market there is highly competitive, and the customer switchover rate is also very high. On the other hand, France and several parts of the US, like Illinois and Minnesota, have regulated retail pricing. The third category would be to expose the retail customer to the wholesale price. Let us do an exercise on this topic.

Exercise 2.18 While Retailer X is allowing a transfer of wholesale prices to its retail customer, the customers will not sign op to the deal since the prices could increase as well as decrease. Therefore, Retailer X needs to take the risk on itself to prevent its customer from being exposed to high/low price fluctuations. Retailer X's risk department calculated this risk as 5% of the selling price. It has a profit target of $250 000, and its overheads are $600 000/yr if the profit target and overheads increase by 3% every year.

1) Make an assumption (e.g. number of customers that Retailer X would be signing up every year, the volume of sales, etc.) and estimate the options for one-time signup costs per customer that are required for Retailer X to meet its GM targets.
2) What other business models can Retailer X come up with?

2.4.3 Local Energy Market

Increasing distributed generation (which is generally small and volatile) in the distribution grid is bringing interesting challenges and opportunities. Local energy markets (LEMs) are potential solutions. These markets can be realized behind the meter or away from the grid, or several of these markets can form an ecosystem that interacts with wholesale and retail markets.

   LEMs can be used as a mechanism for activating flexibilities from the distributed generators (DGs) and preventing grid congestion at the local level. LEMs can be operated by a distribution operator or by an independent entity. They can be run at a community level or a system of such markets could be introduced using a layered approach. Different models of such markets have been proposed. One end of the spectrum is peer-to-peer (P2P), where each prosumer (a consumer who produces as well as consumes energy) could communicate with another and buy/sell independently. At the other end of the spectrum is a central market operator which runs such markets (e.g. using similar principles to a wholesale market). Most of these local markets are currently at the conceptual stage. Let us take an example of the economics of one such market design. This is a forward market, intraday, or day(s) ahead with a double-sided auction. Such mechanisms can be formulated as an optimization problem to minimize costs or maximize social welfare when subjected to power balance constraints (power demand should match the supply), physical constraints (e.g. the number of times the batteries can be charged or discharged), flexibility needs from the distribution operator, etc.

Exercise 2.19

1) Explore the status of LEMs in your country.
2) Do you see increases in distributed energy resources (DERs)?
3) Could LEMs be a potential solution?
4) What are the advantages and disadvantages of an LEM/P2P trading in countries/regions in the global south?
5) If your region already has a feed-in tariff, would an LEM/P2P mechanism be beneficial? Why (not)?

2.5 Visioning

In this chapter, you learned economics concepts to advocate for an energy policy or to put together a business case. Foundational economic concepts like discounting and cashflows, present net values, levelized costs, market operations, ED, and UC, etc. were discussed. Foundational economic concepts like discounting and cashflows, present net values, levelized costs, market operations, ED, and UC, etc., and their applications in electrical engineering, were discussed.

Case 2.1 *Buying a Car* You want to buy a car immediately after getting a job. You are looking for petrol (gasoline), diesel, and electric vehicles. You will drive around 10 000 km/yr. Given a discount rate of 8%, the cost of gasoline to be $2/gal, diesel to be $3/gal, and the cost of electricity 15c/kWh. Additionally, assume the following:

	Gas	Diesel	Electric
Purchase price ($)	25 000	27 000	29 000
Resale value after 10 years	2000	2500	3000
Km per gallon or kWh	29	36	3.4
Lb CO_{2e} per gallon or kWh	20	22	1.5

1) For gasoline, diesel, and EV:

a) What are the total present costs?
b) What are the levelized costs per year?
c) What are the levelized costs per mile per year?

2) If a carbon tax of $50/ton CO_{2e} is applied for the emissions:

a) Calculate total present costs, levelized costs per year, and levelized costs per mile per year for gasoline, diesel, and EV.
b) Are the levelized costs per mile per year for the EV still more significant than the levelized costs for gas or diesel?

3) If the government policy encourages EVs, what kind of subsidies can it consider? How would Question 2 change if you could only afford to put 10% of the purchase price as a down payment and for the rest had to make monthly payments at an interest rate of 4% a year?
4) What happens to the answers in Question 3 above if the loans for the EVs were interest-free?

Case 2.2 *Paper Towels vs. Electric Dryers*   You are tasked to decide whether to replace paper towels with electric dryers in your department restrooms. You must consider two unisex restrooms and one parenting room with two devices in each restroom. Additionally, consider that the devices will be used for seven years. If additional assumptions are required, clearly outline your assumptions, and identify your sources.

1) Simple payback is defined as a ratio of investments and annual savings. It gives the number of years when money saved after the renovation will cover investments. For the electric dryer, what is the simple payback period?
2) ROI is the profitability indicator. It is defined as the ratio of annual savings and investments and expressed as a percentage. What is the ROI for an electric dryer?
3) Compare options using NPV.
4) What are the limitations of each of the approaches above?
5) Which options (paper towel or electric dryer) would you prefer based on the analysis above and why?

Case 2.3 *Understanding the For-profit Business Metrics*   You are working in a technology company. You are bidding on a tender for installing parts of a 10 MW microgrid. You have two models: a technology provider and energy as a service company. As a technology provider you will only provide the technology (like SCADA software system, etc.) for operating the microgrid. The customer pays for the technology. As an energy service provider, you install everything and sell energy (not just a technology). As a service provider, you recover your costs, over a longer period, by selling energy. Civil works and generator installation will be done separately by the end client. You are just bidding for the SCADA and optimization (remember control hierarchy from Chapter 1). You must pay license costs of $50 000. You are required to purchase third-party products (like servers, racks, networking equipment, and software) for $100 000. The delivery project is estimated to last for 12 months. You require 1× full-time project manager, 1× full-time project lead, and 0.5× project engineer. From experience, you need to maintain a risk and warranty reserve of 5% of the end price. Your GM is 20%. Post-delivery of the system, you need to provide support. The yearly services costs will be 10% of your delivery costs.

1) What are your total costs? What is the customer price? Hint:

   Price = Total costs/(1 − Gross margin)

2) What happens if the customer is only willing to pay 10% upfront and the rest upon delivery? (Hint: explore cashflow.)

3) What happens if the customer is willing to pay the price as the costs are incurred?
4) Assuming the microgrid operates for 10 years, how much would you charge (per month or per year) to the customer for the energy as a service model? Please make additional assumptions and note the source if additional assumptions are required.
5) How would the model in Question 4 look like assuming 20 years of operation?

Case 2.4 *Solar Lamps or Kerosene Lights?* You are looking to promote solar lamps over kerosene lights in a remote region. Given the following data:

Type	Cost
Kerosene lamp	$5
One day kerosene cost USD	$0.25
Solar lamp	$13
One-day solar	$0

1) Given that your total cost for manufacturing each solar lamp is $10, what is your GM?
2) From the user perspective, how many days will the total costs of a kerosene lamp be higher than those of a solar lamp? Could you use this information for creating your marketing material?
3) If you introduce the rental costs of a solar lamp to be half of the daily kerosene costs:
   a) How does your GM per lamp look at the end of year 1?
   b) What is the payback period for the kerosene lamp?
   c) What is the payback period for the solar lamp?

References

1 Ritchie, H., Roser, M. (2021). Emissions by sector. https://ourworldindata .org/emissions-by-sector#per-capita-greenhouse-gas-emissions-where-do-our-emissions-come-from (accessed 11 December 2021).
2 The World Bank (2021). Carbon pricing dashboard. https:// carbonpricingdashboard.worldbank.org/map_data (accessed 11 December 2021).
3 Lazard (2020). Levelized cost of energy, levelized cost of storage, and levelized cost of hydrogen 2020. https://www.lazard.com/perspective/levelized-cost-of-energy-and-levelized-cost-of-storage-2020 (accessed 11 December 2021).

3

Modern Electrical Engineering Systems, Current Events, Crises, and Tradeoffs

3.1 Introduction

We stress throughout the book the tradeoffs in system types regarding cost, reliability, and energy poverty that may come during the transition to a carbon-free and just energy world. In this chapter, to illustrate the interconnectedness of electrical systems with other systems, we discuss current events and highlight the linkage between weather/climate events and the electrical grid, geopolitical energy source issues, market manipulation in the free market, energy poverty issues, tradeoffs between green choices of generation and others, humanitarian needs, regulation and deregulation, reliability, and costs. These tradeoffs are made every day during the planning and operation of modern electrical systems.

Global energy crises have plagued (like outages, system failures, etc.) modern electrical systems. This issue of "getting out of equilibrium" may become increasingly common as we enter a new world attempting to deal with sometimes contrary political, economic, and technical forces. Although they represent unfortunate situations, the issues and crises highlighted in this chapter are selected as examples of a few timely events that should be referenced in case studies to highlight the tradeoffs, trends, and current events related to these crises. Since such issues are recent and ongoing, plenty of debate and information should be available.

These crises are primarily two types: grid crises, which come up every so many years in power grids around the world (e.g. the Texas power crisis of 2021), and systematic and serious issues that have plagued some of the power grids (e.g. power cuts in developing nations, carbon and other emissions, geopolitical energy source issues, high costs, energy poverty, etc.).

These issues illustrate the value of a portfolio of generation and storage assets with backup and different attributes like a reliable power grid to balance the costs, reliability, and security.

Modern Electricity Systems: Engineering, Operations, and Policy to address Human and Environmental Needs, First Edition. Vivek Bhandari, Rao Konidena and William Poppert.
© 2022 John Wiley & Sons Ltd. Published 2022 by John Wiley & Sons Ltd.
Companion website: www.wiley.com/go/bhandari/modernelectricitysystems

Some Western European countries made a past decision to eliminate nuclear power. This was made up in part by natural gas via long-term contracts through a pipeline from Russia. This became a geopolitical issue with severe cost and humanitarian ramifications when Russia invaded Ukraine.

In the case of Texas, the tradeoff that seems to have taken place between natural gas as a heating fuel and as a generation source is dramatic. Regarding market issues in wholesale markets, the astronomical (positive or negative) spot prices for electricity are a topic unto themselves. Yet another example, unreliable and inadequate supply, is another unfortunate issue in African and other economics.

Such examples make one wonder how fragile the modern electrical systems we rely on really are. Can we strengthen developed grids and the new ones we will provide worldwide?

In summary, our readers will likely take their energy access for granted. However, having access to secure, clean, reliable, and affordable energy requires many technical, social, environmental, and economic processes to work in the background seamlessly. For most of us, it is magic, and we detangle this magic in several other chapters (e.g. Chapters 7 and 8). Here we give real-life examples and shed light on the tradeoffs made in this seamless process.

After reading this and other chapters, a graduate student, engineer, nonengineer, decision-maker, and members of the informed public will be:

- Acquainted with the concept of current trends and tradeoffs related to modern electrical systems. It also highlights the level of importance of energy/lack thereof, e.g. familiar with people's energy dependency and the social ramifications of outages.
- Acquainted with the general engineering, scientific, social, and economic principles that guide this approach.
- Able to summarize and use the knowledge in this book to make an informed decision to understand better and proactively solve these ongoing events and crises.

3.2 Current Events, Crises, and Tradeoffs in Modern Electrical Systems

In this section, we investigate some of the current events and phenomena by taking examples from (i) extreme weather and climate events, (ii) wholesale market manipulation, (iii) systematic energy crises, (iv) increasing dependence on a single resource, and (v) the pandemic and its impact on the electrical system. After that, we investigate some of the tradeoffs in modern electrical systems. Modern electrical infrastructure is very costly to build, operate, and maintain. Tradeoffs are made every day during the planning and operation of modern electrical systems. Such tradeoffs sometimes regard cost and reliability and other times regard energy poverty. Understanding such tradeoffs and their systematic analysis would help us to better manage the crises and unfortunate events that are happening now because we are transitioning to a carbon-free energy world. Considering this, let us start by exploring some of the recent unfortunate events and investigate some capstone exercises. These events are regrettable because they have affected our society. For example, lack of energy could increase chaos, violence, theft, looting, etc. For instance, during the power crisis in Venezuela in 2019 [1], people were deprived of basic needs like food and healthcare, and some resorted to taking extreme measures, like looting stores. If such outages become systematic, it can not only create a short-term social crisis but also even push back the growth of an entire region/nation for decades. For example, Nepal has faced decades of systematic energy crisis. Its energy usage per capita or energy usage per GDP (gross domestic product) is also significantly lower than similar countries. Per-capita energy usage or, more precisely, per-GDP energy usage is a direct reflection of the economic activities of any country. This systematic problem has pushed Nepal's economic development back for several decades.

Exercise 3.1   What is the electrical energy situation in your country? Have you experienced short-term social ramifications (like looting, social unrest, etc.) due to a sudden lack of access to electrical energy? Or do you have a systematic energy access situation? Why? How could this potentially be fixed?

3.2.1   Europe and Natural Gas: Increasing Dependence on a Single Resource: Policies to Achieve Clean Energy Targets Are Not Simple

Another type of crisis is seen in a power system with an overreliance on a single technique, resource, method, etc. Let us explore an example of a problem involving reliance on a single resource in a predominant geographic source. In 2021 and beyond, European countries are more reliant on natural gas than ever before. It is a cleaner transitional fossil fuel intended to replace dirtier energy sources. The European countries decided first to shut down coal and then prematurely shut down nuclear [2, 3]. The Grafenrheinfeld nuclear power plant (Figure 3.1) was taken offline six months before it was scheduled to close on mid-2010s.

Renewable output is variable because we can't control when the wind blows or when the sun shines. Gas alone can't fuel the postpandemic economic recovery, keep everyone's homes warm, and simultaneously support electricity production.

Further complicating the situation, a large portion of the supply comes via pipeline from Russia. This became a geopolitical issue with severe cost and humanitarian ramifications when Russia invaded Ukraine. Because of the heavy usage and limited supply, gas prices and resultant electrical prices have soared over 500% in 2021-22! After all, the markets are susceptible to price swings because of changing supply-and-demand conditions. The natural gas situation is affecting the supply and demand of the European power system, resulting in an

Figure 3.1 Grafenrheinfeld nuclear power plant in Germany. Source: Christian VisualBeo Horvat/Wikimedia commons/CC BY-SA 3.0.

energy crisis. This has an impact in Europe and globally. Becoming dependent on an energy type or location where one has little control creates dependency. In this case it limits Western Europe in its ability to sanction Russia for its political or humanitarian actions.

As a further example of energy source tradeoffs, gas buyers such as factories may pass the costs to the end-customers by increasing the prices of their products. Other buyers, like some governments, may not be willing to pass them to the end-users, e.g. they may even decide to use coal against the expensive natural gas. Like the EU, shortages of fossil fuels are taking place in China, India, and many other countries. In fragile economies, like Bangladesh, this economic crisis is even more devastating primarily because of their already limited access to resources and dependence on countries already suffering from internal problems.

Policies to achieve clean energy targets are not simple to create. Single energy source failure should, at all costs, be avoided to ensure clean, reliable, and secure modern electrical systems.

Exercise 3.2 Do you think governments are making hasty decisions to quickly achieve a clean energy future? Why (not)? What could have been done better in your country/region?

3.2.2 Extreme Climate and the Grid Needs: Texas Power Crises

In 2021, many Texans were left without electricity due to a rare winter storm (Figure 3.2). We have seen many similar crises caused by extreme weather events, like long-term outages due to yearly wildfires, rolling blackouts to avoid damage, or power outages caused by storms and floods.

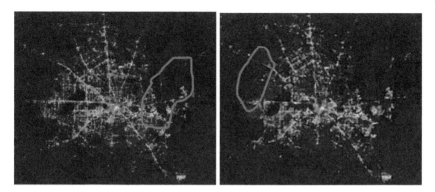

Figure 3.2 Satellite pictures of Texas by NASA: before the storm (left) and after (right). The dark patches show the outages (some of them are marked). Source: NASA.

Another example is outages caused by how extreme temperatures affect grid reliability. The Texas power crisis of 2021 was an example of this. As the temperatures dramatically dropped in Texas, the demand for gas surpassed its prediction and outpaced power generation. Electric Reliability Council of Texas (ERCOT) is the grid operator in Texas. The situation worsened when the weather reduced the ability to supply natural gas to the power plants and caused the wind generators and coal piles to freeze. Texas already had limitations to import adequately from adjacent states as an electric island. This was because of commercial arrangements and physical limits that curbed Texas's electricity import and export with adjacent grids. This resulted in a lack of adequate supply and an increase in demand.

As illustrated in Chapter 2 section 2.3.8, according to the law of supply and demand, an increase in demand and lack of supply in Texas pushed the natural gas prices to their highest. The impact was also seen in the wholesale prices that surged. Wholesale prices below $50 before the storm had surged to tens of thousands of dollars per megawatt-hour!

The grid operator (in this case, ERCOT) had to balance the supply and demand to ensure the grid's reliability. So, when there was more demand and less supply, the only thing they could do was to reduce the demand. Hence, they started load-shedding or planned power outages.

Unfortunately, planned power outages are a norm in fragile economies (see an example from Nepal in Sections 3.2 and 3.3). However, when it happens in places like Texas (where the common public and industrial/commercial systems are not used to working with such outages), the outcome is even worse. No power and unpreparedness meant that all Texan economic activities relying on reliable power were halted.

Such failures should no longer be treated as once-in-a-lifetime events. They are happening more often and are our new normal. Unless we take dramatic steps

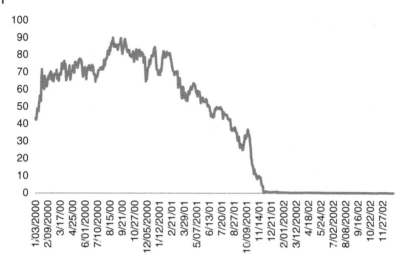

Figure 3.3 Enron's stock price in US dollars. Source: Modified from [4].

to alleviate/mitigate the crisis now, they will repeatedly happen, once every few years. The whole world is expected to see more and more extreme weather events that occur more frequently and stay longer. Hence, they cause more devasting impacts on the grid. Sometimes it is because of cold weather, e.g. Texas. It is also because of hot weather, e.g. California wildfires or Australian bushfires. That is why there is a need for a more resilient grid capable of recovering very quickly from such disasters.

Exercise 3.3 Have you experienced any extreme weather-related blackouts and brownouts in your region? Create a plan, a desktop study, for creating a resilient grid capable of bouncing back after such periodic events.

3.2.3 Market Manipulations: Big Banks to Wall Street Darlings

Chapters 1 and 4 discuss wholesale markets. The market settles based on buy/sell bids in a wholesale market. As trading is involved, companies often use manipulative bidding strategies. Enron is one such example. Enron's case was in the early 2000s. The Enron scandal is a great example. It also had the longer-term effect of limiting the deregulation underway in the US power market.

Once a darling on Wall Street, Enron rose to drastic heights only to experience a dizzying fall (Figure 3.3). At a given height, Enron was trading at $90+ per share [5]. In no time, the shares fell to below $0.50 per share.

Enron's fall was primarily caused by its manipulation in the California wholesale electricity market. Enron's energy traders artificially took powerplants for maintenance during the days of peak demand.

The prices increased due to a lack of supply (see Chapter 2 for a discussion of the law of supply and demand). Enron traders could sell electricity from their available plants at a premium price, many outside California. They also overbooked transmission lines to create artificial congestion, then played in a way that would require the usage of that line. In this way, they could claim congestion fees for illusory congestion. They gamed the market. They eventually went bankrupt and signed billions of dollars in a settlement with several companies and utilities in California.

Let us take another example from the 2010s. JPMorgan Chase reached a $400+ million settlement for market manipulation. The Federal Energy Regulatory Commission (FERC) investigated and found that the traders gamed a complex web of rules to set the prices in the market. This was related to a concept called make-whole payments or uplift payments. Such payments are made by an independent system operator (ISO) or system operator (SO) to the market resources who then experiences a shortfall between their offering and the actual need of the market, for example if an SO commits to buying $60/MWh from a resource but during actual market clearing the price turns out to be $20/MWh. The SO makes additional compensation for fulfilling its prior commitment. In JPMorgan's case, the alleged manipulative bidding strategy yielded tens of millions of dollars from uneconomic plants [6, 7]. One thing to note is that such manipulations (alleged or otherwise) generally require more than a single generator. These manipulations are generally systematic and are done by entities that have market power.

Despite such infrequent mishaps, the wholesale electricity market has benefited the participants and society in general. For example, California ISOs claim over $1.5 billion in benefits from the Energy Imbalance Market (EIM) for their participants. The goal of a market is to satisfy customers' demand for electricity at a minimum cost, in a reliable, cleaner, and secure manner. A good market design should continually evolve, identify, and address critical issues as simply as possible. The billions of dollar in savings from the EIM directly translate to the integration of major renewable energy projects, lesser renewable curtailments, and a significant reduction in costs for the customers.

Exercise 3.4   Does your country/region have a wholesale market? Are there cases of market manipulations? What is market power? Do companies have market power in your region/country? Create a summary report, a desktop study, detailing how such manipulations can potentially be avoided. Who has the responsibility to make the call? Who can make such investigations?

3.2.4   Systematic Energy Crisis: Nepal's Energy Poverty

Power crises in fragile economies are more systemic and long-lasting than in richer economies. We discuss some of this in Chapter 10. Nepal is an example of a fragile economy in need of expensive infrastructure.

Figure 3.4 An example residential inverter/ battery system used in Nepal.

Nepal had been suffering decades of power outages. From the 1980s to the 2010s, urban areas were powered using electricity from the Nepal Electricity Authority (NEA). NEA is a government-owned monopoly.

Rural regions were powered by using community-owned and operated micro-hydropower and alternative energy sources. Foreign countries and non-governmental organizations primarily donated these sources. In both cases, urban or rural, the power was unreliable. The rural electrical system systematically lacked reliability and power quality. The electrical system in large urban areas systematically suffered from 1 to 16 hours of power outages. Please refer to Figure 3.4 to see a residential inverter/battery system that is typically used in Nepal to overcome the impacts of such power outages. People's lives were centered around the power cut.

Let us take some examples. Our energy everyman Jay (protagonist of our book who sporadically appears throughout the chapters) has a sister and other family members. We have been introducing them throughout the book. In this case, his sister lives in Kathmandu. Her day starts in the middle of the night. She sets alarms to wake up in the middle of the night (during the scheduled time when the electrical power is available) to start electric motors that store water for the coming day. She then goes back to sleep and starts her regular life. She is a moderately well-off person. As needed, she could use little luxuries like ironing, cooking in induction and electric cookers, putting clothes in the washing machines, etc. However, she must do these based on the power cut schedule. Given this situation, she seeks alternative energy sources. She relies on gas and kerosene for most of her heating needs. She contemplates installing a small diesel generator or a solar system with a battery in her house. Her workplace has a reliable diesel generator that allows her to send emails and make Internet calls freely. She can write on the computer. When she returns home, she is likely to do so in darkness. She cooks using a gas or kerosene cookstove and handheld emergency lights charged when the electricity is available. This is not just a hypothetical or isolated case; everyone in the country planned their lives around the scheduled power cuts. Electricity was a privilege, and only moderately prosperous people could afford it (e.g. by installing on-premises generators or a battery system).

In rural areas, the electricity from micro-hydropower or other alternative means was highly unreliable. The quality of the available energy was a disaster. For example, the authors witnessed hydropower operating at 25 Hz before the underfrequency relay tripped the machine. The nominal frequency in Nepal is 50 Hz, and systems only work at ± few Hz (max, even in the worst case). Under no circumstances is any residential or industrial equipment designed to operate outside that frequency range.

Therefore, electricity with such a lousy quality could not be used to supply industries like a sawmill, a tin factory, a garment factory, etc. Households only used it for residential purposes. It was only typically used to charge the batteries of the emergency lights or to charge the batteries of cellphones (cell phones were very common during the 2000s – even in rural areas). Such a systematic and decades-long energy crisis plagued the country. There were several reasons for this crisis. Nepal and close to 20% of the world's population need to add reliable electrical capacity to better their health, wealth, and wellbeing.[1] See Chapter 10 for discussions on energy poverty.

Exercise 3.5   Given the background above and the tools you learned in the other chapters. Create a plan, a desktop study, for a successful energy transformation in Nepal. What would the silver bullet look like? What should it include? When and how should it be enacted?

Exercise 3.6   Jay is working for the World Bank and has the opportunity to provide electricity for a region of a country that has no regular power. Funds are limited. How can Jay maximize the resources of the new system that will be designed and built? How can he balance human safety with environmental issues? What if there is access to less-expensive natural gas in that region? Consider the tradeoffs of rationing if they provide more power to industry for jobs and less

---

1 After looking at Section 3.3, we revisit this topic, to better understand Nepal's current outlook, in Section 3.5.

to homes. For further discussion and a specific view, see the editorial from the President of Uganda, Yoweri K. Museveni, in the *Wall Steet Journal* (24 October 2021) [8].

3.2.5 Pandemic's Impacts on Electrical Systems

The COVID 19 global pandemic that started in 2019 brought the entire world to a halt and prompted huge change in the way the world worked and consumed energy over the next two years. During the pandemic, most of the economic activities were stopped. Shops were closed and industries were shut down. The electricity demand fluctuated globally like never before. For example, in China, demand dropped with confinement measures and quickly rebounded after the restrictions were loosened. From April 2020, electricity demand in China was consistently higher than in 2019. Like China, countries have finally started to open after lingering in a (full or partial) lockdown for two years. This has led to increased economic activity and a change in global electricity demand.

Electricity demand has generally started to soar as these countries have emerged from the pandemic. In the case of China, around the same time as the pandemic the government introduced policies to regulate the electricity prices seen by end-users and slowly showed interest in moving away from coal. However, the coal prices were not similarly regulated (the prices hit a record high due to other market conditions), hence the coal power plants were unwilling to produce as (i) coal was expensive and (ii) the electricity price was low and regulated. It was less profitable for them.

More than 50% of China's electricity still comes from coal. As the coal power plants are unwilling to cooperate, the only way to manage reliability and security is to enforce widespread electricity rationing and outages. Such rationing is expected to last for months or longer. Rationing in one of the largest economies could lead to worldwide shortages of Chinese goods and negatively impact the global economy.

Like China, such global pandemics have impacted other power systems. Generally, the pandemic has curbed investments and threatened the expansion of clean energy projects [9]. It has also curbed emissions because of a global economic halt on the positive side. Unfortunately, we can't celebrate as this comes on the back of a crisis and trauma.

Exercise 3.7 Has there been a global pandemic like COVID-19? Or a local one in your region? How did the pandemic impact your electrical system? What happened to the renewable projects? What happened to the demand? What happened to the prices? How can a power system be pandemic resilient?

(See Figure 3.5.)

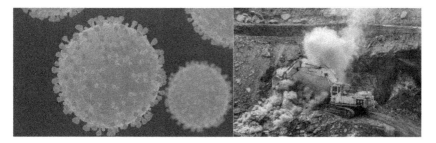

Figure 3.5 Left: COVID-19 virus model (Source: Felipe Esquivel Reed/Wikimedia commons/CC BY-SA 4.0). Right: coal mine. Source: TripodStories AB/Wikimedia commons/CC BY-SA 4.0.

3.3 Tradeoffs

As highlighted in Section 3.2, modern electrical systems undergo unprecedented and expensive changes. The challenges with these changes are generally to meet a wide array of social, economic, physical, and environmental goals. On several occasions, there are tradeoffs to be made to balance the aspirations of competing plans. The topics of tradeoffs might be as contentious as religious or political affiliations. Hence, a thorough understanding is needed. This section highlights some of the tradeoffs in modern electrical engineering systems.

3.3.1 Green Energy Choices vs. Conventional Energy Choices

Some of you may be involved in forecasting and analyzing energy needs and appropriate technologies. Combating climate change requires a significant investment in clean energy technologies. This is a perfect opportunity to select the best and most cost-effective technology to achieve this. But which is the best technology? Wind? Solar? Nuclear? Hydrogen? Storage? Or is it a combination of these? Should emerging technologies be considered? Why? Why not? These are not easy questions to answer. Let's take an example. Country X can have all solar energy, but they need to realize that traditional solar might not be as dispatchable as a traditional generator. Countries like Country X need a dispatchable and flexible resources – like hydro, coal, storage, or natural gas – to balance the energy supply. It is essential to know that even hydro, coal, and gas have limitations. What happens to hydro during a dry year if the coal pile freezes or a natural gas pipeline blows up? In any of those situations, what is our proposed solution? Run diesel generator sets? Brace for blackouts? Light candles? Spend R&D funds in search for newer generation technology? Can Country X afford this? How about the time that this technology takes to hit the market? The answer is not simple. There is always a tradeoff, and decision-makers should know these tradeoffs.

Exercise 3.8 Look at Table 3.1. Can you put together a qualitative metric to compare different technologies. This table also appears in Chapter 7. Can you change the metrics to visually illustrate the information for decision-making in terms of most favorable vs. least favorable? As an example, most favorable is fully shades ● and least favorable is all clear ○.

Table 3.1 Generation sources used in the modern electrical systems and their techno-commercial and environmental attributes.

Table 3.1 Crisis and trade-offs mapping.

Crisis	Tradeoff
Texas power crisis of 2021	Applies to green energy choices vs. conventional energy choices and regulation vs. deregulation tradeoff
	Texas had wind plants that were not winterized. So, the tradeoff could be to use green energy like the wind but winterize it
	Compared to the California crisis example, there was no market manipulation on the electricity side
	Gas regulators did not implement North-American Electricity Reliability Commission (NERC) lessons learned from the winter event. So, the tradeoff is, perhaps, better regulation.
California market manipulation crisis	Applies to regulation vs. deregulation – the tradeoff is market manipulation led to more regulation on market monitoring
	We leave you to explore market monitoring and market power mitigation topics
Nepal's energy crisis	This crisis and its solution relate to resource scarcity and regulation vs. deregulation
	The tradeoff is some regulations brought competition, but more regulation is needed. There were other tradeoffs for supplying reliable power to industrial vs. residential customers and coping with the restricting of the long-serving governmental monopoly
	We would invite you to explore these topics further
Europe overreliance on a single resource: natural gas	Applies to green energy choices vs. conventional energy choices
	The tradeoff is that one can use nondispatchable green energy
	Still, they must be deployed with dispatchable and flexible sources. One can decide to switch off specific sources (e.g. nuclear); however, the consequences need to be thoroughly understood.
China's energy policy and coal price misalignment	China is heavily reliant on coal power plants
	The misalignment between the coal price and the end-user price shows interactions between the tradeoffs made in different markets (regulation/deregulation in the coal market impacting the electricity market). Another tradeoff is a country's choice to protect health vs. letting the market run as usual; and the resultant impact on the electricity sector
	This is a topic we want you to explore potentially as a capstone project

	PV	Onshore Wind	Offshore Wind	Hydro Power	Nuclear	Coal	Natural Gas	Combined Heat and Power	Diesel Generator	Biomass	Thermal Storage	Electrical/battery Storage	Hydrogen	Electric Vehicles (G2V and V2G)	Virtual Power Plant	Microgrid
Renewable																
Carbon Release																
Supply Energy																
Supply Capacity[1]																
Supply Flexibility																
Burns Fuel																
Capital Costs																
O&M Costs																
Salvage Costs																
Risk to Environment																
Risk to investment																
Conventional or New																
Requires access to robust T&D																

[1] The intermittent generation sources like wind/solar can provide energy but not capacity or flexibility primarily because of unpredictable and intermittent nature.

Least Favorable ◄─────────────────► Most Favorable

Another way to make these assessments is using the lifecycle analysis (LCA) technique. It is a cradle-to-grave analysis technique that determines the holistic quantitative economic and/environmental impact of technology [10]. Decision-makers typically decide based on qualitative measures (like the ones shown in the table) or LCA quantitative techniques. For example, Germany's decision to shut down all six of its nuclear power plants by the end of 2022 as part of a plan set out in 2011 will have involved a process of considering a number of tradeoffs. Without such analysis, the decision would be a catastrophic one.

Exercise 3.9 What energy sources are used in your region/country? Who makes the decision? Are green energy choices political or apolitical decisions? How would you analyze future energy needs versus economic, environmental, or quality attributes? What would you do to balance the need to act soon with the need to act wisely on significant long-term infrastructure decisions?

3.3.2 Regulation vs. Deregulation

Electricity is deemed a commodity in most countries. It is generated using power producers and transmitted using transmission and distribution lines (see Chapter 1). We also understand that electricity cannot typically be stored (hence the need for transmission and distribution systems). The investments required in the modern electrical systems are much more significant, and the implementation timeframes are much longer. As a result, the complexity of providing electricity is not as simple as delivering other commodities like food or water.

Regulation is defined as rules or directives made and maintained by the concerned authorities to be followed and enforced by certain actors. It is a means to control a business (or its segments) through laws passed by the government to protect the interests of consumers. Conversely, deregulation is the elimination of those laws and rules.

We can look at various business segments in modern electrical systems to understand these concepts.

Electricity generation, transmission, distribution, and retail are example segments. Other features would be extraction, transmission, distribution, and actual fuel usage (like coal or natural gas or hydropower). Let's look into the first set of example segments: generation, transmission, distribution, and retail.

The regulation that we refer to in this segment set is a mechanism by which the government puts a set of rules to guide the working of the vertically integrated utilities. Vertically integrated utilities own generation, transmission, and distribution systems. They own the generators, transformers, switchgear, poles, and wires. They also typically own an exclusive right to use these assets. The end-customer would have to be served by these utilities and would not have other options (as they would in a free market). Hence, the government regulates the rates at which these utilities charge the end-customer, e.g. Minnesota Public Utility Commission (PUC) regulates Xcel Energy in Minnesota, USA.

The PUC has a tough job. On the one hand, it ensures a monopoly. On the other hand, it should only allow for reasonable returns. If the PUC lets the utility loose, the electricity prices may skyrocket. If the PUC strictly curbs the returns (knowingly/unknowingly), the utility might go bankrupt and the customers lose access to electricity.

In a deregulated distribution/retail market, the distribution utilities are required to divest their ownership in a generation, and transmission, e.g. as a part of the ongoing deregulation of the utilities, in the late 1990s, Southern California Edison Company sold 10 power plants (~7.5 GW) in just a year or so [11].

There could also be another set of entities called the retailer in a deregulated market. The customers can choose to buy electricity from any retailer, e.g. Victoria in Australia.

In a regulated market, the utility must show a return on its investment. Generally, a regulatory commission (like the PUC) carefully evaluates if the investment is reasonable, justifiable, and prudent. It allows (or does not allow) the costs to be passed to the end-customers. In return for a regulator regulating the profits, the regulated utility generally enjoys a monopoly, i.e. the regulator will not allow other utilities to compete in its (regulated utility's) area.

Deregulation allows for competition and power of choice, i.e. the end customer can choose from many providers or resellers. Generally, deregulation seems better. However, some examples tell otherwise, e.g. a Strong reason for the California energy crisis of the 2000s was market manipulation caused by poor-deregulation. Hence the choice of regulation vs. deregulation is not simple. It involves tradeoffs. And, when a market goes from regulation or deregulation or vice versa, there are always winners and losers.

Exercise 3.10 Are you being served by regulated or deregulated energy markets? What would you prefer? Why (not)?

3.3.3 Reliability vs. Costs

We recently saw announcements from Pacific Gas and Electric (PG&E) in the US that they had started rolling blackouts to prevent costs from Californian wildfires. They had a dilemma. If overhead transmission lines were not shut off, they could cause additional wildfires. But shutting off the overhead transmission lines meant customers in those areas had to rely on backup power for days instead of hours. Moreover, there is a high additional cost if such lines are placed underground instead of overhead. For PG&E, there was a tradeoff required between costs and reliability.

Exercise 3.11 Chapter 2 introduces the concept of levelized costs. What are the levelized costs per mile of building a transmission line in your region?

Here is another example. Western Power, which operates an electrical system for most of Western Australia, wrote an article vouching against 100% reliability. Their reliability was 99.93%, ~3 hours of a power outage for an average customer per year. They suggested against further improving reliability as it would create an unreasonable cost burden for customers. While discussing their outage and reliability matrices, they said that it is never possible to ensure the same level of reliability at the same price at the same time. This example indicates tradeoffs related to reliability vs. costs that the utilities make in their everyday operations. Sometimes there are more innovative ways to deal with reliability vs. costs.

Let's take another example from Western Power in Australia. Western Power in 2020 ran a program using standalone power systems (SAPS) of 1000 microgrids like solar, batteries, and even hydrogen electrolysis instead of maintaining long transmission and distribution systems that served remote customers to achieve anticipated savings of A$259 million [12].

Exercise 3.12 As shown by the case of Western Power, should reliability and costs in modern electrical engineering systems be mutually exclusive? Why (not)? On the one hand, Western Power says it is impossible to achieve this. On the other hand, the 1000 microgrids could serve as a counterexample. Is it? Why (not)?

Exercise 3.13 We show the reliability vs. costs tradeoff using the California PG&E example. Among other things, this case showed that underground transmission lines have high reliability but come with high costs compared to overhead transmission lines with low reliability and low costs. PG&E's case unfolded very dramatically with bankruptcy filings and so on. Please conduct a desktop study on this topic. One set of students could write an opinion piece (one or two pages) taking PG&Es side, and another group of students could write an opinion piece (one or two pages) from the regulator's (and hence end users') perspective.

3.4 Crises and Tradeoffs Mapping

To summarize, each of the five crises discussed previously might have been avoided if the tradeoffs had been adequately understood.

Let us look at an example from Nepal where the tradeoffs were adequately understood and the decade-long power crisis was solved. Things changed dramatically from 2017 to 2018, especially in the NEA's urban areas and areas supplied. The domestic power cuts were stopped after several decades. NEA that was almost always a loss-making company started reporting profits. It was able to reduce the system losses by over 10%. It was done by lowering electricity theft and line losses.

This change, however, was achieved by longer-term policy and system transformation, and change in the leadership of NEA. NEA understood the tradeoffs. For example, the Electricity Act in the 1990s allowed independent power producers (IPPs) to generate, thereby increasing the supply. The Emergency Action Plan in the mid-2010s led to the formation of a new transmission and grid company, a generation company, and a power trading company which immensely helped in this transition. The restructuring of a large governmental monopoly and allowing IPPs, on the one hand, could have brought a lot of uncertainty and reduced organizational productivity.

On the other hand, it could have fostered dynamics and promoted free-market principles. Similarly, rerouting power from industrial to residential customers starved some industries of a reliable electricity supply and made them furious. It could have negatively impacted the economy. It could also have significantly impacted the decades-long residential power outages. Each decision had both pros and cons that had to be thoroughly evaluated. Nepal did just that. Hence, such restructuring, reduction of system losses, rerouting of power, and action by the new dynamic leadership of NEA jointly led to this final push and ended several decades of prolonged energy crisis.

Given this impressive turnaround, Nepal has a long road ahead. Only a fraction (<3% [13]) of its 43 GW hydropower potential has been realized, and the current consumption of electricity is primarily for residential purposes (>80%). NEA does not cover the entire country (mainly its rural areas). Nepal awaits another dramatic transformation into evolving as a country that could supply reliable, clean, and secure electricity to its people and even export its excess electricity (as electricity or even packaged as hydrogen and other packageable forms) to countries around the world.

Further, please see Table 3.1, which links crises with tradeoffs.

Exercise 3.14 In hindsight, how could these crises have been avoided? What are some steps that could have been taken to avoid the issues? Please explore this further.

3.5 Visioning

In this chapter, you learned to investigate different and sometimes conflicting issues and tradeoffs surrounding modern electrical engineering systems in the form of case studies. We have provided other toolkits throughout the book to analyze, understand, and better address these issues. After reading this chapter and this book, we expect you to advocate for an energy policy or put together a business case. It is essential to understand the tradeoffs.

Foundational engineering and market concepts (e.g. market operations discussed in Section 3.2.2), foundational economic concepts (e.g. supply and demand topics in Sections 3.2.4 and 3.2.5), and foundational market social concepts (e.g. energy poverty topics in Section 3.2.3) need to be understood to manage crises and tradeoffs. This is a common theme that we stress throughout this book.

Tradeoffs are made every day during the planning and operation of modern electrical systems. Such tradeoffs sometimes concern cost and reliability and at other times energy poverty. At other times, they are about the choices of energy sources. Understanding such tradeoffs and their systematic analysis will lead to the better management of the crises and unforeseen events that are happening now and will happen in the future. Tradeoffs are not new. Tradeoffs are not new. They have always been part of energy management. It is more urgent we understand them as we transition to a carbon-free" energy supply for a world that is suffering an increasing number of energy crises, in order for us to better manage the transition and better deal with these crises.

A particular case for the tradeoffs made regarding energy poverty needs to be stated. Up to 25% of the world's population does not have regular access to electricity (see Chapter 10). This is startling to many in the twenty-first century. The tradeoffs that have or have not been made to supply electricity from any available source involve environmental and cost issues discussed throughout this book. As we go forward to address this social justice issue, it will include calculating the human cost of lost healthcare, industry, education, and life safety. This direct human energy cost needs to be balanced with global warming and energy source economic tradeoffs.

We set out in this book to help you, the student/practitioner, create a toolkit of knowledge to address the new challenges in modern electricity. In this chapter, you saw some real-life examples of this. We are hopeful that in the next 20–30 years the electrical grid will become cleaner, more technologically advanced, and more available to all of humanity. As you now understand the fundamentals and the tradeoffs, we hope you can help to facilitate this change.

Case 3.1 *Regional Hot Topics/Current Events and Tradeoffs* This chapter presents case studies giving examples of crises that highlight the importance of tradeoffs. Find several articles on these topics or other hot topics in your region. As a further exercise, what is a current hot topic for the electrical grid for your region? What are the tradeoffs that are made? If you had decision-making authority, how would you address these issues? Prepare a presentation for your colleagues.

Case 3.2 *Strong vs. Fragile Economies* Electric vehicles (EVs) are a headline in stable, emerging, and fragile economies. In a case study in Chapter 2, we looked at buying an EV in a strong economy. The fragile economies also have their case. Take the example of Nepal. In the past decade, small commercial EVs are becoming prominent (see Chapter 10 for further details). While it was a disruptive, positive local innovation, it also:

- Increased the power requirement of the already power-starved grid.
- Caused more accidents because the government was struggling to categorize it as a licensed vehicle.
- Created traffic jams due to many EVs instead of a few buses.

A similar situation was experienced in the US with electric scooter experimentation. Are EVs a blessing or a curse? What are the upsides and downsides? What are the EV-related tradeoffs in your regions?

Case 3.3 *Clean vs. Reliable Sources of Energy* Do you think there is a tradeoff between clean and reliable energy sources? Why (not)? What form of energy is abundant in your country? Prepare a two-page summary report on whether (or not) your country should opt for increasing the penetration of clean energy resources. If it impacts reliability, how can the problems be mitigated?

Case 3.4 *Sector Coupling* Sector coupling is a way of connecting several sectors to get a holistic view of how the activities of one sector affect another. Are sectors, e.g. transportation and electricity, agriculture and electricity, industry and electricity, heating and electricity, etc., in your region coupled? Prepare a ~10-page presentation on sector coupling in your region and explain it to your colleagues.

References

1 BBC News (2019). Venezuelan crisis: running out of food, power and cash in blackout. https://www.bbc.com/news/world-latin-america-47522208 (accessed 11 December 2021).

2 Roche, D. (2021). Gripped by energy crisis, Europe considers breaking climate promises and turning to coal. https://www.newsweek.com/gripped-energy-crisis-europe-breaking-climate-promises-coal-gas-1637291 (accessed 11 December 2021).

3 Stapczynski, S. (2021). Europe's energy crisis is coming for the rest of the world, too. https://www.bloomberg.com/news/articles/2021-09-27/europe-s-energy-crisis-is-about-to-go-global-as-gas-prices-soar (accessed 11 December 2021).

4 Segal, T. (2021). Enron scandal: the fall of a Wall Street darling. https://www .investopedia.com/updates/enron-scandal-summary/.

5 NPR (2004). Tapes: Enron traders plotted market rigging. https://www.npr.org/ templates/story/story.php?storyId=1920810 (accessed 11 December 2021).

6 Tracy, R., Fitzpatrick, D. (2013). J. P. Morgan settles electricity market case. https://www.wsj.com/articles/SB10001424127887324170004578637662037547582 (accessed 11 December 2021).

7 Financial Times (2013). JPMorgan to settle over power rigging claims. https:// www.ft.com/content/aa6ea7ae-f886-11e2-92f0-00144feabdc0 (accessed 11 December 2021).

8 Museveni, Y.K. (2021). Solar and wind force poverty on Africa. https:// www.wsj.com/articles/solar-wind-force-poverty-on-africa-climate-change-uganda-11635092219?st=3lacf5xiztqfbwf&reflink=desktopwebshare_permalink (accessed 11 December 2021).

9 IEA (2021). Covid-19 impact on electricity. https://www.iea.org/reports/covid-19-impact-on-electricity (accessed 11 December 2021).

10 ScienceDirect (2017). Life cycle assessment. https://www.sciencedirect.com/ topics/earth-and-planetary-sciences/life-cycle-assessment#:~:text=Life%20cycle %20assessment%20is%20a,manufacture%2C%20distribution%2C%20and%20use (accessed 11 December 2021).

11 NS Energy (1998). US utilities rush to sell generation assets. https://www .nsenergybusiness.com/news/newsus-utilities-rush-to-sell-generation-assets/ (accessed 11 December 2021).

12 Energy Mix (2021). Western Australia plans 1000 new microgrids combining solar and storage and hydrogen. https://www.theenergymix.com/2021/03/03/ western-australia-plans-1000-new-microgrids-combining-solarstoragehydrogen (accessed 11 December 2021).

13 The World Bank (2019). Power-less to powerful. https://www.worldbank.org/ en/news/feature/2019/11/25/power-less-to-powerful (accessed 11 December 2021).

4

The Influence of Wholesale Energy Markets in Policy and Pricing Discussions: Part 1

4.1 Introduction

Why should readers care about the influence of wholesale energy markets in policy discussions? There is no doubt in the world that an energy transformation is taking place. Energy is transformed from large unit sized fossil-fuel-based generators to small unit sized renewable distributed generators. Energy policy must keep up with this transformation. Specifically, readers should understand the US market developments to apply lessons learned for other international markets, and vice versa. Energy markets are exciting because energy policies are always trying to catch up with customer developments. This lag in energy policy is true in both the US and other international settings.

   In this chapter, readers would appreciate the push and pull between markets and policy drivers. Carbon-free goals drive policy in many international markets. In the US, environmental goals drive policy at the state level. The lack of a comprehensive federal law complicates policy for readers. Some states have organized markets, and some don't. Some states allow the retail choice of electricity, and some don't. Hence this chapter brings together the influence of organized markets on energy policy.

4.1.1 Market-based Solutions to Enable Energy Transformation

On the one hand, wholesale energy market proponents see energy markets as the true solution for society's benefit. Because the evidence is clear that carbon emissions are down in organized markets, it shows that a market-based approach works. (PJM shows 10 million tonnes of carbon emissions reduction since the PJM market was established [1]). On the other hand, state regulatory construct proponents do not want an energy market without the guardrails, i.e. regulation, because they want to influence the process. Both market participants and state regulators come together and discuss their proposals at market committee meetings.

Modern Electricity Systems: Engineering, Operations, and Policy to address Human and Environmental Needs, First Edition. Vivek Bhandari, Rao Konidena and William Poppert.

4.1.2 It Is Energy Markets, Not Electricity Markets

Whenever a commodity is bought or sold in bulk in the marketplace, "wholesale" markets apply. This is true for energy as well, hence the term "wholesale energy markets" applies to buying and selling large quantities of energy. And specifically, the term "energy" applies to this context, not "electricity," because the word "energy" is more comprehensive than "electricity," and we are talking of energy transactions across multiple regions in a country and across multiple countries in a continent.

Hence, this chapter focuses on the influence of wholesale energy markets in policy discussions, not the influence of electricity markets – even though we talk about electricity and discussing retail electricity in other chapters.

4.1.3 The United States Regional Transmission Organization Developments

In 2020, New Jersey thought about leaving the Pennsylvania–New Jersey–Maryland Interconnection (PJM) capacity market as among the options to consider in meeting their state's carbon goals because membership of a regional transmission organization (RTO) is voluntary in the US.[1] RTO membership is dependent on each electric transmission utility decision. And the kind and type of market they want depends on their state's electric authority approval. The utility's state regulator approves the RTO membership because state regulators pass on RTO membership costs to the consumers.

At the time, New Jersey was thinking about leaving the PJM capacity market utilities, in the southeast US utilities were thinking of forming a framework called the Southeast Energy Exchange Market (SEEM), which is neither an RTO nor an energy imbalance market (EIM) found in the US's western interconnection (see Section 4.2.2) [2]. Even though the word "market" is in SEEM, the SEEM is not designed along traditional energy market lines, such as having an independent board of directors.

4.1.4 International Energy Market Developments

These market developments are not unique to the US. China tried to have a countrywide energy market. Vietnam is taking a phased approach to establishing a robust energy market. The Philippines has a decade-old wholesale electricity

---

1 Even though there is an effort underway to make RTO membership mandatory. See CLEAN Future Act, Section 217 c. https://energycommerce.house.gov/sites/democrats.energycommerce.house.gov/files/documents/0128%20CLEAN%20Future%20Discussion%20Draft.pdf (accessed 12 December 2021).

spot market [3]. Australia has a regional wholesale electricity market. Before the current President took over, Mexico had a free market regulated by the energy regulator Comision Federal de Electricidad (CFE) operating without government intervention. But under Mexican President Andrés Manuel López Obrador's proposed energy reforms, CFE will be eliminated, leading to more state control over energy markets and power purchase agreements [4]. A small central Asian country, the Kyrgyz Republic, is convinced of the benefits of reducing state-owned generators' monopoly power [5].

In most international settings, where there is a lack of independent power producers (IPPs), different countries explore an energy market framework to provide a service for distributed energy resources (DERs) such as rooftop photovoltaic (PV) or provide transparency in the buying and selling of electrical energy.

Exercise 4.1   As you know, Jay is an international energy consultant. He is in your country. In this chapter, you will present/prepare and explore several items around wholesale markets for Jay.

Explore the wholesale energy markets operational in five different regions. Does your country have a wholesale energy market(s)? If not, do you think it should? If yes, what can be learned from the five regions that you have selected? Prepare a two- to three-page research brief outlining your findings for Jay.

4.1.5   Don't Expect a Policy to Lead Energy Markets

Policies are rarely ahead of markets for goods and services. Look at the penetration of distributed solar in wholesale energy markets. There are no regulations that uniformly apply and govern DERs across the board at US RTOs. Yes, there is a Federal Energy Regulatory Commission (FERC) Order 2222 on DER Aggregation, but that is relatively new.

"Aggregation" means bundling more than one resource to participate in energy markets or other ancillary markets. For example, if it does not make economic sense for a small 50 kW rooftop solar to bid into a market, one 50 kW unit could be aggregated with another 50 kW unit, and the resulting 100 kW unit could meet the size threshold for market participation. That is how the new FERC Order 2222 on DER Aggregation works. Similar aggregation rules apply in other regions and territories that are seeing the rise in DERs.

4.1.6   Finally, Energy Markets Are Fascinating and Complex

There is another reason for the reader to understand the complexity of wholesale energy markets and energy policy. And that is the energy markets' depth at some RTOs.

Take the example of the PJM capacity market. PJM capacity market regulations are complex because they are mandatory and look ahead three years compared to the Midcontinent Independent System Operator (MISO) capacity market, which is voluntary and looks ahead a year. But the PJM capacity market attracts higher prices compared to MISO's capacity market. Hence, even though not all US energy markets are equal even under the same federal energy regulator, they attract different market participants based on market incentives, leading to complex rules.

Take the case of a system operator (SO; sometimes also known as an independent system operator or ISO) that operates under different market rules. Australian Energy Market Operator (AEMO), the ISO in Australia, runs the Wholesale Electricity Market (WEM) for western Australia and the National Electricity Market (NEM) for eastern Australia. Though the ISO is the same, the rules are different in these markets.

If energy policy and regulations are of interest to someone, there is no escaping from the complex rules and acronyms. This density is the primary reason this chapter is relevant to energy policy discussions. If the reader understands the history, context, and potential implications of energy policy regulations from across the world, they will be better informed to represent their position, whatever that might be. In this book and this chapter, we help you to simplify this complexity.

Exercise 4.2 Did your country/region explore (or have) a wholesale energy market? Explore the history and context. Why was the market in place? Why is it (not) envisioned? What are the benefits and costs?

4.2 Do Energy Markets Influence Policy?

Energy markets don't drive policy: they inform energy policy discussions. For example, PJM capacity market rules inform New Jersey's offshore wind procurement policy. And the caution is, there is no perfect energy market. What works for Texas does not work for PJM states, because Texas does not have a capacity market and does not come under FERC rules but PJM does have both.

What works in Western Australia might not work in the Northern Territory (the Northern Territory in Australia currently does not have an organized market and is evaluating its options to run such a market). Similarly, what worked in California ISO before may not work after the EIM took off in 2014 because California ISO must now work with other states that are not in a market.

Also, there are countries and regions with a comprehensive energy policy but no market yet. The primary reason Vietnam pursues energy markets is to reduce the state-owned generation's influence in the energy business. This reason is the same for the Kyrgyz Republic and China, hence energy policy is driving the need for energy markets in these countries.

4.3 How Does Policy Benefit Market Operations?

FERC Order 841 mandates electric storage resource participation in US energy markets. To show how policy benefits market operations, let us take an example of storage technology, like a battery, that participates in the California ISO market. In 2020, California ISO had 1300 MW participating as stand-alone storage. The California ISO market benefited from the FERC storage policy because there is clear evidence that storage can provide frequency regulation at a low cost compared to other market resources. "Battery resources can contribute toward meeting both regulation and mileage requirements at relatively low cost compared to other resource types" [6].

Storage is not the only example of a policy that benefits energy market operations. Dynamic line rating (DLR) is another example because a variable transmission line rating compared to a static rating impacts how much transmission capacity is available in the energy market. As the International Renewable Energy Agency (IRENA) brief on DLR indicates, Belgium, Bulgaria and Slovenia, France, Italy, Uruguay, and Vietnam are some international places where the grid operators have adopted DLR [7].

Exercise 4.3 Pick an energy policy from your country or region. Find out how this policy originated. Did an event take place on the electric grid that initiated this policy? What are the benefits and costs of this policy? After the enactment of this policy, did the challenge go away or lead to another challenge?

4.4 Joining an Energy Market Is an Important Decision

The infrastructure costs of setting up an energy market are at least several hundred million US dollars with functions such as a control room, a backup control room, a physical workspace for all the engineers to work together, and the communications equipment needed to dispatch supply and demand. This market setup cost is not to be taken lightly. In many jurisdictions, convincing the stakeholders of the energy market benefits is one task, but the challenge of managing the costs of an energy market startup is an entirely different task.

If the costs are high, the savings are even higher. The first step of quantifying the savings starts with conducting a benefit-to-cost study. For all the potential RTO members, "What are the projected benefits?" is the first question they have.

4.4.1 Benefit-to-cost Studies

Benefit-to-cost studies compare the projected benefits of a utility inside an RTO to outside an RTO. A business case must be made that the transmission system will be more reliable with an RTO than without an RTO. At most RTOs, the first step is to turn over the transmission assets to an independent grid manager who impartially manages access to the transmission system., meaning the RTO manager does not favor a transmission owner to access the transmission system. Their main job is to ensure that every transmission owner has equal access to the transmission system.

In some cases, the reason for joining an RTO can come from regulators who are concerned primarily about transmission system access. In the south, US authorities received complaints from IPPs on the incumbent utility. Look at the November 2012 statement from the Department of Justice in the US, closing their investigation into Entergy after Entergy announced their intention to join MISO RTO [8]. In Entergy's situation, the state of Arkansas asked Entergy to investigate joining Southwest Power Pool (SPP) or MISO. So, even though joining RTOs is voluntary, utility regulators tell their utilities to join an RTO to ensure an independent party manages the transmission system.

4.4.2 Energy Imbalance Markets

Whether instructed by their regulators or on their own, more and more utilities are making the energy market leap. But this is often being done in stages, partly due to the high costs of market startup and partly due to their state regulator's desire to test the waters about RTO membership benefits, such as governance. "Governance" means who gets to sit on the RTO independent board of directors. In some states like California, the governor appoints members to the California ISO board of governors.

One such market staged approach is the energy imbalance market, or EIM. What is an EIM? An EIM is a real-time energy market used for the primary purpose of balancing the energy needs of a specific region.

Western EIM, since its establishment in 2014, is slowly but steadily coming to fruition in the 2020 timeframe. The reason for western EIM is California ISO is looking to expand its footprint, and western interconnection states are analyzing alternatives to large federal-government-owned hydropower companies such as Bonneville Power Administration (BPA) and Western Area Power Administration (WAPA). This reason is similar to Vietnam, China, and other countries looking to reduce reliance on government-owned utility companies.

The California ISO was the first to enter the western EIM in 2014, followed by PacifiCorp, a large investor-owned utility (IOU) serving Oregon, northern California, and southeastern Washington [9]. Arizona utilities joined next in 2015–2016. Others, like Xcel Energy in Colorado, are in the process of joining an EIM.

This western EIM serves as a prime example of staged market entry that other countries and regions could mimic. As Table 4.1 indicates, it took nearly 10 years

Table 4.1 Western EIM benefits range from various sources.

Organization	Benefits	Time	Reference
NREL	$1.46 billion with full EIM	March 2013	NREL report [10]
CAISO	$1 billion surpassed	July 2020	CAISO press release [11]

NREL: National Renewable Energy Laboratory; CAISO: California ISO

for an EIM to take off with more than a billion dollars in savings. Even as utilities in the western US join the western EIM, plans are underway to layer an extended day ahead market (EDAM) on top of the EIM [12].

4.4.3 Value Proposition Studies

A value proposition study shows RTO membership benefits. Since RTOs provide reliability, planning, and operational benefits, a value proposition study puts a dollar value on those benefits to remind the current RTO members and regulators about the membership benefits of staying in the RTO. This membership retention is especially true where RTOs compete for memberships, e.g. mainland US.

For example, Table 4.2 is derived from PJM's value proposition study [1].

The table indicates that PJM RTO provides an aggregate of $3.2 billion of benefits at the low end and $4 billion at the high end. The benefit of deferring generation investments ("with RTO") if not for the market ("without RTO") is on the top at $1.2 billion. At $1.1 billion, the second biggest benefit provider is the generator interconnection queue. The remaining benefits of energy production cost savings due to market operations – keeping the lights on (i.e. control room operations) and reducing emissions by dispatching least cost units – add another $1 billion in benefits.

Table 4.2 PJM value proposition table.

PJM benefits	Low end	High end
Generation investment	$1 200 000 000	$1 800 000 000
Integrating more efficient resources	$1 100 000 000	$1 300 000 000
Energy production costs	$600 000 000	$600 000 000
Reliability	$300 000 000	$300 000 000
Emissions	$10 000 000	$10 000 000
Total	$3 210 000 000	$4 010 000 000

Source: Based on [1].

An MISO value proposition study similarly calculated $3.6 billion in benefits to its members [13]. Considering the MISO's operational cost of $296 million, MISO RTO's benefit to cost ratio would be 12:1 in 2020 [12]. At SPP RTO, the benefit to cost ratio is 14 : 1 [14].

Exercise 4.4   Owing to blackouts in Texas during the cold winter week of 15–19 February 2021, there is a debate in US electricity markets that the Electric Reliability Council Of Texas (ERCOT) should join either MISO or SPP. Can you put together an essay on why ERCOT should or should not remain independent?

4.4.4   Regulatory Compliance and Audits

In addition to the market benefits, there are compliance benefits of joining an RTO. In the US, most of the RTOs are tied to the balancing authority (BA) compliance function, and this is how RTOs fall under the NERC jurisdiction in addition to FERC.

Some RTOs explicitly list NERC and FERC compliance benefits, whereas some with the staged approach of entering energy markets do not have all the compliance benefits articulated yet. MISO RTO claims that there are $96–$133 million in compliance benefits to its members [13].

In the author's experience with NERC and FERC audits, NERC audits are conducted periodically every three years with a set start and end date. But the FERC audit of RTOs is not known in advance and does not have a set schedule. As the example of a compliance audit finding report indicates [15], NERC and FERC staff accompany regional entity audit teams when reviewing utilities.

Exercise 4.5   As you know, Jay is an international energy consultant. He volunteered to serve as an expert on a trip to Rwanda, Africa. This US Agency for International Development (USAID), US Energy Agency Partnership (USEA), and Power Africa-led workshop has representatives from all the countries in the Eastern Africa Power Pool (EAPP). Power Africa is a US government-led partnership coordinated by the USAID. This workshop brings together stakeholders from Kenya, Uganda, Tanzania, Rwanda, DR Congo, and Ethiopia.

Prepare a 10-slide presentation on why USEA/USAID should put out a bid/request for a proposal on behalf of EAPP for a market integration study. That market integration study should include benefit to cost, value proposition analysis, and benefits to EAPP from energy markets and regulatory compliance. The first 2–3 slides of your presentation should be "high level" – for executives of USEA/USAID/EAPP. Can you do this for Jay?

4.5 States with Multiple RTOs

Because RTO membership is voluntary in the US, there are states with multiple RTOs, each with their stakeholder committees creating stakeholder meeting fatigue for the state regulators with limited staff. A case in point, a state like Iowa must follow both MISO and SPP stakeholder meetings because both MISO and SPP have members in Iowa.

Sections 4.5.1–4.5.5 contain examples from the US indicating it is possible to have countries and regions in organized markets, each with regional independence.

4.5.1 The Three RTOs of Texas

Texas Has Three RTOs: Electric Reliability Council of Texas, Southwest Power Pool, and Midcontinent Independent System Operator A Republican state like Texas does not have a policy favoring fossil-free generation, unlike California. But Texas is known for wind generation, and Competitive Renewable Energy Zones (CREZ) was a model for building transmission to integrate wind. And ERCOT does not fall under the FERC jurisdiction. The Public Utility Commission (PUC) of Texas is the sole regulator for ERCOT. The PUC of Texas chairperson sits on the ERCOT board as an "unaffiliated" member [16].

ERCOT is the only ISO in the US that compensates for primary frequency response in the ancillary services market. Another unique aspect is that ERCOT does not have a capacity market. And MISO, which has members in Texas, has a voluntary capacity auction. Both East Texas and Northeast Texas Electric Cooperatives are "transmission using members," as per SPP bylaws [17]. Hence Texas has three RTOs within its state boundary.

Exercise 4.6 Owing to the pandemic, Jay was not traveling internationally in February 2021. As a result, he experienced firsthand blackouts in Texas during the cold snap in the week of 15–19 February 2021.

Jay watched with fascination how the PUC of Texas, ERCOT, and the Texas Railroad Commission (the regulator for natural gas) were mentioned in the news. Jay wants to write a paper on lessons learned on the policy that applies to your country/region. What should Jay include in this policy paper? What lessons learned from the 15 February event apply to your country/region's regulators? In your country, who manages electricity and primary fuel? What happens if there are blackouts in your region? What happens if your location experiences sudden shifts in weather? If you are in a cold region, what happens if you suddenly experience hot spells? If you are in a region like Texas, where summers are hot, what happens if snow falls?

4.5.2 Missouri: SPP and MISO

The RTO members in the state of Missouri drive the energy dialogue for the most part. Ameren Missouri, a subsidiary of the Ameren utility, is a MISO member. St Louis, the major load center in Missouri, is in MISO, but the rest is in SPP RTO.

In addition to a major IOU like Ameren, Missouri has many municipal members who formed the Missouri Joint Municipal Electric Utility Commission (MJMEUC). MJMEUC is relevant when discussing Missouri's RTO participation at both MISO and SPP because MJMEUC has 70 members spread all over Missouri [18].

Additionally, the Missouri Public Utility Alliance (MPUA) is a qualified transmission owner and active at SPP and MISO RTO stakeholder committees [19].

4.5.3 Illinois: PJM Interconnection and MISO

As Ameren Missouri is a MISO member in Missouri, Ameren Illinois is a member of MISO in Illinois. Unlike in Missouri, Ameren serves the rest of the Illinois state outside of the major load center, Chicago, which is in PJM RTO. This splits matters for two reasons. First, Chicago, served by the utility Exelon, follows mandatory PJM capacity market rules. In contrast, the rest of Illinois follows MISO voluntary capacity auction rules.

Second, Illinois is the only state in MISO that allows aggregators of retail customers (ARCs), while the rest of the MISO states do not allow ARCs. By permitting aggregators such as ARCs, Illinois allows utilities, including Ameren Illinois's, demand response programs to participate in MISO and PJM markets.

4.5.4 States with Multiple RTOs Creates "Seams" Issues

Seams within the RTO context are the fictitious border between two RTOs. Having multiple RTOs serving a single state also creates so-called seams issues. These seams can be frustrating for utility commissioners who are new to the electric utility industry.

These seams can also lead to complaints at the FERC. Northern Indiana Public Service Company (NIPSCO) filed a complaint at FERC on both MISO and PJM that they are not planning for transmission that would decrease the congestion in the northern Indiana region due to the seams issue [20]. As a result of this 2013 NIPSCO compliant, FERC ordered in 2016 that both MISO and PJM develop a coordinated system plan study.

A major source of frustration for regulators in a state with multiple RTOs is that each RTO allocates costs for transmission a bit differently. Cost allocation is discussed in Section 4.10.

4.5.5 Joint and Common Market: PJM Interconnection and MISO Effort

Because of the proximity of MISO and PJM, both RTOs have an operating agreement that documents cooperation for reliability, planning, and markets. This cooperation also grew out of the 2003 blackout when it was widely believed that PJM could see a problem on the MISO side, but there were no RTO communication protocols in place for information sharing.

The Joint and Common Market (JCM) effort has its website[2] and stakeholder working groups staffed by PJM and MISO RTO experts, including stakeholders.

4.6 Other Organized Wholesale Markets

Like some states in the US, countries around the world are experiencing renewable penetration, a rise in DERs, and a need for ancillary services. Hence international institutions such as USEA, USAID, NARUC International, World Bank, ISO/RTO consortium GO-15, and EPRI International, to name a few, are all involved in sharing best practices across these countries.

This subsection identifies other world regions where wholesale markets exist, e.g. Australia and Germany. We eliminated Mexico from this discussion due to President Andrés Manuel López Obrador AMLO's election stalling Mexico's energy market reforms.

We also introduce other regions that are looking into creating such markets, e.g. Vietnam. Finally, Nepal and African countries are introduced to showcase the potential of such markets in developing nations.

Exercise 4.7   As you know, Jay is an international energy consultant. He just got a project from the Bill & Melinda Gates Foundation. Based on what Jay has learned in this chapter, can you help Jay with this assignment?

Explore the wholesale energy markets operational in India, China, Canada, New Zealand, Poland, and the UK. First, do these countries have a wholesale energy market(s)? If yes, how does the energy market in India (as an example) compare and contrast with ERCOT in the US? If not, what steps should the country take to prepare for an energy market? Prepare a two- to three-page research brief outlining your findings for Jay.

4.6.1 Australia

Australia has two wholesale electricity markets and is currently envisioning having a third market. The first market is called NEM, covering the eastern

---

2 https://www.miso-pjm.com.

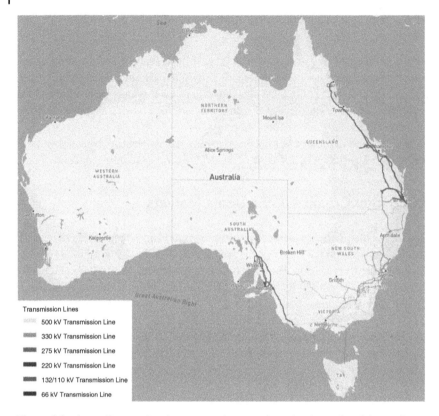

Figure 4.1 Australia map showing west and east regions that have electricity markets (note: only includes certain KV TX lines; and it does not include non-AEMO territories). Source: Modified from [21].

states, including Tasmania (Figure 4.1) [21]. The second market is called Western Electricity Market (WEM) and covers Western Australia. The third market is envisioned to run in the northern states. This existing market is run by the AEMO, who also acts as two separate SOs for these regions. Since AEMO runs both markets, the products and processes are very similar, and the differences primarily account for regional intricacies.

On the one hand, Australia sees an unprecedented rise in DERs, and on the other, the market currently is relatively old to account for these changes, e.g. the wholesale markets are regional (there is one price for one state/region vs. nodal). But both NEM and WEM systems have issues with system strength.

System strength is a complex, relatively new, and evolving topic. It is provided primarily by spinning machines like hydro, coal, gas generators, and synchronous condensers. The rotating mass in these machines helps to keep the voltage waveform strong simply because they are connected to the system and are spinning in synchronism. If there is a fault or a disturbance, these machines can provide short bursts of energy and stabilize the system. With DERs, renewables, and inverter-based loads (e.g. modern fridges and washing machines), the system is losing this rotating mass and hence becoming weaker. Thus, newer changes to wholesale market rules and newer technology are needed to maintain the system's integrity. The Australian market is thus waiting for an overhaul.

4.6.2 Germany

Germany is the largest economy in Europe. And it is also famous for its energy transition "Energiewende," which calls for the closing of all nuclear plants in Germany by 2022. That nuclear plant closing decision came after the Fukushima nuclear plant incident due to a tsunami in March 2011. High German retail energy prices have been making news lately partly due to German reliance on natural gas imports from other European countries and partly due to an increase in renewable energy such as solar.

Germany is committed to renewable energy, as evidenced by the data in Table 4.3 that shows Germany has more than 50% of its energy coming from renewable energy [22]. Germany wants to be net-zero by 2045 [23].

Part of the solar energy success in Germany is because of the number of policies. Currently, there are 10 policies in effect for solar energy in Germany, according to the International Energy Agency (IEA) policy database [24]. For example, the latest policy, "Package for the future: Expansion of renewable energies," eliminates

Table 4.3 Percent generation by fuel type in Germany in 2020.

Fuel type	Percentage generation (2020)
Wind – RE	27%
Brown coal	16.8%
Uranium	12.5%
Gas	12.1%
Solar – RE	10.5%
Biomass – RE	9.3%
Hard coal	7.3%
Hydropower – RE	3.7%
Others (including oil)	0.8%
Total	100%

Source: Based on [22].

the cap for solar PV, among other things like increasing the target for offshore wind from 15 to 20 GW by 2030 [25].

There are nine policies in effect for wind energy. Offshore wind is a major fuel type for Germany to achieve 65% of electricity consumption from renewable energy by 2030 [26]. Currently, German renewables account for 45% of consumption [27].

Since the Electricity Market Act was passed in 2016, the German electricity market design is in its second iteration. In the Electricity Market 2.0, Germany is focused on "security of supply" by guaranteeing free price formation, monitoring the security of supply, upholding balancing group commitments, and prolonging the grid reserve [28]. Additionally, Germany eliminated "avoided grid charges" starting in 2021 and introduced a capacity reserve to address potential capacity shortfalls due to reliance on renewable energy.

4.6.3 Vietnam

Vietnam has a staged market approach with many hydropower plants and a heavy reliance on state-owned (Vietnam Electricity EVN) plants. Vietnam aims to start a retail market for electricity in 2021 [29], but this might be delayed due to the pandemic [30].

Vietnam is focused on transitioning from coal (Figure 4.2 shows the growth in historical coal production) to solar energy procurement. The latter is facilitated by enabling corporate buyers to enter into a Direct Power Purchase Agreement (DPPA) pilot program with the utilities. This DPPA pilot mechanism enables direct customer participation from the spot market and lays the foundation for future retail electricity markets. The latest stakeholder meeting notes suggest the size of this DPPA pilot is 1000 MW in Vietnam [31].

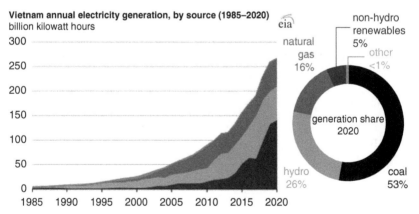

Figure 4.2 Vietnam annual electricity generation, by source (1985–2020). xxxvi / U.S. Energy Information Administration / Public Domain.

Vietnam's Wholesale Electricity Market Reform (VWEM) implementation plan looked like this:

- Pilot VWEM Step 1 (Paper Market) – 2016
- Pilot VWEM Step 2 – 2017–2018
- Full VWEM (MOIT Decision No: 8266) – 2019
- Long Term VWEM – 2019 onwards

From our experience with change in political leaders, we are not sure this electricity market reform is progressing as planned. But the key features of the mandatory cost-based gross pool are:

- All generators (≥30 MW) must directly participate in the market (as sellers) for selling their energy.
- Build, operate, transfer (BOT) power plants must participate in the VWEM directly or via a BOT trader (under EVN).
- Strategic multipurpose hydropower plants (SMHPs) must participate in the VWEM directly or via an SMHP trader (under EVN).
- All retailers (five power corporations) must buy energy from the market for meeting their customers' demand.

The plan was to have a gradual transition from a cost-based to a price-based pool model. Vietnam's top energy regulator, the Electricity Regulatory Authority of Vietnam (ERAV), focuses on the DPPA pilot program, which was anticipated to start in the third quarter of 2021. The USAID Vietnam Low Emission Energy Program (V-LEEP) is supporting ERAV in setting up this DPPA pilot [31].

4.6.4 Nepal (Potential)

Nepal is a landlocked country sandwiched between India and China. It has no known deposits of oil, gas, or coal. Therefore, biomass, oil products, and hydro are the main sources of energy. Of them, hydropower is primarily used for generating electricity. With a commercially exploitable hydrogeneration capacity of 42 GW, it was only generating ~0.8 GW.

Until the 1990s, all the hydropower generation was the responsibility of Nepal Electricity Authority (NEA), a Nepali government-owned company. The private sector could generate hydropower starting in 1992. In 2016, ~38% of the installed capacity was private sector generation.

NEA is also responsible for planning, developing, implementing, and operating transmission. It is also the de facto monopoly for electricity distribution. This monopoly had led to huge inefficiencies, losses, and lack of access to reliable, quality electricity for productive use.

Debundling and breaking the NEA monopoly is the key to reforming the electricity sector of Nepal and reducing inefficiencies, losses, and improving accessibility. Since the systematic restructuring, Nepal has magically addressed a lot of its electricity-related issues.

In 2021, Nepal entered into bilateral agreements to purchase from the Indian energy market. As there is a limit to the amount that can be imported, its benefits are limited too. However, should Nepal realize its full hydropotential, a surplus could be sold to the neighboring markets like the Indian energy market or packaged as transportable fuel like hydrogen. Given today's situation, the former seems more plausible.

Following this, the way for Nepal to leapfrog is to start moderate investments into minigrids, localized energy markets, and wholesale energy/flexibility markets. This restructuring will help Nepal build reliably and quickly from the bottom while waiting for trillions of investment dollars to build the electricity infrastructure and adjacent policies it needs. For further details, please refer to Chapters 3 and 10.

4.6.5 Africa (Potential)

Some parts of Africa already have power pools. Currently, EAPP and South Africa Power Pool are active in various parts of Africa. But as Table 4.4 indicates, more solar generation is planned for eastern Africa, and this trend of more renewables would continue in other parts of Africa, hence a market mechanism is needed. Markets provide a market signal to share renewable energy. Power pools have good potential for evolving as a market.

As energy market experts, the authors believe that Africa has much potential as an energy market. Energy markets grew out of power pools in the United States, such as the New England Power Pool (NEPOOL), formed in 1971, which turned over the day-to-day operation of the New England markets to the New England ISO in 2005 after setting up markets in 1999 [38]. Southwest Power Pool (SPP) is an RTO, which has its roots as a power pool all the way back to 1941 [39]. PJM has its roots as a power pool when it had established an administrator called "PJM Interconnection Association" back in 1993 [40].

Similarly, the power pools in Africa EAPP and SAPP can individually set up energy markets and form ISO-EA and ISO-SA in the future.

Exercise 4.8 As you know, Jay is an international energy consultant. He volunteered to serve as an expert on a trip to Namibia, southern Africa. The Bureau of Energy Resources at the US Department of State, the National Association of Regulatory Utility Commissioners (NARUC), led workshop has regulators from Namibia, Botswana, and Zambia in the Southern Africa Development Community (SADC) region. Regional Electricity Regulators Association of Southern Africa (RERA) is the partner for this workshop.

Table 4.4 Installed and expected generation in eastern Africa in megawatts. Numbers in the bracket are the capacities of the announced plants.

	Hydro	Geothe-rmal	Ther-mal	Natural Gas	Coal/Peat	Biomass	Wind	Solar	Methane	Electricity Access (%)
Kenya [32]	826	828	749			28	336 (450)	51 (250)		74%
Uganda [33, 34]	927		103			96		50 (40)		28%
Tanzania	574		89	893			200	(150)		33%
Rwanda	125		50		15			8 (30)	26	52%
South Sudan [35]			131					(100)		24%
Democratic Republic of Congo	2542		135					(250)		9%
Ethiopia [36, 37]	3817		7.3				426	(1000)		44%
Somalia			100					6		15%
Total wind announced							450			
Total solar announced								1820		
Total wind + solar announced							2270			

Prepare two 15-slide presentations, with appendices if needed but not exceeding five slides, on renewable energy technologies (first topic) and energy storage (second topic). In each topic identify (i) the technical requirements in connecting the technology to the grid, (ii) the limitations of distributed and utility-scale technology, and (iii) the amount of technology that can be successfully integrated with the existing grid and what grid infrastructure improvements are needed for a greater degree of technology penetration. The first two or three slides of your presentation should be for executives of the US Department of State, NARUC, and RERA.

4.7 Leaving Energy Markets Is a Decision Not to Be Taken Lightly

Leaving the energy market means there are exit fees for transmission-owning and nontransmission-owning members. In addition to exit fees, MISO transmission

Table 4.5 Reliability must run contract value at US RTOs.

RTO	RMR contracts value [41]
CAISO	$62 738 163 in 2018
NYISO	$0
SPP	$0
MISO	$16 247 431
PJM	$37 592 050
ISO-NE	$0

owners First Energy and Duke Energy Ohio found that they must pay transmission costs allocated to them even if they left MISO for PJM.

Additionally, grid operators pay reliability must run (RMR) units not to retire if they are critical for grid reliability. MISO, the grid operator in Indiana, paid almost $50 million for RMR contracts in 2014 and $16 million in 2018, shown in Table 4.5 [42]. All the ISO/RTO performance reports are posted on FERC's site [43]. Since most US grid operators are not for profit, this money is essentially paid by market participants.

Hence, leaving an energy market should be considered critical because it means paying exit fees plus transmission project costs in addition to losing revenue for a critical unit such as RMR.

4.7.1 First Energy and Duke Energy Ohio Left MISO

If we look back into the history of energy markets in the US, Midwest utilities leaving their grid operator for PJM was a major milestone. Both First Energy and Duke Energy Ohio (Duke Indiana stayed in MISO) left MISO in 2009 because they didn't think their capacity was valued much [44]. PJM's RPM offered a better capacity price for their units. Hence, they left MISO because membership in RTOs is voluntary. This First Energy and Duke Ohio example illustrates that utility membership is voluntary at RTOs.

4.7.2 New Jersey Threatened to Leave PJM's Capacity Market

Due to a minimum offer price rule (MOPR) pricing rule at PJM, FERC reacted in March 2016 to a complaint filed by the natural gas generators in the PJM market [45]. As a result of that complaint, which alleged that natural gas generators were at a disadvantage of not being dispatched in PJM's market due to the state policy interventions on renewables, FERC initiated these MOPR proceedings.

The main sticking point in this controversial MOPR proceeding is that FERC rejected a PJM proposal in 2018 because the Republican chairperson at FERC felt that state policies were suppressing energy prices.

The state of New Jersey was exploring the option to leave PJM's capacity market because the state did not feel PJM's capacity market valued state policies enough [46]. With a Democrat chairperson at FERC in January 2021, once President Biden took over at the White House, New Jersey did back out of this idea to leave PJM's capacity markets [47]. This New Jersey example shows states have the option of asking their utilities to leave RTOs, specifically capacity markets, because states have that ultimate authority to ensure resource adequacy.

4.8 States or Countries without RTOs

In addition to states with RTOs, i.e. organized markets, there are states without organized markets, such as Alabama, Georgia, South Carolina, Florida, and Tennessee [48]. In some countries there are states that do not have organized wholesale markets, for example the Northern Territory of Australia.

A state or a country joining an energy market is a complex process. It takes years to bring all the stakeholders together and at least a year or two of market trials before the actual energy market is up and running. So, naturally, the question is: "Who runs the show before energy markets?" The short answer is that the energy regulators do in some situations, and the energy companies do in others.

Exercise 4.9 Jay is now living in your state. What kind of wholesale electricity market does he have? Does he have a centralized operator? Or is the electricity traded without such an operator? Who is responsibility for maintaining reliability? Who is responsible for setting the policies? How old is the current market design? What reforms are being envisioned and why?

4.8.1 Who Maintains Reliability?

The energy regulator is generally responsible for the essential goal of keeping the lights on. This responsibility differs from one jurisdiction to another, from one country to another. For example, at a state level in the United States, the Public Service Commission (PSC) or the (PUC) are responsible for reliability in states that do not have organized markets.

And at a country level, it is the federal energy regulator or the ministry of energy responsible for regulating electric utilities. In the case of Australia, reliability is AEMO's responsibility, which is regulated by the Australian Energy Market Commission (AEMC).

A key point to note is that a regulator watches over the electric utilities to comply with the reliability standards. This regulator is most often responsible for all utilities such as telecommunications, water, and in some US states the insurance industry [49]. Hence, even without an energy market, there is always a regulator responsible for the electric system's reliability. And with an energy market, the state regulator still retains that authority for its state's capacity needs and transmission and distribution reliability.

4.8.2 How Are Capacity Needs Assessed?

In a state without a market, the state PUC has regulatory authority for looking over the utility's shoulders for future capacity needs. This framework is known as integrated resource planning (IRP). A state IRP process is sometimes contentious because IOUs responsible for complying with IRPs downplay renewable capacity if not for the pressure from environmental advocates.

The reader can research recent utility IRPs and learn how different state processes capture future capacity needs on the system [50, 51]. Not all states have a consistent IRP schedule. In some states, like Minnesota, utilities must submit IRPs every three to five years, whereas in others, like South Dakota, the utilities can make a regulatory filing whenever they choose to [52]. Hence, a regulator is always responsible for the electric system's capacity needs even without a capacity market.

4.8.3 How Are Transmission Needs Assessed?

With an energy market, states depend on the market operator to perform the regional transmission planning function, while their transmission utilities continue to perform local planning obligations. Before energy markets, transmission owners and utilities transacted energy over the telephone. If a utility needed capacity during an anticipated peak demand time, they called their neighbor and contracted for that capacity. The amount of support from their neighboring utility was limited by the transmission capacity available on the interconnecting facilities.

To quantify the support needed in megawatts, the Loss of Load Expectation (LOLE) model helped identify how much support was needed to maintain reliability. That interconnection support paved the way for transmission planning and designing higher ratings of transmission capacity for power transfers.

LOLE studies determine the amount of capacity benefit margin (CBM) needed to maintain the "one day in 10 years" reliability standard.

FERC defines CBM as:

> Capacity Benefit Margin or CBM means the amount of TTC preserved by the Transmission Provider for load-serving entities, whose loads are located on that Transmission Provider's system, to enable access by the

load-serving entities to generation from interconnected systems to meet generation reliability requirements, or such definition as contained in Commission-approved Reliability Standards [53].

That reliability standard says the power system is designed not to lose load of more than 1 day in 10 years, which translates into 0.1 day per year. One-tenth of a day is 2.4 hours. But not all hours in a day are created equal, and losing power for 2.4 hours in the 6 p.m. to 9 p.m. peak demand time is not the same as losing it during the 1 a.m. to 4 a.m. off-peak time. Hence that one day in 10 reliability standard is mostly for "planning reserves" on the system, not operating reserves.

Planning reserves are decided by looking ahead a year or more on the horizon, whereas an operating reserve is concerned about the next day or week, up to two weeks. Anything beyond two weeks is a "planning time horizon" [54]. So, without an energy market, the regulator responsible for the electric system's reliability oversaw transmission projects needs to maintain planning reserves.

4.8.4 Capacity Benefit Margin Is Relevant in States Without RTOs

A capacity benefit margin (CBM), the capacity on the transmission lines or many transmission facilities called "flowgates," is needed to reserve transmission capacity for energy purchases in states without RTOs, but it is not needed within a market because the RTO manages transmission access.

Locational capacity needs assessment replaced CBM. As Figure 4.3 indicates for the state of California, a local capacity needs map would indicate how much capacity must be within a particular area because of that area's contribution to peak load for reliability purposes [55].

For example, in California, the Los Angeles Basin area is defined as an area with local capacity needs because that area contributes 40% of the peak load to California's overall system peak. This locational capacity needs concept acknowledges that not all areas within a system contribute to reliability challenges in an energy system.

4.8.5 Relevant Concepts for Areas without a Market

Some concepts like available flowgate capacity (AFC), available transfer capacity (ATC), and energy accounting (EA) are still relevant in regions with grid operators. However, these concepts apply to imports and exports from the market to the nonmarket region [56], because in a market region all resources are deemed deliverable (there is enough transmission capacity to deliver the unit output), and hence they have network access. And market-to-market coordination of flowgates does take place [57].

Local capacity areas

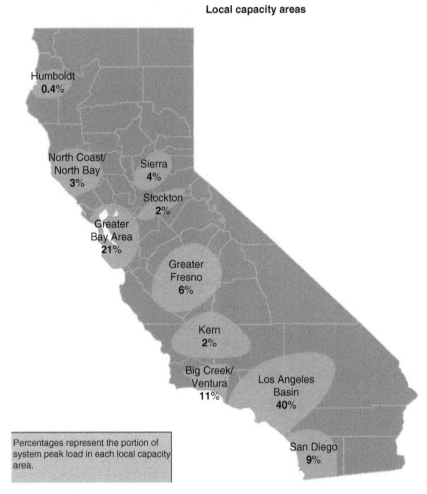

Figure 4.3 California ISO locational capacity needs map. Source: [51]/California Independent System Operator Corporation.

LOLE analysis feeding CBM and calculation of ATC explain how transmission planning concepts are relevant in capacity markets today. The reader should understand which concepts are still relevant in the market to nonmarket. New concepts like transmission reservations are added with energy markets.

4.8.6 Energy Markets and the Importance of Transmission Investments

The reader should know the difference between firm and nonfirm access on the transmission system, which will explain the capacity market and registration process at RTOs. When new or even existing resources have generator interconnection agreements to connect at a node on the power system, they get firm transmission access for the capability of their resource. But this transmission reservation is not firm all the time. Sometimes this transmission access is non-firm and is known as an energy resource interconnection service (ERIS). A firm transmission service would be a network resource interconnection service (NRIS).

NRIS assures firm transmission reservation on the system. When resources are studied for interconnection and network upgrades are identified for firm access and the interconnection customer does not want to pay for those upgrades, they can also have nonfirm ERIS as an option. This ERIS option allows the interconnection customer to take on the risk of nonfirm access. Additionally, under energy emergency alerts, nonfirm access resources are first curtailed.

How important is model building for detailed transmission studies? A case in point is the wording here:

> PJM uses the TARA commercial software package to create AFC cases for analyzing monthly, weekly, daily, and hourly transmission service requests. These cases serve as the base case models for the AFC calculation for a specific period. Solved base case models for monthly, weekly, daily, and hourly time frames are developed multiple times each day.

Exercise 4.10 Jay is mentoring engineering students in your state. Jay must give a lecture to engineering, policy, and legal students during a visit to the university campus.

Can you help Jay put together this hour-long lecture? Jay's presentation must include the engineering fundamentals so that both policy and legal students can understand. Jay must talk about how historical transmission planning has evolved today under energy markets in the region. Since policy and legal students are in the audience, Jay should take the opportunity to suggest possible policy changes needed in the future and call out what laws/regulations need to be changed. Jay should not lose this opportunity to influence future policymakers.

4.8.7 Transmission Planning and Capacity Markets Are Intertwined

In states or regions without organized markets, we have established that the state PUC is responsible for maintaining reliability. And they have a process for future capacity needs assessment, namely the IRP process.

That IRP process shows the future capacity needs of the system. But how to bring that capacity to meet load? Hence a parallel process for transmission planning exists to hold public utilities accountable for their future transmission plans to bring this capacity to load. The transmission interconnection capacity needed to meet load is quantified in LOLE studies, and concepts like CBM, AFC, and ATC apply. Those concepts are not relevant in a market construct, because the

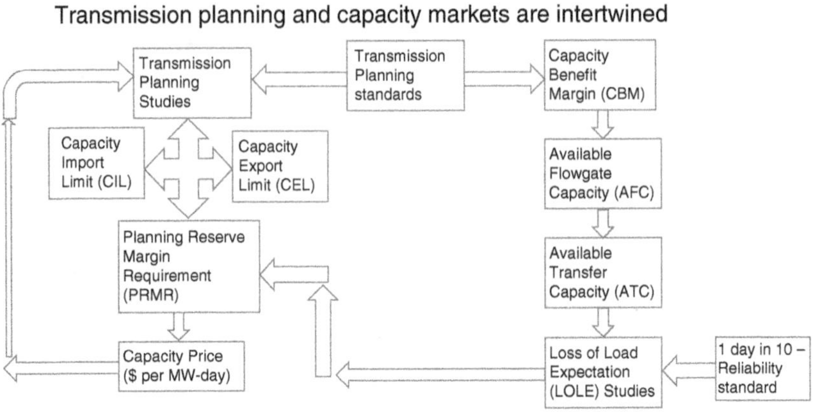

Transmission planning and capacity markets are intertwined

Figure 4.4 Transmission planning and capacity markets are intertwined.

market operator conducts studies to deliver those capacity resources to load, hence "deliverability."

In a market, it is the responsibility of the market operator to qualify the capacity resources and the demand resources. The market qualification process for new resources and existing resources and the planning studies to determine how much planning reserves are needed are all inputs into the capacity market, as shown in Figure 4.4.

Once the capacity market runs and shows market prices for future capacity, IPPs and other resource owners plan to meet those capacity obligations. As soon as one auction is run, the planning process for another auction starts with the planning studies. Hence these market processes are sequential.

To summarize, transmission planning is very complex and, at the same time, rewarding for planning engineers. Several software tools are needed to run these complex calculations. Some discussion and mention of these modeling tools can be found online [58, 59].

If planning transmission is complex, negotiating and convincing the load to pay for transmission is equally complex. Hence Section 4.10 provides a template for allocating costs for transmission based on their perceived need on the transmission system.

4.9 Cost Allocation of Transmission Projects

In states that have organized markets for regional transmission projects, the first step is the independent market operator board approving transmission projects. The board approves transmission projects because they are part of the regional planning package. After board approval, the utilities seek state approval. This state process is usually called a certificate of public convenience and need (CPCN) filing. In states not part of organized markets, the transmission utilities directly go to the state approval process.

A necessary step before utilities starts building transmission is cost allocation. What is cost allocation? Since we have already established that transmission projects are needed to maintain reliability, determining which utility pays for transmission is cost allocation.

Transmission projects can be classified by project needs, such as reliability, economic, or public policy. They can also be classified according to the voltage class, for example 345 kV and above. What voltage is considered transmission can be a tricky question in some markets. For example, 34.5 kV is considered transmission voltage in Iowa, whereas in most regions a voltage of 100 kV and above is a transmission system.

Based on market and data availability, cost allocation of transmission projects increases in complexity. As a simple matter, transmission utilities pay for transmission. The transmission owners recover these costs from transmission rates that they charge the transmission customers. As the voltage level increases, the benefits of transmission increase. Hence, who pays for transmission depends on who benefits the most, which is the general principle on which US transmission projects cost allocation works.

Because an independent transmission company (ITC) earns revenue when T is built, they may propose only T solutions for an identified need on the T system. Similarly, an investor-owned utility (IOU) revenue model only works if they invest in capital projects around generation, especially if they cannot own transmission. Hence an IOU would propose only generation projects.

Even with FERC Order 1000, there is a feeling that not enough transmission has been built in the US. Others believe lack of competition in lower voltage transmission is helping incumbent transmission owners, not independent transmission developers.

We will take the example of a large reliability project, e.g. building transmission lines. The need for a reliability project is to comply with NERC standards and address simple things like load growth, asset replacement due to age, natural disaster, or other acts of God.

4.9.1 Reliability Project Cost Calculation

Each transmission project allocation might be different, but we are taking MISO as an example here to provide a reference to understand how cost allocation works.

One hundred percent of the reliability project cost is assigned to the local transmission pricing zone (TPZ) at MISO. The entire MISO region is split into 12 cost allocation zones (CAZ), and in each CAZ, there are several local TPZs based on the number of transmission owners (TOs) [60].

In CAZ 1, there are seven TPZs, each corresponding to the TO:

- DPC – Dairyland Power Cooperative
- GRE – Great River Energy
- MDU – Montana Dakota Utilities
- MP – Minnesota Power
- NSP – Northern States Power
- OTP – Otter Tail Power
- SMP – Southern Minnesota Municipal Power.

Assigning 100% of the reliability project cost to the proposing transmission owner makes the most sense because it would benefit the incumbent TO. And the TO recovers these costs from transmission customers in their area.

Section 4.9.2 discusses the cost allocation of a large economic project. The need for an economic project is transmission congestion.

4.9.2 Economic Project Cost Calculation

An economic project cost allocation is complex compared to a reliability project cost allocation. Continuing our MISO example, economic projects that specifically address the transmission congestion and make the market more efficient by removing that congestion are called market efficiency projects (MEPs), and hence the reason behind having an energy market which makes all the difference in congestion data availability.

The specific criteria for an economic project to qualify as an MEP at MISO include these three conditions [61]:

- Only 230 kV and above voltage class projects.
- The total project cost has to be more than $5 million.
- The benefit to-cost ratio should be greater than 1.25.

One hundred percent of the MEP costs are distributed to the CAZ appropriate with the benefits. For example, Huntley-Wilmarth is an MEP in MISO Transmission Expansion Plan 2016 (MTEP16) shown in Figure 4.5. This Huntley-Wilmarth 345 kV T line connects Huntley substation in north Iowa close to the Minnesota border to Wilmarth substation in Minnesota.

The length of this 345 kV project is 50 miles, and two MISO TOs are involved. On the Iowa side, the TO is ITC-Midwest, and on the Minnesota side, the TO is Xcel Energy. After running production cost analysis, as shown in Table 4.6, MISO found that 65% of the MEP costs should be allocated to Iowa CAZ, 34.5% allocated to Minnesota CAZ, and the remaining 0.5% allocated to Illinois CAZ.

MEPs primary measure of economic efficiency is
adjusted production cost (APC) savings

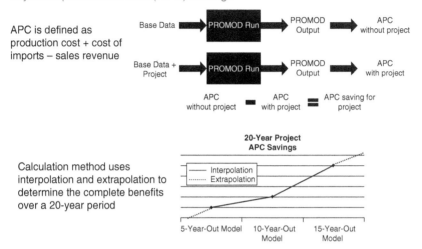

APC is defined as
production cost + cost of
imports – sales revenue

Calculation method uses
interpolation and extrapolation to
determine the complete benefits
over a 20-year period

Figure 4.5 Adjusted production cost (APC) savings calculation for a MEP. Source: [61]/Midcontinent Independent System Operator, Inc.

Table 4.6 Huntley-Wilmarth market efficiency project (MEP) cost allocation Illustration.

MISO CAZ	% of benefit share
1	34.5%
2	0%
3	65.0%
4	0.5%
5	0%
6	0%
7	0%

The production cost analysis for finding who benefits from a transmission project is an engineering analysis that involves model building, analysis, and reporting on the model results. In the specific case of MEP, a model run is first made without the project. And then, a model run is made with the project.

The difference in the adjusted production cost (APC) metric between the model run with the project and without the project is APC savings, i.e. how much this

MEP saves in APC. To confirm these savings, and to acknowledge that T projects are not built overnight, and an asset life of T line is usually 40 years, this model runs are conducted for plus 5, plus 10 and plus 15 years from the in-service date (ISD) of the MEP as shown in Figure 4.5 [61].

4.9.3 Adjusted Production Cost Calculation

An APC calculation is important to assess the benefits of economic and public policy projects. To calculate which CAZ benefits from an MEP, an APC calculation is run for each TO member of MISO. Realizing that transmission projects could benefit both MISO members (intrapool) and non-MISO members (interpool), the APC formula is:

An hourly company APC = production cost + intra pool cost + inter pool cost

The economic transactions between MISO members are intrapool because they are in the same market. And the economic transactions between MISO and non-MISO members are in the interpool because they reflect market to nonmarket transactions. Interested readers should refer to the MISO white paper on this topic [62].

4.9.4 Public Policy Project Cost Calculation

Owing to public policy goal, the need for a transmission project is usually around meeting renewable goals or moving to a carbon-free electric system.

If 100% of the reliability project cost is assigned to the local TPZ at MISO, and 100% of the MEP is distributed to the CAZ appropriate with the benefits, 100% of MVP costs are postage stamped to load. The term "postage stamp" in transmission cost allocation means that, since the T project's benefits are all over the region, the entire load not a specific zone, pays.

The specific criteria for a T project to qualify as a public policy project type called multi value project (MVP) at MISO includes these three conditions [60]:

- Only 100 kV and above voltage class projects.
- The total project cost has to be more than $20 million.
- Meet one of the three criteria below.
    - o Criterion One – Public policy goal: The T project should reliably and economically deliver energy to meet state, federal, or any other regulatory body's requirements.
    - o Criterion Two – Economic goal: The T project should have a benefit-to-cost ratio of greater than or equal to one.
    - o Criterion Three – Reliability goal: The T project should address "one transmission issue associated with a projected violation of a NERC or regional entity standard and at least one economic-based transmission issue that provides economic value across multiple pricing zones."

Table 4.7 MISO and CAISO transmission project benefit categories.

MISO MVP benefit metrics	CAISO transmission economic assessment methodology [63]
Congestion and fuel savings	Production benefits
Reduced operating reserves	Capacity benefits
Optimized wind turbine investment	Public policy benefit
Avoidance of future transmission investment	Renewable integration benefit
Reduction of planning reserve margin	Avoided cost of other projects
Reduced transmission line losses	

Table 4.8 MISO and SPP transmission project cost allocation.

Who pays?	MISO	SPP [64]
Entire region pays, i.e. postage stamped	100 kV and above if MVP	For all 300 kV and above projects
Only local TO/load pays	For all reliability projects, the local TO pays	For all 100 kV and below projects, local load pays
Split between entire region and zone	Voltage must be 100 kV and above for MVP, and 230 kV and MEP	Above 100 kV and below 300 kV – the SPP region pays 33% and local zone pays 67%

Even though a project type is MVP, the word "economic" figures prominently in each of the three criteria similar to an MEP. Hence the reader should understand the complexity of MVP project cost allocation arising from the economic benefits aspect.

To illustrate how similar and different MISO and California ISO transmission project cost allocation categories are, please consult Table 4.7.

As we can see from the MISO and California ISO table, production cost savings, public policy benefits, and renewable integration are similar. And MISO specifically called out T line losses, and California ISO called out capacity benefits.

Another example is comparing MISO and SPP cost allocation based on transmission project voltage type, shown in Table 4.8.

Exercise 4.11 Jay is now mentoring master's and PhD students in your state. Jay must give a lecture to engineering, policy, and legal students during a visit to the university campus. Several university professors and representatives from think-tanks are also in the audience. Some of the financial investors have been invited as well.

Can you help Jay prepare a presentation on transmission projects' cost allocation in your state/region? All the transmission project types should be considered. Any suggested future improvements should be written down, including opportunities to invest.

4.10 Visioning

This chapter introduces different energy market elements to equip the reader with the tools to navigate complex energy policies. We believe in the power of wholesale energy markets in reducing carbon. With moves afoot to reduce humanity's carbon footprint by interconnecting renewables, energy policymakers must look at market-based solutions in addition to regulatory mandates.

Based on the information in this chapter, the reader can work on emerging technologies such as energy storage and distributed energy in the US and internationally. The reader should now understand how to transition between market constructs with an established value proposition and nascent markets with foundational concepts such as CBM.

In the future, cost allocation for new and emerging technologies and their role in transmission will be critical. There are new transmission technologies on the market as well. With the cost allocation concepts from this chapter, the reader should navigate the complexity of interconnecting both utility-scale and distributed-scale renewable resources.

Exercise 4.12   As you know, Jay is an international energy consultant. He just got a project from the Bill & Melinda Gates Foundation. Based on what Jay has learned in this chapter, can you help Jay with this assignment?

Explore the wholesale capacity markets operating in India, China, Canada, New Zealand, Poland, and the UK. First, do these countries have a wholesale capacity market(s)? If yes, how does the capacity market in India (as an example) compare and contrast with PJM in the US? If not, what steps should the country take to prepare for a capacity market? Prepare a two- to three-page research brief outlining your findings for Jay.

Scaling Up Renewable Energy (SURE) II Experience Matrix Excerpt, cited here as an example of market requirements without calling it an energy market. SURE is one of the initiatives under USAID for advancing partner countries to develop the technical capabilities needed to add more renewable energy [65].

This SURE case study illustrates a pathway toward creating the conditions present for forming an energy market. If we read the subtasks under each significant task in this table, we notice that a market operator performs those functions. For example, the transmission planning department at a grid operator performs the development of demand forecasts and involves stakeholders to discuss those demand forecasts. For each case, select one or more subtask/illustrative activities applicable to your country/region and create a two-page summary report on what has been done in this regard.

Case 4.1   SURE – *Strategic Energy Planning*

Task 1: Strategic Energy Planning	Market Operator Function	Specifics
Illustrative activities		
Capacity-building/training of stakeholder groups involved in power sector planning at the national or subnational level	Y	
Development of tailored, country-specific planning processes (IRP or IRRP)	Y	Capacity Market
Development of demand forecasts	Y	
Development and validation of critical data for power sector planning, such as generation inventories, hourly demand data, critical meteorological information	Y	
Training of country counterparts on power sector modeling and joint analysis	Y	
Power system resilience assessments	Y	
Perform multicriteria analysis to identify renewable energy zones in countries	Y	
Development of country-specific renewable energy zones (REZs) process and legislation	Y	
RE resource mapping (wind, solar, etc.) or resource data improvements and validation	Y	

Case 4.2   SURE – *Competitive Procurement of Renewable Energy*

Task 2: Competitive Procurement of Renewable Energy	
Illustrative activities	
Provision of training and workshops on techniques and best practices for competitive procurement/reverse auctions in both utility-scale and distributed generation	Y
Support in the design and adoption of regulatory and legal frameworks such as public/private partnership laws, independent power producer laws	Y
Development of documentation including schedules, auctions rules, auction manuals, model PPAs, and other documentation based on best practices and tailored to country needs	Y

Development of procurement and transfer of auction bid model/platforms	Y
Development and review of prequalification documents, bidding framework, and guidelines	Y
Development of environmental and social impact assessment frameworks	Y
Technical reviews of legal documents, tariff design, PPAs, and other relevant documents such as guarantees and political risk insurance	Y

PPA: power purchase agreement.

Case 4.3   *SURE – Grid Integration of Renewable Energy*

Task 3: Grid Integration of Renewable Energy

Illustrative activities

Grid integration studies

Convene modeling working groups and technical advisory committees to perform grid integration studies	Y	
Data collection and validation for grid integration studies	Y	
Training, capacity building, and joint analysis on grid integration modeling for country counterparts, such as production cost modeling	Y	
Procurement of power system modeling software for use by in-country stakeholders	Y	
Data quality assessments and identification of data gaps	Y	
Data collection and data procurement to fill necessary gaps for performing in-depth grid-integration studies	Y	
Development of grid integration roadmaps based on study outputs	Y	

Grid integration pilots

Design, implementation, and assessment of grid integration pilots in partnership with other actors, leveraging 1 : 1 investment, in areas such as • ancillary services • demand response as flexibility	Y	Ancillary Services

Flexible system operations and markets

Analysis of balancing area cooperation opportunities, including impacts on reserve sharing, coordinated scheduling, and consolidated operations	Y	
Design of or improvement of energy and power markets, including ancillary services markets	Y	
Training of SOs inflexible operation of power system	Y	

Regulatory and policy improvements

Review and revision of regulations, codes, and interconnection standards to reflect best practices for improved VRE integration and system flexibility, both at utility-scale and distributed PV	Y

Wind and solar forecasting

Designing, collecting, and evaluating metrics for forecasting vendor trials	Y
Analysis of the impact of forecasting on power system reserves	Y
Creation of meteorological datasets	Y
Training of SOs on the integration of day ahead, hourly, and subhourly solar and wind forecasts into system operations and market dispatch	Y

Case 4.4 *Four Questions*

1) How is an EIM different from SEEM? What are the pros and cons of SEEM? In which international settings would SEEM work?
2) Can utilities participate a la carte in energy, capacity, and ancillary services markets? For example, can a utility participate in the MISO energy and PJM capacity markets? Explain your reasons.
3) How does the relationship between a PUC and RTO work? What happens in situations when the RTO does not administer a capacity market? What influence does RTO transmission planning have in front of the PUC commissioners? What about states/regions and countries without RTOs?
4) After the transmission planning process, who initiates the buildout of transmission? Can utilities object to the need for new transmission to keep their costs low?

References

1 PJM (2021). PJM value proposition. https://www.pjm.com/about-pjm/~/media/about-pjm/pjm-value-proposition.ashx (accessed 12 December 2021).

2 Morehouse, C. (2020). Duke, Dominion, Southern file SEEM proposal with state regulators, plan to file with FERC by end of year. https://www.utilitydive.com/news/duke-dominion-southern-file-seem-proposal-with-state-regulators-plan-to/592072 (accessed 12 December 2021).

3 Philippines Electricity Market Corporation. (2021). About PEMC. https://www.wesm.ph/about-us/about-pemc (accessed 12 December 2021).

4 Baker, A. (2022). Mexico's AMLO asks Congress to eliminate energy regulators, limit power sector competition. https://www.naturalgasintel.com/mexicos-amlo-asks-congress-to-eliminate-energy-regulators-limit-power-sector-competition/ (accessed 23 February 2022).

5 USAID (2020). Central Asia energy regulatory partnership. https://www.usaid.gov/central-asia-regional/fact-sheets/central-asia-energy-regulatory-partnership (accessed 12 December 2021).

6 Department of Market Monitoring (2021). CAISO 2020 annual report on market issues and performance, p. 49. http://www.caiso.com/Documents/2020-Annual-Report-on-Market-Issues-and-Performance.pdf (accessed 12 December 2021).

7 International Renewable Energy Agency (2020). Dynamic line rating, Table 2. https://www.irena.org/-/media/Files/IRENA/Agency/Publication/2020/Jul/IRENA_Dynamic_line_rating_2020.pdf (accessed 12 December 2021).

8 United States Department of Justice (2012). Justice Department Statement on Entergy Corp's Transmission System Commitments and Acquisition of KGen Power Corp's Plants in Arkansas and Mississippi. https://www.justice.gov/opa/pr/justice-department-statement-entergy-corp-s-transmission-system-commitments-and-acquisition (accessed 12 December 2021).

9 Western Energy Imbalance Market (2021). About. https://www.westerneim.com/Pages/About/default.aspx (accessed 12 December 2021).

10 National Renewable Energy Laboratory (2013). Examination of potential benefits of an energy imbalance market in the Western Interconnection. https://www.nrel.gov/docs/fy13osti/57115.pdf (accessed 12 December 2021).

11 California ISO (2020). ISO announces the Western EIM surpassed $1 billion in benefits. http://ISO-Announces-Western-EIM-Surpassed-1Billion-Benefits.pdf (accessed 12 December 2021).

12 Utility Dive (2020). The three key challenges to expanding the West's real-time energy market to day-ahead trading. https://www.utilitydive.com/news/the-3-key-challenges-to-expanding-the-wests-real-time-energy-market-to-day/578390 (accessed 12 December 2021).

13 Midcontinent Independent System Operator (2022). Value proposition. https://www.misoenergy.org/about/miso-strategy-and-value-proposition/miso-value-proposition (accessed 22 February 2022).

14 Southwest Power Pool (2021). 14 to 1: the value of trust. https://spp.org/documents/58916/14-to-1%20value%20of%20trust%2020190524%20web.pdf (accessed 12 December 2021).

15 WECC (2016). Compliance audit report. https://www.nerc.com/pa/comp/Audit%20Repots%20DL/2016_public_wecc_pac.pdf (accessed 12 December 2021).

16 ERCOT (2021). Board membership. http://www.ercot.com/about/governance/directors (accessed 12 December 2021).

17 Southwest Power Pool (2021). Members and market participants. https://www
.spp.org/about-us/members-market-participants (accessed 12 December 2021).

18 Missouri Joint Municipal Electric Utility Commission (2021). MJMEUC member map. https://cdn.ymaws.com/mpua.org/resource/resmgr/images/maps/
mjmeuc_map1.png (accessed 12 December 2021).

19 Missouri Public Utility Alliance (2021). RTP activity. https://mpua.org/page/rto
(accessed 12 December 2021).

20 Federal Energy Regulatory Commission (2021). Northern Indiana Public
Service Company v. Midcontinent Independent System Operator, Inc. and
PJM Interconnection, LLC.: FERC docket: no. EL13–88. https://www.ferc.gov/
sites/default/files/2020-05/E-4_83.pdf (accessed 12 December 2021).

21 US Energy Information Administration (2017). Australia. https://www.eia.gov/
international/analysis/country/AUS (accessed 12 December 2021).

22 Burger, B. (2021). Public net electricity generation in Germany 2020: share
from renewables exceeds 50%. https://www.ise.fraunhofer.de/en/press-media/
news/2020/public-net-electricity-generation-in-germany-2020-share-from-
renewables-exceeds-50-percent.html (accessed 12 December 2021).

23 Patel, S. (2021). Germany shifts net-zero target to 2045, sets tougher limits for
energy industry. https://www.powermag.com/germany-shifts-net-zero-target-to-
2045-sets-tougher-limits-for-energy-industry (accessed 12 December 2021).

24 International Energy Agency (2021). Policies database. https://www.iea.org/
policies?country=Germany&technology=Solar%20PV&status=In%20force
(accessed 12 December 2021).

25 International Energy Agency (2021). Package for the future: expansion of
renewable energies. https://www.iea.org/policies/13506-package-for-the-future-
expansion-of-renewable-energies?country=Germany&status=In%20force&
technology=Solar%20PV (accessed 12 December 2021).

26 International Energy Agency (2021). Offshore Wind Energy Act (Amendment):
Increase of Expansion Target. https://www.iea.org/policies/11508-offshore-
wind-energy-act-amendment-increase-of-expansion-target?country=Germany&
status=In%20force&technology=Wind (accessed 12 December 2021).

27 Umweltbundesamt (2021). Renewable energies in figures. https://www
.umweltbundesamt.de/en/topics/climate-energy/renewable-energies/renewable-
energies-in-figures (accessed 12 December 2021).

28 Federal Ministry for Economic Affairs and Climate Action (2021). Electricity
market 2.0. https://www.bmwi.de/Redaktion/EN/Artikel/Energy/electricity-
market-2-0.html (accessed 12 December 2021).

29 National Association of Regulatory Utility Commissioners (2021). International: Asia: Vietnam. https://www.naruc.org/international/where-we-work/
asia/vietnam (accessed 12 December 2021).

30 IHS Markit (2021). Vietnam's transition away from solar tenders is likely to start between 2025 and 2027. https://ihsmarkit.com/research-analysis/vietnams-transition-from-solar-tenders-likely-start-2025-2027.html (accessed 12 December 2021).

31 Viet Nam Energy Partnership Group (2021). Technical working group 3: energy sector reform, report of the sixth meeting. http://vepg.vn/wp-content/uploads/2021/05/VEPG_TWG3_6thMeeting-Report_fin.pdf (accessed 12 December 2021).

32 USAID (2021). Kenya: Power Africa fact sheet. https://www.usaid.gov/powerafrica/kenya (accessed 12 December 2021).

33 Monitor (2019). Solar power capacity to hit 50 mW. www.monitor.co.ug/News/National/Solar-power-capacity--hit-50MW/688334-4928366-j1li0rz/index.html (accessed 12 December 2021).

34 GL Africa Energy (2021). News of GL Africa Energy. https://www.glaenergy.com/news/GLAE-Celebrates-Seventh-Anniversary.html (accessed 12 December 2021).

35 Bellini, E. (2019). South Sudan to get 20 MW/35 MWh solar-plus-storage plant. https://www.pv-magazine.com/2019/12/05/20mw-35mwh-solarstorage-plant-to-be-built-in-sudan (accessed 12 December 2021).

36 World Bank Group (2021). Ethiopia. https://www.scalingsolar.org/active-engagements/ethiopia (accessed 12 December 2021).

37 Ethiopia Electric Power Company (2021). Power generation. www.eep.com.et/en/power-generation (accessed 12 December 2021).

38 New England Power Pool (2021). About NEPOOL. https://nepool.com/about-nepool (accessed 12 December 2021).

39 Southwest Power Pool (2021). About us. https://www.spp.org/about-us (accessed 12 December 2021).

40 PJM (2021). PJM history. https://www.pjm.com/about-pjm/who-we-are/pjm-history (accessed 12 December 2021).

41 Midcontinent Independent System Operator (2020). FERC docket no. AD19-16-000 report. https://cdn.misoenergy.org/Tab%20A%20-%20FERC%20Common%20Metric%20Final%20Spreadsheet%204840-2212-6800%20v.2489262.pdf (accessed 12 December 2021).

42 Midcontinent Independent System Operator (2020). FERC docket no. AD19-16-000 report. https://cdn.misoenergy.org/Tab%20A%20-%20FERC%20Common%20Metric%20Final%20Spreadsheet%204840-2212-6800%20v.2489262.pdf (accessed 12 December 2021).

43 Federal Energy Regulatory Commission (2021). Electric: performance metrics. https://www.ferc.gov/industries-data/electric/electric-power-markets/rtoiso-performance-metrics (accessed 12 December 2021).

44 FindLaw (2017). Midcontinent Independent System Operator Inc v. Duke Energy Ohio Inc Duke Energy Kentucky Inc FirstEnergy Service Company. https://caselaw.findlaw.com/us-6th-circuit/1865203.html (accessed 12 December 2021).

45 Advanced Energy Economy (2021). FERC expands minimum price rule, hurting advanced energy and consumers in nation's largest market. https://info .aee.net/hubfs/FERC-MOPR-Update-%20brief-Jan.2020%20.pdf (accessed 12 December 2021).

46 Governors' Wind & Solar Energy Coalition. (2018). NJ regulator threatens to pull out of PJM power market. https://governorswindenergycoalition.org/n-j-regulator-threatens-to-pull-out-of-pjm-power-market (accessed 12 December 2021).

47 NJ Spotlight News (2021). NJ changes course on opting out of regional power grid. https://www.njspotlight.com/2021/07/nj-changes-course-on-opting-out-of-regional-power-grid (accessed 12 December 2021).

48 Federal Energy Regulatory Commission (2021). Electric: RTOs and ISOs. https://www.ferc.gov/industries-data/electric/power-sales-and-markets/rtos-and-isos (accessed 12 December 2021).

49 State Corporation Commission (2021). Consumers. https://www.scc.virginia .gov/pages/Home (accessed 12 December 2021).

50 Arizona Corporation Commission (2021). Docket details. https://edocket.azcc .gov/search/docket-search/item-detail/22167 (accessed 12 December 2021).

51 Minnesota Public Utilities Commission (2021). eDockets: search documents. https://www.edockets.state.mn.us/EFiling/edockets/searchDocuments.do? method=showeDocketsSearch&showEdocket=true&userType=public (accessed 12 December 2021).

52 Michigan PSC (2021). IRP requirements for MISO states. https://www .michigan.gov/documents/mpsc/IRP_Requirements_for_MISO_States_554264_7 .pdf (accessed 12 December 2021).

53 Federal Energy Regulatory Commission (2021). References to capacity benefit margin in FERC orders. https://www.nerc.com/pa/Stand/Project%20200607 %20MODV0Revision%20DL/Comment_Form_MOD-004_Attachment1.pdf (accessed 12 December 2021).

54 Midcontinent Independent System Operator (2022). MISO transmission owners agreement: appendix B: section II, MISO planning staff and the planning advisory committee. https://cdn.misoenergy.org/MISO%20TOA%20(for %20posting)47071.pdf#doc16294 (accessed 22 February 2022).

55 California ISO (2020). 2019 annual report on market issues and performance. http://www.caiso.com/Documents/2019AnnualReportonMarketIssues andPerformance.pdf (accessed 12 December 2021).

56 California ISO (2021). Available transfer capability implementation document: MOD-001-1a. https://www.caiso.com/Documents/AvailableTransferCapabilityImplementationDocument.pdf (accessed 12 December 2021).

57 Midcontinent Independent System Operator (2021). MISO coordinated flowgates. http://www.oasis.oati.com/woa/docs/MISO/MISOdocs/MISO-COORDINATED_FLOWGATES.pdf (accessed 12 December 2021).

58 PowerGEM (2021). PAAC: PowerGEM AFC/ATC calculator. https://www.power-gem.com/PAAC.html (accessed 12 December 2021).

59 PowerWorld (2021). Available transfer capability (ATC). https://www.powerworld.com/products/simulator/add-ons-2/atc (accessed 12 December 2021).

60 Midcontinent Independent System Operator (2021). MISO regional and interregional cost allocation, p. 8. https://cdn.misoenergy.org/MISO%20Regional%20and%20Interregional%20Cost%20Allocation%20Reference90295.pdf (accessed 11 December 2021).

61 Midcontinent Independent System Operator (2021). MISO project types and cost allocation methodologies. https://cdn.misoenergy.org/20210222%20Transmission%20Cost%20Allocation%20Workshop%20Item%2002%20525025.pdf (accessed 12 December 2021).

62 Midcontinent Independent System Operator (2021). MISO adjusted production cost calculation white paper. https://cdn.misoenergy.org/MISO%20APC%20Calculation%20Methodology125160.pdf (accessed 12 December 2021).

63 California ISO (2017). Transmission economic assessment methodology (TEAM). http://www.caiso.com/Documents/TransmissionEconomicAssessmentMethodology-Nov2_2017.pdf (accessed 12 December 2021).

64 Southwest Power Pool (2021). SPP 101. https://spp.org/documents/31587/spp101%20-%20an%20introduction%20to%20spp%20-%20all%20slides%20print.pdf (accessed 12 December 2021).

65 USAID (2021). Scaling up renewable energy. https://www.usaid.gov/energy/scaling-renewables (accessed 31 January 2021).

5

How to Put Together a Regulatory Policy by Following a Process

5.1 Introduction

What should readers expect to learn from this chapter? Readers are not thinking about the regulatory policy when looking at an industry report. Even though a utility strategy might be right in their face, they may or may not realize its impact on the industry.

For example, suppose a utility announced 100% carbon-free goals by 2050. In that case, the reader may not realize the impact of that announcement on distributed renewables (most likely the utility plans utility-scale projects more than a distributed generation), or even the reader may not understand if this goal is achievable by 2050. It's not aggressive at all. If the goal is reaching carbon-free energy by 2030 or 2035 when you are in 2020 that would be an aggressive goal.

In this chapter, after discussing the different flavors of regulatory policy, readers are introduced to the five-step process for regulatory policy. These five steps are: (i) understanding the customers of the process, (ii) understanding the output from the regulatory process, (iii) understanding the regulatory process itself, (iv) understanding the inputs to the process, and (v) understanding whom the stakeholders are providing input to this regulatory process.

The regulatory policy is often seen as a second priority to business development if sales and business development are strong for an organization. Additionally, legislative affairs are confused with regulatory policy. In this chapter, the reader understands how regulatory policy is different from legislative affairs and how different regulatory strategies occur due to the business drivers.

Additionally, since people are central to executing any policy, the chapter discusses communicating with a specific policy goal. Not all communications are created equal: some report on key initiatives, whereas others have a specific compliance objective. Readers could get frustrated if they attended a compliance-driven meeting with a business development learning objective.

Modern Electricity Systems: Engineering, Operations, and Policy to address Human and Environmental Needs, First Edition. Vivek Bhandari, Rao Konidena and William Poppert.
© 2022 John Wiley & Sons Ltd. Published 2022 by John Wiley & Sons Ltd.
Companion website: www.wiley.com/go/bhandari/modernelectricitysystems

Hence this chapter gives the reader insights into researching these communications and the look of successful meetings.

5.2 What Is a Regulatory Policy?

A policy is a long-term vision or goal. A policy is written, displayed, and publicly announced.

A regulatory policy has an intended regulatory outcome. This regulatory outcome, in the United States (US), can be a rate increase at a state's public utilities commission (PUC) or a Federal Energy Regulatory Commission (FERC) Order 841 on electric storage resources. In the former, the investor-owned utility (IOU) would benefit, and in the latter, a storage developer would benefit.

For example, in the US, Xcel Energy announced achieving 100% carbon-free by 2050 as a policy goal [1]. This announcement is Xcel Energy's policy toward meeting investor and customer expectations. Xcel's regulatory strategy would vary state by state where it serves those customers to meet this corporate goal.

Xcel Energy, doing business as (d.b.a) Northern States Power (NSP) Minnesota, must work under the Minnesota PUC regulatory construct, and Xcel's utility in Colorado, Public Service Colorado (PSCo), must work under Colorado PUC rules. The plant workers' union structure and the bargaining power might constrain Xcel in Minnesota versus Colorado. Hence, Xcel's strategic approach toward coal plant retirements and the impact of job loss in Minnesota would be different from Colorado's. So, the strategy varies, but the policy remains the same for Xcel Energy.

Take another example: Google's commitment to operating on 24/7 carbon-free energy in all data centers and campuses worldwide by 2030 [2].

Since Google is not an IOU, the regulatory constraints under which Google operates are much different from those of Xcel Energy. But, if Google has data center operations and renewable energy projects under power purchase agreement (PPA) at three different regional transmission organizations (RTOs), it must have a regulatory strategy for federal energy because RTOs work under FERC direction. And how Google approaches RTOs would vary based on the intended outcome to achieve that 100% carbon-free goal for their operations. Hence the strategy varies, but the policy remains the same for Google.

Moving from US examples to Uganda, the regulatory policy for an economy like Uganda's must be aggressive compared to that of the US.

In a recent US Energy Agency (USEA) project, from the Uganda Energy Mix Diversification Strategy point of view [3], Uganda has an integrated resource plan (IRP) that brings together multiple governmental agencies and works with the technologies that are available today versus technologies that might be economical in the future like nuclear power. Uganda Electricity Generation Company,

Ltd.'s (UEGCL) regulatory policy is worth mentioning here because Uganda has an installed capacity of 1300 MW. But Uganda's National Development Plan goal calls for 40 000 MW of installed capacity by 2040 [3]. Aspiring to go from 1300 MW installed capacity to 40 000 MW in 20 years is an aggressive goal.

Shifting gears from Uganda to Europe, Europe had a regulatory policy in a grant for utilities to test new technologies. It was called the Seventh Framework Programme for Research and Technological Development (FP7) for the 2007–2013 period [4], and later Horizon 2020 EU for 2014–2020 [5]. These European grants (FP7 €50 billion, Horizon 2020, €80 billion) were funding between 50% (for companies) and 100% (for universities) of the participants' costs to support taking risks in testing unproven technologies. This policy is successful as it decreases the cost of testing/improving technologies only if deployed afterward. However, this is not always the case. European countries have similar programs to support research.

Another form of support in Europe is provided by regulators (national regulatory authorities) when they allow the transmission system operator (TSO) to depreciate the whole amount invested (the 100%) even if the TSO only spent 50%. Also, regulators incentivize the TSO by allowing an additional benefit for its shareholders based on the grant being successfully collected.

And yet another form of support from regulators is to allow their TSO to test technologies even if not permitted by the current regulation, e.g. Terna (Italy) has been allowed to deploy batteries to test the ability to provide flexibility to the grid even though TSOs aren't authorized to own and control batteries (except for backup power at substations) [6].

Any entity seeking to connect its generators to the grid must put together a regulatory policy. Here is another example of an international mining giant. Rio Tinto, a mining giant, wants to achieve net-zero emissions by 2050 and reduce emissions intensity by 30% by 2030 [7]. It operates in many different countries, including Australia and the Americas. Among other things, Rio installs renewables and/or buys power from the renewables. Rio Tinto, depending on its mine site, must have a regulatory strategy in place. This example is how the regulatory strategy differs from the policy.

Exercise 5.1 As you are aware, Jay is now working internationally. He is now in your country. He needs to understand and look for regulatory strategies for your utility. Based on the publicly available information, can you help summarize for him the regulatory strategies of your local utilities?

5.2.1 What Does a Successful Regulatory Policy Look Like?

Success for one person (or group) might be a challenge for another. Keeping that in mind, let us look at some examples related to renewables. In the US, the stock market and state regulators reacting favorably to a utility's carbon-free goal announcement in a progressive state (states that favor renewables over fossil fuels) is a success for an IOU that operates in multiple states.

In other parts of the world, like in Saudi Arabia, the government generally manages coal, oil, and gas. It is their staple income. Therefore, a pro-renewable policy would be deemed successful if the message is – renewables save internal fuel consumption (so that it can be exported or saved) or renewables are showcase projects to enhance the country's image globally.

Yet, in other countries, e.g. Morocco or Nepal, a successful regulatory policy would be to frame renewables to reduce dependency on foreign fossil fuels, whereas, in other cases, success could be defined as the side that won the argument with the regulator to restrict residential solar deployment, which is entirely plausible if the majority party in that state is not in favor of renewable energy (like so-called nonprogressive US states or some parts of the Middle East).

Going back to the Xcel Energy example from the US, Xcel, in October 2020, received approval from the Department of Energy (DOE) for a $10.5 million grant for a pilot hydrogen project at one of its nuclear plants in Minnesota [8]. For Xcel, this is what a successful regulatory policy looks like because Xcel only pays $2 million for testing a new technology, which is a good deal for Xcel and its customers and investors.

The state regulators are happy because ratepayers do not have to pay for this pilot project. The investors, the hydrogen and related equipment vendors, and many environmentalists are happy because Xcel is leading the industry with new technology. So, Xcel is a good investment. This example is a win–win situation for many of the stakeholder groups.

Exercise 5.2   Are your local utility (utilities) regulatory policies pro-renewables or against renewables or agnostic to them? What are the drivers and barriers?

5.2.2   Influence of Energy Policy and Regulatory Actions in Fragile Economies

Some of the same policies and programs detailed in this section apply to less developed and more fragile world economies but are often severely limited by the availability of funding, traditions, and the legacy administrative infrastructure. This limitation is why approximately two billion people are not served with electricity globally. See Chapter 10, which discusses the issue of global energy poverty in much more detail.

An example of policy driven by international agencies is worth examining. Several international agencies can help advance access to energy, and more specifically clean energy, in fragile economies. International agencies like the World Bank, various United Nations departments, and nongovernmental organizations (NGOs) with a mission of technological and humanitarian outreach can be instrumental in furthering the progress of energy policies.

The United Nations High Commission for Refugees (UNHCR) is a specific case of energy policy influence. Displaced populations or refugees are some of the most underserved people in the world. They are very limited in their infrastructure and amenities, such as clean water and energy. Displaced populations often are foreign nationals that have fled natural or manmade disaster situations. A host country may accept them within its borders, but they lack the resources or citizenship to receive the basic amenities available to the home country's citizens.

For this reason, an already impoverished group has an even greater need. Less expensive and environmentally disadvantaged energy sources, such as diesel generators, provide limited electrical power for refugee camps and settlements' administrative buildings. Little or nothing is provided for people's homes or for street lighting safety.

UNHCR has recently begun to initiate global policy programs to help bring financial resources, local regulations, technological resources, and administrative or structural resources to refugee camps. A task force called the Clean Energy Challenge Action Group is currently in place and will likely be rolling out these types of needed programs for several years to come.

Exercise 5.3   As you know, Jay is an international energy consultant. He has a consulting assignment to serve as a grid expert trainer for developing and conducting a five-day training course for staff in XM, Colombia's system operator; UPME, the energy planning agency; the Ministry of Mines and Energy; CREG, the regulatory board; and other electricity-sector organizations. This training takes place in Bogota, Colombia.

Prepare a 20-slide presentation with appendix but not exceeding five slides on an overview of distributed energy resources (DER). This presentation should cover the following topics: DER in front of and behind the meter, what resources are included in DER vs. DG (distributed generation), policy and regulatory issues, and DER rate structures.

5.3   Different Types of Regulatory Policy in the Electric Utility Industry

Readers should note that the policy varies according to the business need, whether called a regulatory policy or regulatory affairs. In this section, we explore the

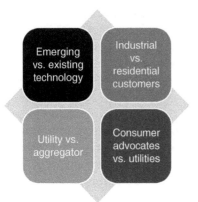

Figure 5.1 Different flavors of a utility's regulatory policy.

potential combinations observed in the utility industry starting with the utilities shown in Figure 5.1.

5.3.1 Utility's Regulatory Policy for Emerging vs. Existing Technology

Demand-side technologies such as light-emitting diode (LED) lighting and air conditioner (A/C) saver switches are well known as demand side management (DSM) technologies in the utility industry. Compared to new programs, existing DSM program savings are well documented. The regulatory strategy for a utility in a state regulatory proceeding for these known and existing technologies is different from, for instance, a new and emerging technology such as a microgrid. The utility may ask for pilot project approval on a microgrid rather than a request to approve a ratepayer-funded DSM program.

For example, let us compare the strategy for demand response against electric vehicle (EV) infrastructure technology. Does the IOU contract a third-party provider of electric vehicle supply equipment (EVSE), or does the IOU own the EVSE infrastructure? That answer leads to the utility strategy at the state PUC. This question is a $400 million question in California, which has the highest EV adoption in the US [9].

The state of California has announced a policy goal of ending all gasoline-powered cars by 2035. As a result, the state is pushing for more EV charging infrastructure, commonly referred to as EVSE. Each of the top three IOUs in California is spending millions of dollars on EVSE to achieve that state transportation goal.

Exercise 5.4 How is the current penetration of EVs in your country? What does the future look like in your region? You can use the *Global EV Outlook 2020* report for reference [10].

5.3.2 Utility's Regulatory Policy for Industrial vs. Residential Customers

The interplay between utility and the industrial customer becomes important if utility revenue is mostly from the industrial customer, especially if the industrial customer has major energy costs. For example, Minnesota Power in the US derives >60% of its revenue from industrial customers [11]. So, Minnesota Power's regulatory policy for industrial customers differs from Xcel Energy in Minnesota, which serves mostly residential customers.

If Minnesota Power is dealing with industrial customers and their associations, Xcel Energy is dealing with residential consumer advocates such as Citizens Utility Board (CUB) Minnesota.

5.3.3 Regulatory Policy for a Utility vs. Aggregator

A large IOU typically has multiple states, regional, provincial, or national jurisdictions; thus, its utility regulatory strategy in a state like Colorado (without a wholesale energy market) is different from that of Minnesota (with the energy market operator MISO). And this IOU policy is different from a community choice aggregator (CCA), which focuses more on their limited service area within a state like California.

But that does not mean the IOU and CCA are competitors, because they can be partners in addressing a community resiliency need. As an illustrative example, Pacific Gas and Electric (PG&E) is the IOU, and Redwood Coast Energy Authority (RCEA) is the CCA. They have partnered with Redwood Coast Airport and Humboldt county in California to keep the airport open even during natural disasters [12].

So, it is not always true that utility and aggregator business interests are the opposite, even though their regulatory strategies are different. How are they different? The IOU and CCA regulatory strategies differ in their scope. IOUs are focused on states and regions; CCAs are focused on the communities they serve.

5.3.4 Regulatory Policy for Consumer Advocates vs. Utilities

Readers should seek and understand different sides of an argument in the industry. The reader does not have to go to multiple places to check out these diverse perspectives. If they look at a utility Integrated Resources Planning (IRP) they can see that the consumer advocates' policy is mostly at odds over the utilities.

Some utilities say it was not economical for distributed-scale solar and storage even in 2020, and they want to wait for 5–10 years when the technology is proven. But the consumer advocates want the utility to include more distributed-scale solar and storage now because they see it is included in other utility resource plans.

5.3.5 Regulatory Policy in Regions with and without Organized Markets

Absent market barriers, it is equally important to realize that markets provide additional revenue streams for any technology in a regulatory context. As a result, a utility's energy efficiency program can bid into the wholesale capacity market administered by an independent system operator (ISO) like PJM but not in a state like Georgia that does not have an organized wholesale energy and capacity market. There are several parts of the world where such markets exist and places where they don't exist.

For example, in Australia, two wholesale electricity markets are operated by the same system operator but under slightly different mandates. Further, in the third region in Australia, the Northern Territory, such markets do not exist. Hence, there are three flavors of the regulatory policy process.

5.3.6 Regulatory Policy for an ITC vs. TO

At places where wholesale energy markets exist, the energy-efficiency (EE) program could find revenue by bidding into the capacity market. The regulatory strategy for an independent transmission company (ITC) is unlike an incumbent TO in a state that allows utility participation in organized markets.

Some US states like Minnesota have the right of first refusal (ROFR) for their incumbent TOs. The transmission owner can choose to reject the proposal from an ITC like LS Power. This Minnesota ROFR law is why LS Power has petitioned the Supreme Court of the United States (SCOTUS) to overturn the law [13]. But SCOTUS has refused to hear the case.

5.3.7 Regulatory Policy for Supply-side such as Nuclear Fuel

According to the US Energy Information Administration (EIA), as of 1 May 2020, 57 nuclear plants are operating in the US [14]. Not all of them are economical due to renewable energy penetration in the market. Recent nuclear plant retirement announcements and bail-out packages by states such as Ohio show that the utility (First Energy) regulatory strategy includes a specific fuel choice (nuclear energy).

There are two pathways for a utility that owns nuclear plants in states that have organized markets. (i) Does the utility take a proactive approach and retire uneconomic nuclear plants in meeting 100% carbon-free goals by a certain date in the future (Exelon in Chicago, Illinois, seems to be pursuing this strategy) or (ii) is the utility going to argue that nuclear energy is without carbon emissions and hence the PUC should provide incentives for keeping the nuclear plant running, including ratepayer funds for storing spent nuclear fuel, even if it is uneconomic given the market prices for renewable energy such as solar and wind? Xcel Energy in Minnesota may be following this path.

In a state like Georgia that does not have an organized market, there is not much regulatory policy needed for Georgia Power, the IOU in Georgia, if the state utility commission favors continuing the support for nuclear power.

5.3.8 Regulatory Policy for Demand-side such as Demand Response

As mentioned earlier in Section 5.3.1, a regulatory strategy around demand-side options such as demand response is an ongoing utility journey.

Demand response could utilize concepts like demand charges, time of use charges, smart use of building management systems, microgrid management systems, virtual power plants, and EV charging systems to control the onsite loads, including the storage systems. There might be new technologies added to the mix. Please refer to Chapter 7 on how the power system works for further details on some example schemes.

An example of a failed DSM program is the emergency generator response program operated by utilities in the US 20+ years ago. The utilities correctly ascertained that many emergency generators were in commercial and industrial buildings. These provide fire protection, elevators, and emergency lighting backup.

Formal programs were developed by the utilities where a significant rate break was given to the customer in return for utilizing their generator during peak periods upon a signal or phone call from the utility. But as consciousness about air quality and global warming grew, it was discovered that the electrical peaks to be shed were on warm or even air inversion days in many areas. Further, these customers were often clustered in downtown or industrial areas. A significant number of diesel generators under these conditions caused a measurable drop in outdoor air quality. Ultimately, this was the downfall of the programs.

5.3.9 A Regulatory Policy with Compliance Purpose

There are governmental, quasigovernmental, or independent bodies in countries with the mission to assure an effective and efficient reduction of risks to the reliability and security of the electrical grid. North American Electric Reliability Corporation (NERC), in the US, is such a body. The reader should note and distinguish utility policy around compliance with NERC and NERC regional reliability organizations. This policy can be as simple as the utility advocates to relax compliance obligations when testing new technology in the pilot phase.

5.3.10 Regulatory Policy for Technology Provider vs. National Laboratory

There is another perspective that the reader should note while observing government-funded national laboratories' role. By design, a national lab might have a testing site that is handy for a technology provider. By providing this test site, the national lab might be ensuring its relevance in the industry. But the national lab does not have the same pressure as the technology provider to succeed in the marketplace.

5.3.11 Regulatory Policy Drives Partnerships

Not everything is about an organization taking on the entire task. An ITC might find a partner as part of its regulatory policy to introduce itself to the marketplace. An IOU might partner with a CCA on a specific community resiliency project, as seen with PG&E (Humboldt County, California's utility) and Redwood Coast Energy Authority (RCEA), which administers Humboldt County's Community Choice Energy program. A technology provider might partner with a consumer advocate for expanding the solar market in a state. So, individual business interests drive potential partnership opportunities.

5.3.12 How Do We Know the Regulatory Policy Is Working?

An organization needs to take a step back and assess if the regulatory policy set at an offsite is still working before going into another annual offsite.

In the case of Xcel and other utilities, renewable portfolio standards (RPS) (percentage mandates) have successfully provided a substantial penetration of wind and other renewable energies in the grid from the 2000s to the present. Any additional doubts gave way to what proved to be a win–win situation for the customers, the utilities, and the environment. Production tax credits and accelerated depreciation aided this.

In Australia, the target to meet 33 GWh by renewables by 2020 was met in September 2019 [15]. This was a win–win situation for utilities, retailers, and consumers. Now that electricity produced by renewables is increasing, the government focuses on reliability and price reduction. Being a highly regulated market, the actors in this market are now complying to support this.

Finally, we know FERC Order 841, on opening up markets for energy storage, is working, by looking at Figure 5.2 that shows more than 40 000 MW of storage capacity in the PJM queue as of September 2021.

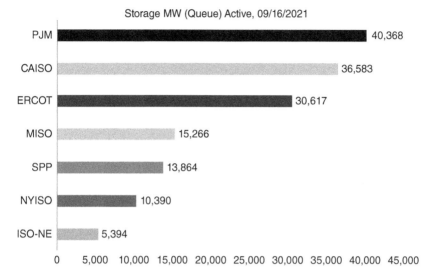

Figure 5.2 Storage megawatt capacity in US RTO/ISO interconnection queue.

Exercise 5.5 As you know, Jay is an international energy consultant. He has a consulting assignment to serve as a grid expert trainer for developing and conducting a five-day training course for staff in XM, Colombia's System Operator; UPME, the energy planning agency; the Ministry of Mines and Energy; CREG, the regulatory board, and other electricity-sector organizations. This training takes place in Bogota, Colombia.

Based on his experience in the North American market prepare a 25-slide presentation, for Day 2, with some backup or appendix slides.

Title - Integrating DER into the electricity markets.

Topics to be covered:

- DERs and the XM market status
- International experience
    - DER management systems
    - Business models
    - Interaction requirements between ISO, DSO, and DER aggregators at high penetration levels of DER
    - DER measurement, evaluation, and verification
    - DER supply and demand in the electricity markets
    - Examples of current DR programs in North American wholesale markets
    - DER integration and DER providers

- o Outlook of DER integration (in front and behind the meter) and the future role of ISOs.
- Integrating DER into XMs market – lessons for XM.

5.4 There Are Five Steps in Any Regulatory Policy Process

Readers are familiar with process definitions by now in their academic and possibly work experience. As in a study report where it makes sense to start with the conclusions and summary before providing a background, strategy discussions should start with the customers in a regulatory process. Who are the customers for the principal organization putting together this regulatory policy? As we have already mentioned, customers for a utility are different from those for a consumer advocacy group.

This process has five steps (modeled after a process-based tool called suppliers inputs process outputs customers, or SIPOC[1]), shown in Figure 5.3 and described in this section. However, we do not intend to suggest that the process is linear, because it is not. For example, our experience suggests the outcomes are predetermined, and then the study of the customer is conducted. At other times, the regulatory process is fully understood, and then the outcomes are predetermined, then the other steps are performed.

5.4.1 First, Understand the Customers of this Process

An engineer needs to understand the customers because senior executives at an organization are the internal customers, whereas the consumers and regulators, i.e. the stakeholders, are the external customers. As a reader, this internal and external customer concept is hard to conceptualize. But this concept is essential if one is to understand the people involved in this regulatory process. A fresh-from-college utility engineer or an analyst is not sitting in front of a PUC commissioner at a rate case. Their supervisor, the director of regulatory policy/strategy, is in front of the regulator.

This focus on the customer is important because they drive the value creation of the utility business model. Without customers, there is no revenue for the utility. For example, a transmission-dependent utility (TDU) is focused on the state PUC for approval of a generation plant, whereas an ITC is focused on the RTO

---

1 SIPOC stands for suppliers, inputs, process, outputs and customers. This is a process based tool that shows how most major processes have customers, and suppliers who provide inputs and depend on outputs of the process.

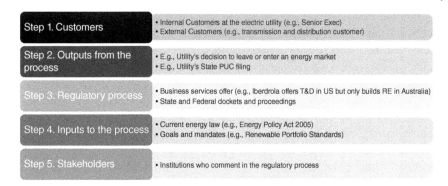

Figure 5.3 Five-step process to understanding regulatory policy.

for approval of a transmission project. The ITC wants to seek revenue to satisfy its transmission customer, i.e. the TDU, whereas the TDU seeks to create value for its distribution customer. In both instances, utility and ITC are looking for a favorable rate of return from state and federal regulators. The external customers are the state commissioners for the TDU and the RTO board for the ITC.

This focus on customers earlier in the process enables building intentional relationships. In some instances, it is prohibited to have communications after an RFP is out or when a case is open before the utility commission unless the communication happens in the open docket proceeding.

In the earlier example of a utility versus a transmission company, the utility regulatory policy is spent developing relationships with the utility commission staffers because the utility commission is the regulator approving the utility resource plans. In contrast, the transmission company regulatory policy is attending the RTO stakeholder committees and developing relationships with the RTO staffers because the RTO is the FERC jurisdictional entity approving the transmission company plans.

5.4.2 Second, Understand the Output from the Regulatory Process

In the second step of this process, it is important to understand the expected deliverables from this strategic process. Again, if the audience is internal executives, the strategic document might contain confidential business details, which won't be shared in the external-facing strategy document or job descriptions.

For instance, an internal "visioning exercise" enables an honest discussion of the detractors, whose voices are heard the most, and the key constituents. This internal visioning exercise might serve its purpose of bringing together executives and line managers toward sharing the common vision and retaining utility members for a grid operator.

But that internal policy document put together due to multiple discussions is only an external one-line statement regarding the grid operator's value proposition. Hence readers need to understand that there is more to the external-facing policy documents and announcements. They must dig deeper to understand policy or strategy.

One way is researching what the grid operator member executives are communicating in the investor presentations. Another way is looking at the grid operator's board meeting materials. Looking at what their market monitor suggests in terms of market improvements is another way to dig deeper. And finally, key external-facing stakeholder committee meetings contain a vast amount of information on the future direction of the grid operator.

This policy document lists out action items and milestones to be met during the year. In most cases, these milestones become part of the goals document with compensation tied to achieving those goals for the organization and the individual responsible. These are commonly referred to as short-term incentives, or STIs for short.

In any for-profit organization, executive bonuses are often tied to profitability. Therefore, successful regulatory policies that result in more profit would contribute to one's annual rewards and bonuses.

It is also likely that a resource gap is identified due to these internal policies or strategy sessions. Or a decision is made to "internalize" the external legal consulting costs associated with legal support.

It is important to realize that some things are outside your control at this juncture of the strategic process. If a technology provider is geared up to enter the US market from Germany, for instance, and finds out a new regulatory mandate on distributed energy resources makes their product more relevant now to what they thought about six months ago that changes the whole investment and marketing strategy of the provider in their discussions with the utility customers.

The reverse could be true. The technology provider might have to scale down their plans to enter the US if an announcement from the White House restricts foreign providers supplying electric utility equipment [16] or levies tariffs on all foreign goods and services, including solar panels [17].

5.4.3 Third, Understand the Regulatory Process

The third step in the five-step process is understanding the regulatory process because individual state and country processes are different. Experience in the state utility regulatory process of New Mexico is not the same as the experience in Florida.

Additionally, to put together regulations for newer demand-side technologies, one needs to understand the current rules accommodating demand response programs. The experience with DSM programs prepares institutions for DERs because the US FERC definition for DERs says any resource connected to the distribution grid behind a customer meter is a DER.

If we step back and think about some of the drivers behind the state processes, there is a federal driver behind rules governing interstate transmission and sale of energy. Hence it makes sense to pay close attention to the federal energy regulatory process.

Another reason is that a big multinational company's regulatory policies change according to the country. For example, EDF would change tactics in France versus Germany. Another example is that companies operating in "flexibility markets" are more interested in energy markets in Australia and Europe but not in the US because flexibility is not yet compensated.

Another example is that Iberdrola does T&D in the US and Europe but only builds renewables in Australia. The strategies are tied to forecasted profitability in the business segments. But these profits are ultimately tied to what is happening concerning regulation in these countries.

Like the US, different Central American countries are in various stages of DER adoption. Pressure from the poultry industry in Panama, the desire to maintain the green canopy in Costa Rica, public lighting under the local government control in El Salvador (not under the traditional distribution utilities), workshops with significant users in Guatemala, large industrials consuming 70% of Mexico's energy, and qualified consumers installing self-generation in Honduras are all drivers behind the interest in DER regulatory policy for regulators in Central America.

5.4.4 Fourth, Understand the Inputs to the Process

The fourth step in the regulatory process is identifying all the different inputs into the process. A regulatory policy for an ISO might include inputs from the independent market monitor (IMM) and the board of directors, whereas the regulatory policy for an electric cooperative would include inputs from their member cooperative elected officials. And this stakeholder input influences what the ISO and the electric cooperative are planning to do in the next couple of years.

In the case of a state PUC, when the people elect a governor and appoint the PUC chair who drives the agenda, one could argue the people had a voice in the PUC appointment. But the outcome of any regulatory proceeding is not known in advance, just by appointing a utility commissioner. No one knows how they vote on a specific utility docket proceeding when the time comes.

The inputs to the regulatory process are right in front of our eyes. For instance, many utility commission filings are public, including the testimonies provided in support of a utility policy. Readers don't look at these documents, because they are unaware of their informational value in public utility commission dockets. In these cases, the docket number or the case number becomes important to focus on a specific topic such as storage as a transmission only asset (SATOA), FERC docket no. ER20-588.

Searching for that specific docket number on the FERC eLibrary pulls up the entire list of documents filed by MISO, MISO TOs, and other interested parties on the topic of MISO's SATOA FERC filing. This process helps a reader if they want to research the role of energy storage as a transmission-only asset. The average industry follower does not know where to look on any given topic, especially on new and emerging regulatory topics.

This docket number is critical for the state regulatory filings as well. For example, if we are searching for docket no. E002/RP-19-368 on the Minnesota PUC eDocket website [18], which the Department of Commerce manages, we will find Xcel Energy's IRP. In Arizona, the Arizona Corporate Commission (ACC) functions like the Minnesota PUC. So, if the reader wants to pull up the Arizona utilities IRP filings, they must search for docket no. E-00000V-19-0034. That ACC link shows all the parties who filed comments on the Arizona utilities IRP filings [19].

Someone interested in researching how Xcel Energy in Minnesota is preparing for its commitment to a 100% carbon-free goal can compare that approach to Arizona IOUs such as Arizona Public Service (APS) – the case number is the best way to look at raw data straight from the IOUs.

5.4.5 Fifth, Understand the Stakeholders in this Regulatory Process

The fifth and final step is focusing on the people leading the steps until now. By focusing on the people, the reader can learn how leaders' experiences early on in their careers led them to execute a policy successfully or unsuccessfully. Because it is quite likely that if a policy calls for smoothing out relations with utility commissioners some personalities are doing the opposite.

And the problem for readers is that hardly anyone might take them into their confidence to explain why some worked in a role while others did not. It is left up to the imagination of the observer.

5.4.6 Applying the Five Steps to the Dynamic Line Rating Policy Context

FERC has issued a notice of proposed rulemaking (NOPR) on transmission line ratings that includes incorporating dynamic line rating (DLR). Hence, Figure 5.4 shows how to take the five steps and apply them to a regulatory policy that mandates the inclusion of DLR in transmission line ratings in the US.

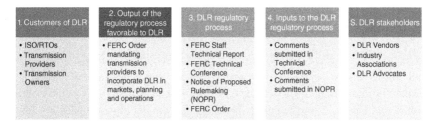

Figure 5.4   DLR regulatory policy five-step process example.

Figure 5.5   Australia regulatory structure five-step process.

5.4.7   Applying the Five Steps to the Australian Policy Context

If we take all the five steps learned here to understand the regulatory policy and apply that to an international country such as Australia, shown in Figure 5.5, we realize that there are common elements to any regulatory policy.

Exercise 5.6   Jay is in Australia and has a consulting assignment to serve as a grid expert. He has read this chapter and needs to apply/adapt the five-stage process to the National Electricity Market in Australia.[2] Please adapt and expand the process as appropriate.

5.5   How Does Regulatory Policy Drive Legislative Affairs?

Companies employ individuals as lobbyists in the countries that permit lobbying. Such hiring is done to influence legislators to advance the policies in the interest of the lobbyist's employer. Before a regulatory policy becomes law, legislators and their policy advisers reach out to industry associations and lobbyists of several relevant organizations to "take their pulse" over the proposed legislation. The ruling party and the minority party may gather these opinions to know whether the proposed legislation would have enough votes to pass in the legislative body. In

---

2 Please note that there are currently two wholesale electricity markets in Australia National Electricity Market (which covers eastern states of Australia including Tasmania) and Western Electricity Market (which covers Western Australia). Australia is looking into formulating a third market in the Northern states.

some cases, lobbyists might help frame amendments to a bill so that the legislation works for their organization also.

Exercise 5.7   As you know, Jay is an international energy consultant. After Australia, now he has a consulting assignment to serve as a grid expert trainer for developing and conducting a five-day training program for staff in XM, Colombia's system operator; UPME, the energy planning agency; the Ministry of Mines and Energy; CREG, the regulatory board; and other electricity-sector organizations. This training takes place in Bogota, Colombia.

Prepare a 25-slide presentation with appendix if needed but not exceeding five slides on Planning the bulk power system for high DER penetrations. This Day 3 presentation should cover the following topics: introduction to planning for high DER penetrations; load forecasting in different time horizons, from long term to real-time, with DERs; hosting capacity, feeder and system level; impact of DERs on bulk power system reliability and stability; special considerations regarding the impact of behind-the-meter DER on system reliability; load flow and dynamic studies for assessing the impact of DER on the bulk power system; and modeling DER and use of the simulation models to assess the reliability of DERs on bulk power system reliability.

5.6   Additional Examples of Regulatory Policy Driving Regulatory Success

A regulatory policy (see Section 5.2 for a definition) should drive regulatory success. But business development generally comes first and often without thought to potential regulatory barriers, many of which will be hard to overcome or take time to negotiate further down the road when a business's systems are already in place, making it harder for the business to change its direction. For example, imagine that an aggregator can operate in different states/regions of a country, and this country has multiple markets, e.g. the US.

In the US, a case in point is the lack of success for third-party aggregators in states like Minnesota or regions that have opted out of their demand response program participation in wholesale energy markets. If aggregators know upfront which state allows third-party aggregators (like Illinois) and those that don't (like Minnesota), they could focus their resources on states that allow their business model.

5.6.1   Salvation Army's Heat Program

As an example of a long-functioning humanitarian energy policy by an NGO, Heat-Share is a utility bill (heating and electric) program developed by the Salvation

Army [20]. The Salvation Army is a longstanding faith-based organization that assists on a secular basis. This program in the US has provided close to $50 000 000 worth of utility bill assistance to over 150 000 households since 1982.

Residential customers apply directly to the Salvation Army to determine eligibility for assistance. In addition to financial assistance with energy bills, monies can be provided for equipment efficiency improvements and repairs. It is used by low-income seniors, disabled people without a steady income, and others experiencing financial emergencies. The average household receives approximately $400.

Salvation Army or its equivalent operate in various countries, and they typically support the people who face difficulties paying their bills. However, depending on the regional needs, their policy focus might be slightly different. Hence, the flavor of regulatory success could be different.

Nevertheless, coming back to our US example, energy utilities such as Xcel, among others, have made HeatShare a policy by encouraging their total customer base to voluntarily fund the program by electing regular automatic donations to the program via their utility bills. It is a private program, and families must first prove they've been denied government assistance.

In most cases, these regulatory successes originated out of industry communications, which we discuss in Section 5.6.2.

5.6.2 Example of IOU Listening to Stakeholder Comments in IRP Proceedings

In another example of regulatory success driven by citizen and community engagement, Xcel Energy in Minnesota in June-end 2021 decided to abandon its plan to put together a big natural gas plant and instead planned to operationalize small natural gas units.

CUB [21], Vote Solar [22], and several other organizations had commented at the Minnesota PUC that natural gas was not needed given the strides made by renewable energy sources such as solar and energy storage.

This "Alternate Plan" of Xcel [23], which Minnesota PUC must approve, has 81% carbon-free resources by 2030 compared to the 70% carbon-free resources in an earlier plan submitted in June 2020.

Even though it is a common understanding that Xcel Energy listened to clean energy stakeholders in putting together this alternate plan, distributed generation stakeholder constituents are not satisfied with Xcel's new resource plan.

The reader should note that international experiences differ a lot. Unlike in the US, where an IOU must apply at the state PUC for a resource plan, it works the other way in, for example, Belgium.[3] The Belgian government must make a

---

3 According to Christophe Druet, email dated 19 July 2021.

strategic plan regarding energy by consulting the energy sector and use that information to grant permits to those interested in developing power plants.

5.6.3 Citizen and Industry Group Influence on Renewable Standards

In developed countries, there are several policies and programs to promote cleaner energy, aid with utility costs, achieve greater efficiency, and promote new technologies. These involve participation by government agencies, NGOs, and citizen groups.

Citizen, trade, and environmental groups have been instrumental in cleaner energy and new energy technologies. This group involvement has been a worldwide trend for the last 30+ years. The examples are numerous. They include fully independent organizations, organizations that are partially government-supported, and many more.

The US Environmental Protection Agency (EPA) categorizes Green Power Partnership-related programs and organizations into four functional stakeholder groups [24]:

- organizations
- certification programs
- trade organizations
- government.

To see the effect of some of these groups on policy, consider the case of RPS. An RPS indicates a goal, either optional or government-mandated, for a percentage of renewables or a defined level of clean energy. In effect, renewable standards vary greatly from country to country and even from region to region within a country. This varied effect is due to their grassroots nature where many advocacy groups (such as Natural Resources Defense Council [25]), local government [26] and regulatory bodies have created these from the bottom up.

It is also observed that electrical generation is becoming more decentralized with the advent of distributed generation technologies and merchant and user power production. The environmental movement has been characterized by strong citizen participation, including professional advocacy groups and trade associations that represent businesses (such as Advanced Energy Economy [27]) and organizations engaged in this industry.

Two researchers at the University of Arizona conducted a study of how stakeholders participated in the decision-making process around renewable standards [28]. They discovered that stakeholder knowledge makes their participation more meaningful, as do relationships with decision-makers and policymaking agencies. They also find that many governmental entities have developed formal and informal opportunities for participation in electric sector decision-making. This type of open process gives citizens a democratic voice. It can be diminished by the fact that multiple voices can make the decision process more cumbersome. Additionally, less informed voices do not necessarily lead to a better process. We hope that this book's toolkit of information helps our readers be informed participants in policymaking processes.

The example given above seems somewhat applicable to Western countries. However, in contrast to the above, in countries like China, the renewable standards are generally set at the central level, and the citizen and industry groups are at the receiving end of a successful (or unsuccessful) renewable regulatory policy.

Exercise 5.8   Jay is continuing his assignment from Exercise 5.7. He has a consulting assignment to serve as a grid expert trainer for developing and conducting a five-day training program for staff in XM, Colombia's system operator; UPME, the energy planning agency; the Ministry of Mines and Energy; CREG, the regulatory board; and other electricity-sector organizations. This training takes place in Bogota, Colombia.

Prepare a 25-slide presentation with appendix if needed but not exceeding five slides on DER interconnection standards. This Day 4 presentation should cover the following topics: the importance of interconnection standards; IEEE 1547 standard: focusing on aspects relevant for the bulk power system operator; and international experience and standards.

Exercise 5.9   We would ask Jay/the reader to research the RPS or equivalent for the state, province, or country where they currently live. Find at least three active stakeholder groups that fit into the EPA categories listed in Section 5.6.3 and see whether these groups were instrumental in the development of these clean energy standards.

5.7   How Does Regulatory Policy Drive Individual Participation in Industry Communications?

Until the policy is executed, the results are unknown. Otherwise, the policy remains on paper. People execute the policy, and people must participate in various industry communications at all levels. Communications can take the form of a memo, a presentation, a one-on-one meeting, a report, or even going to court.

Picking up on the third-party aggregator's example discussed in Section 5.6 on business development, we don't expect aggregators to show up in nuclear energy communications as an extreme case. Hence, if readers are focused on learning about jobs in the industry and talking to current third-party aggregators, they should research technology-driven, demand-response, or DER-focused communications.

The pandemic increased virtual calls and virtual conferences. This made many organizations and individuals consider the best ways of interacting with people virtually. This prompted many to consider the influence of social media platforms when it comes to the dissemination of industry communications.

Participating in a virtual conference is a skill we all had to learn on the move during the pandemic in 2020. Moving from a virtual room where a small team focused on a subtopic to reporting results in the larger room was seamless once we learned how the organizer's tool worked. The benefit of participating in the virtual conferences is travel cost savings and keeping in touch with our network of professionals.

Regarding the social media platforms' role in industry communications, platforms such as LinkedIn, Twitter, and Facebook have opened real-time updates on events in progress and allowed the users to share, like, and add individual opinions on any topic. This real-time feed is invaluable for someone who cares about a topic. Social media platforms are here to stay. Hence, we encourage our readers to embrace the new technology and tools to advance regulatory policy. Following are some venues:

- Utility-law communications/seminars.
- Technology-driven communications/conferences.
- Markets-focused communications/symposiums.
- Communications focused on utilities by industry groups.
- Consumer advocate conferences.
- Transmission-focused communications/seminars.
- Nuclear-energy-focused conferences.
- Demand-response- and DSM-focused communications.
- Compliance-focused conferences.
- National laboratory events.

5.7.1 How Do You Know You Had a Successful Event?

Readers need to know that successful business-development-focused executives and other industry veterans schedule one-on-one meetings before an event. These senior industry professionals have a calendar packed with meetings before they even set foot outside their office and spend time traveling. They have already researched the companies and individuals they are visiting and why they are talking to them. These meetings are very intentional.

Since most meeting organizers know who attends their meetings, their titles, and which business interests they represent, readers who plan to attend these industry conferences should learn this information before making the financial and time commitment to attend the event. It is important to understand "don't show up and throw up," meaning: don't just attend a meeting and present and leave the meeting. Mingle with people, talk to them about topics they care about and make a personal connection. It's not always about you!

Hence, you know you had a successful conference if you met with people you wanted to meet with, talked with them about what you wanted to talk to them about, and hopefully made a positive impression that they want to work with you in the future.

Event participation is successful if you can establish a contact that later becomes a champion for you. This person, first, has an interest in promoting your cause, second, must have power and influence, third, there has to be something in it for them (e.g. in terms of their career or business growth), and finally they must sell on your behalf when you are not there. Remember, a champion is different from a coach who merely advocates for you but does not have the power to make and sway the decisions. But a conference is only successful once you have identified a connection with a champion.

Exercise 5.10   As you know, Jay is an international energy consultant. He is continuing his consulting assignment to serve as a grid expert trainer for developing and conducting a five-day training program for staff in XM, Colombia's system operator; UPME, the energy planning agency; the Ministry of Mines and Energy; CREG, the regulatory board; and other electricity-sector organizations. This training takes place in Bogota, Colombia.

Prepare a 25-slide presentation with appendix if needed but not exceeding five slides on power system operations with high DER penetration. This Day 5 presentation should cover the following topics: approaches for incorporating DER forecasting into generation scheduling and dispatch for real-time, intraday, and day-ahead time horizons; impact of DER on operational reserves; approaches and best practices for improving ISO control center monitoring and control of DERs; options for managing DERs, such as the use of aggregators; and interoperability with distribution companies.

5.8   Visioning

Our vision about regulatory policy in the future requires a thorough understanding of stakeholders and processes. We have established that regulatory processes are time-consuming, process-driven exercises that set the record and involve the diverse interests of all stakeholders. The regulatory policy comes from environmentalists, customers, government, and business interests like the stakeholders involved. This regulatory process creates a record and shows the diversity of stakeholder interests.

With this chapter, readers and professionals should have the tools to follow the regulatory process with a five-step process to impact energy policy.

Since the regulatory policy is transparent and hidden simultaneously, readers need to know what regulatory success looks like and when something is not working. Since people are the center of institutions trusted with developing energy policies, this chapter also equips the reader with events and conferences where they might find like-minded individuals to bounce ideas off.

Case 5.1   *Regulatory Policy for an Oil and Gas Company to Work in the Solar Industry*

Situation

Let us take a hypothetical example of BP, the multinational oil and gas company. BP wants to invest in renewable fuels and energy because of changing consumer preferences for renewable energy, like solar.

How would you strategize and advise BP? Which markets in the US are worth BP's attention? What specific technology would enable BP to play to its strength? Who should BP partner with, and for how long?

Analysis

Some of the questions to conduct initial analysis and research would include:

- Understanding the current renewable energy market penetration for BP.
- How far along are BP engineers in renewable fuels?
- What is the consumer demand for renewable fuels, and where is it?
- What trading platforms exist for renewable fuels?
- Who are the potential partners for BP's size and scale?

These questions help frame solutions.

Possible Solution

Based on the analysis conducted, a possible solution is for BP to start taking steps toward monetizing renewable fuel trading. This fuel investment will build upon the BP scale and financial situation. Additionally, BP can reduce the cost of distributed solar by investing in aggregator platforms that integrate energy storage and other emerging technologies.

Another advantage BP has, relative to other industry players, is that when BP has an event, it is guaranteed that people will show up. Both international and domestic. Both supporters and detractors. Both regulators and technology providers. Hence, BP must be intentional in taking steps toward the solar industry because there are downstream implications no matter what.

Case 5.2   *Convincing a State to Mandate Certifications for All Solar Installations*

Situation

JaySolar is a Minnesota-based residential and commercial solar installer with a unique patented "no holes" drilled design for rooftop solar. They are also Underwriters Laboratory Inc. (UL) listed for Minnesota. But that UL testing is not a requirement in the state of Minnesota.

How would you put together a regulatory policy to convince PUC Minnesota that the UL listing is essential for rooftop solar in the state? How do you convince JaySolar to hire you as a lobbyist? Who is going to be partners in this advocacy? Who is going on the other side of the argument?

Analysis

Some of the questions to conduct initial analysis and research would include:

- Understanding the market need for a UL listing in the solar market.
- What are all the benefits of a UL listing?
- What are the implications of a mandatory UL listing on solar components?
- What is the transition period for suppliers to catch up with the UL listing?
- Does UL have the resources in Minnesota to certify everyone?
- Does JaySolar have a better chance of convincing a neighboring state like Wisconsin before making the Minnesota pitch?

These questions help frame solutions.

Possible Solution

Based on the analysis conducted, JaySolar could convince the Minnesota regulators of the UL listing's value for solar panels, specifically rooftop solar. JaySolar must show data and evidence where customer satisfaction is high compared to other suppliers under varying weather conditions. This research could mean JaySolar funding an independent research organization to conduct customer surveys impartially.

Case 5.3 *Regulatory Policy for a Regional Company Based in One Region to Operate in Another Region*

Situation

Let us take a hypothetical example of BatteryEd, the New York City-based utility company. BatteryEd Transmission (BET) is a subsidiary of BatteryEd, the utility company. BET successfully bid on independent transmission projects in New England and New York ISO. Now, BET wants to explore the Midwest market for transmission investment opportunities.

How would you advise BET on the MISO market entry? Who are potential partners for BET? Which MISO meetings are worth BET attending? What are the first steps in introducing BET to MISO and MISO regulators? Would your strategy be different for MISO North and South regions? At what point would BET give up and go home?

Analysis

Some of the questions to conduct initial analysis and research would include:

- Understanding the strengths of BET in the New York and New England markets.
- Assess the market needs for new transmission investment in the Midwest.
- Which regulators are in favor of ITCs versus who favor TOs?
- Who are likely partners for BET?
- Is BET a better partner for TOs in the MISO north region or the south?

These questions help frame solutions.

Possible Solution

Based on the analysis conducted, a possible solution is for BET to become a MISO member and gain experience bidding on a competitive transmission project. That experience enables BET to understand MISO processes.

BET would have to decide to exit the market if BET cannot land a transmission project investment in the first five years.

Case 5.4 *Convincing an RTO to Keep the Costs Lower*

Situation

Let us take a hypothetical example of the CUB, the Minnesota consumer advocate. CUB Minnesota is thinking about increasing its voice in the MISO stakeholder process because CUB represents electric utility consumers' interests in Minnesota's entire state.

How would you advise CUB Minnesota? Since there are many MISO stakeholder committee meetings in planning, reliability, and market areas, where would you suggest CUB focus its limited resources? Do you suggest CUB increase its budget? If so, how should CUB convince its funders of additional funds?

Analysis

Some of the questions to conduct initial analysis and research would include:

- Understanding the current budget for CUB.
- Sitting down with the CUB board and the executive director to discuss what outcome they expect to see and when.
- Understanding MISO stakeholder processes.
- What expertise is required to follow the technical arguments at MISO meetings?

These questions help frame solutions.

Possible Solution

Based on the analysis conducted, a possible solution is for CUB Minnesota to dip its toes, first, in MISO's planning stakeholder committees to emphasize regional transmission projects' consumer costs. CUB can take the cue from Consumer Advocates of PJM states (CAPS) and compare notes with other CUB organizations in organized markets.

If CUB learns from CAPS that reliability-justified transmission project costs are a significant driver behind consumer costs, CUB should learn the reliability-justified project cost allocation process at MISO. This focus identifies partners at MISO stakeholder meetings to fill in the gaps for the CUB knowledge base.

References

1 Xcel Energy (2021). Net-zero energy provider by 2050. https://www.xcelenergy.com/environment/carbon_reduction_plan (accessed 11 December 2021).
2 Google (2021). We must help build a more sustainable future for everyone. https://sustainability.google/commitments# (accessed 11 December 2021).
3 United States Energy Association (2021). Uganda energy mix diversification strategy for the Uganda Electricity Generation Company Ltd ("UEG-CL") report. https://usea.org/sites/default/files/Uganda%20Energy%20Mix

%20Diversification%20Strategy%20March%202021.pdf (accessed 11 December 2021).

4 European Commission (2007). FP7 in brief. https://www.ehu.eus/documents/ 2458339/2849729/fp7-inbrief_en.pdf/84bd26a8-ab12-469a-8ed9-475917b36cd1? t=1411241328000 (accessed 11 December 2021).

5 European Commission (2020). What is Horizon? https://ec.europa.eu/ programmes/horizon2020/what-horizon-2020 (accessed 11 December 2021).

6 Energy Storage News (2020). Italy's grid operator Terna "moving in the right direction" to accommodate battery storage. https://www.energy-storage.news/ italys-grid-operator-terna-moving-in-the-right-direction-to-accommodate-battery-storage (accessed 11 December 2021).

7 Rio Tinto (2021). Climate change report 2020. https://www.riotinto.com/en/ sustainability/climate-change (accessed 11 December 2021).

8 Hughlett, M. (2020). Xcel gets $10.5m federal grant for pilot project on hydrogen. https://www.startribune.com/xcel-gets-10-5m-federal-grant-for-pilot-project-on-hydrogen/572754271 (accessed 11 December 2021).

9 St. John, J. (2020). California targets nearly $400m to fill gaps in EV charging infrastructure. https://www.greentechmedia.com/articles/read/california-targets-384m-to-fill-gaps-in-electric-vehicle-charging-infrastructure (accessed 11 December 2021).

10 IEA (2020). Global EV outlook 2020. https://www.iea.org/reports/global-ev-outlook-2020 (accessed 11 December 2021).

11 Minnesota Public Utilities Commission (2021). MP's 2021 integrated resource plan. https://www.edockets.state.mn.us/EFiling/edockets/searchDocuments .do?method=showPoup&documentId=%7b70795F77-0000-C41E-A71C-FD089119967C%7d&documentTitle=20212-170583-01 (accessed 11 December 2021).

12 Scatz Energy (2021). Redwood Coast Airport microgrid. https://schatzcenter .org/acv (accessed 11 December 2021).

13 LS Power (2020). LS Power petitions Supreme Court to overturn anti-competition electric transmission laws that hurt consumers. https://www .lspower.com/ls-power-petitions-supreme-court-to-overturn-anti-competition-electric-transmission-laws-that-hurt-consumers (accessed 11 December 2021).

14 EIA (2020). Frequently asked question: "How many nuclear power plants are in the United States, and where are they located?" https://www.eia.gov/tools/ faqs/faq.php?id=207&t=3 (accessed 11 December 2021).

15 Hilson, Z., Cunsolo, A. (2021). Electricity regulation in Australia: overview. https://uk.practicallaw.thomsonreuters.com/w-010-9549? transitionType=Default&contextData=(sc.Default)&firstPage=true (accessed 11 December 2021).

16 Brasher, L.T., Czarniak, J.A., Estes, J.N. III, et al. (2020). Trump administration limits acquisition. https://www.skadden.com/insights/publications/2020/05/trump-administration-limits-acquisition (accessed 11 December 2021).

17 Office of the United States Trade Representative. (2018). President Trump approves relief for US washing machine and solar cell manufacturers. https://ustr.gov/about-us/policy-offices/press-office/press-releases/2018/january/president-trump-approves-relief-us (accessed 11 December 2021).

18 Minnesota Public Utilities Commission (2021). eDockets. https://www.edockets.state.mn.us/EFiling/edockets/searchDocuments.do?method=showeDocketsSearch&showEdocket=true (accessed 11 December 2021).

19 Arizona Corporate Commission (2021). Arizona utilities IRP filings. https://edocket.azcc.gov/search/docket-search (accessed 11 December 2021).

20 Salvation Army (2021). Keeping the heat on. https://centralusa.salvationarmy.org/northern/heatshare-program/ (accessed 11 December 2021).

21 Citizens Utility Board (2021). Xcel no longer pursuing gas power plant, proposes more renewable power. https://cubminnesota.org/xcel-is-no-longer-pursuing-gas-power-plant-proposes-more-renewable-power (accessed 11 December 2021).

22 Vote Solar (2021). Innovative solar model can help all Minnesota Xcel customers save on electric bills. https://votesolar.org/innovative-solar-model-can-help-all-xcel-customers-save-on-electric-bills (accessed 11 December 2021).

23 Xcel Energy (2021). Delivering reliable, affordable, clean energy: Xcel Energy Upper Midwest plan. https://www.xcelenergy.com/staticfiles/xe-responsive/Company/Rates%20&%20Regulations/Resource%20Plans/Upper%20Midwest%20Energy%20Plan%20-%202021.pdf (accessed 11 December 2021).

24 United States Environmental Protection Agency (2021). Green Power Partnership. https://www.epa.gov/greenpower (accessed 11 December 2021).

25 Narita, K. (2013). State renewable portfolio standards create jobs and promote clean energy. https://www.nrdc.org/resources/state-renewable-portfolio-standards-create-jobs-and-promote-clean-energy (accessed 11 December 2021).

26 United States Environmental Protection Agency (2021). Green Power Partnership top 30 local government. https://www.epa.gov/greenpower/green-power-partnership-top-30-local-government (accessed 11 December 2021).

27 Hanis, M. (2021). Report: five case studies show DER benefits wholesale electricity markets. https://www.aee.net/articles/report-five-case-studies-show-der-benefits-wholesale-electricity-markets (accessed 11 December 2021).

28 Rountree, V., Baldwin, E. (2018). State-level renewable energy policy implementation: how and why do stakeholders participate? https://www.frontiersin.org/articles/10.3389/fcomm.2018.00006/full (accessed 11 December 2021).

6

How Institutions Shape Energy Policy

6.1 Introduction

How institutions shape energy policy is a vast topic because energy policy is huge. Many topics go under the "energy policy" umbrella. In the US, these topics could include all the 18 titles and several subtitles from the US's Energy Policy Act of 2005.[1]

In the US alone, there are several federal and state institutions that influence energy policy. The White House, the Department of Energy (DOE), the US Congress, the Federal Energy Regulatory Commission (FERC), and several national associations, alliances, and technology companies are some of them.[2] The latter are increasingly relevant in energy policy discussions due to the direct impact of social media and consumers' increasing appetite for data that is translating into rising energy demands at technology companies' data centers, as seen in Google's chart in Figure 6.1.

Additionally, both electric vehicles (EVs) – the US is expected to reach 58 terawatt-hours by 2030 [1] – and Bitcoin mining (91 terawatt-hours of electricity annually [2]) have significant energy needs. Many of these same companies and their customers are also helping to lead the conversion to clean energy.

Several institutions influence energy policy, but only one institution is trusted with regulatory and legislative mandates to set energy policy. In the US, the

1 The 18 titles in the Energy Policy Act of 2005 are: (i) Energy Efficiency, (ii) Renewable Energy, (iii) Oil and Gas, (iv) Coal, (v) Indian Energy, (vi) Nuclear Matters, (vii) Vehicles and Fuels, (viii) Hydrogen, (ix) Research and Development, (x) Department of Energy Management, (xi) Personnel and Training, (xii) Electricity, (xiii) Energy Policy Tax Incentives, (xiv) Miscellaneous, (xv) Ethanol and Motor Fuels, (xvi) Climate Change, (xvii) Incentives for Innovative Technologies, and (xviii) Studies.
2 FERC is a subagency of the DOE. For more information, visit: https://www.federalregister .gov/agencies/energy-department (accessed 11 December 2021).

Modern Electricity Systems: Engineering, Operations, and Policy to address Human and Environmental Needs, First Edition. Vivek Bhandari, Rao Konidena and William Poppert.
© 2022 John Wiley & Sons Ltd. Published 2022 by John Wiley & Sons Ltd.
Companion website: www.wiley.com/go/bhandari/modernelectricitysystems

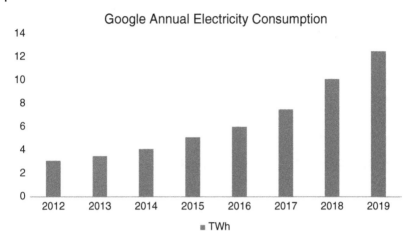

Figure 6.1 Google annual electricity consumption.

US Congress makes all laws, including energy policy law. The federal and state agencies must implement those laws.

Countries like Germany have a comprehensive countrywide energy policy (*Energiewende*), whereas the US does not have a strong federal energy policy.[3] The US states are free to set their statewide energy policy like the renewable portfolio standard (RPS) in the past, and now, 100% carbon-free goal by 2050. In the US, federal policy is centered around enabling energy efficiency and energy efficiency appliance standards. It is less about dictating what states should do.

As we parse through energy policy, it is tough to relate to what works in the US with something closely related to another country or region. Additionally, different regional political interests drive what is topical in energy policy.

For example, in the US, the Trump administration took over in 2017. Since one of their campaign promises was to save coal – the US DOE initiated a proceeding to factor in the reliability of fossil-based units. The FERC commissioners, in a unanimous 5–0 vote, rejected that Trump DOE-led proceeding [3]. There are parallels in other countries. For example, in Mexico, President Andrés Manuel López Obrador (called AMLO for short) rejected the Mexican energy market progress and stopped Mexican federal energy regulator Comisión Reguladora de Energía (CRE)'s market rules.

---

3 We realize this statement depends a lot on where one is located. If you are based in Europe, EU countries can be seen as states as they are bound to implement European Union (EU) regulations, whereas EC would be the equivalent of the federal government. Take, for instance, Germany: there the *landen* have some freedom on how they implement the *energiewende*, just as Germany has to implement the EU regulations but still has some latitude on how to steer things (Christophe Druet, personal communication).

So, politics drives energy policy. However, most energy regulatory institutions and policy-setting institutions are deemed apolitical. As a result, these institutions continue to do their job even under political pressures. This chapter captures those institutions' stories.

This chapter sheds light on institutions that are instrumental in setting energy policy in their countries and regions. For driving a change in society, sometimes we look at historical institutions in a new light. We turn to institutions we trust.

A case in point in most countries today is the rise in distributed solar, i.e. distributed energy resources (DERs). As a result, the forecast for DERs becomes increasingly critical to keep the lights on, literally. Because the electric utilities have to keep up with the increasing demand for DER compensation and forecast where DERs will pop up next in their service area, this increasing demand for DERs forecasting is something that historical load forecasting institutions can handle if given the policy mandate.

And usually, even in partisan political environments, there are trustworthy institutions in the public eye. Even if political leaders' intentions are partisan, in most cases, the data gathering and reporting are done in a factual, impartial manner, leaving the narrative to the political establishment. Because, as the example with DER forecasting suggests, how much DERs get compensated by the distribution utility directly relates to the number of DERs predicted to interconnect in the future.

An apolitical institution would enable robust policy discussions on this DER compensation topic. Because both political leaders and regulatory officials trust these institutions' data, there is less acrimony about the data and more discussion about the policy direction.

There are very few institutions worldwide that have earned the people's trust in the energy policy area. This chapter is their story. By reading this institutional narrative, the graduate student and nongovernmental organization (NGO) would know where to turn for data before setting energy policy. The author hopes that energy policy discussions control society's cost to integrate these new technologies by focusing on the data and assumptions. Otherwise, ordinary people in society may not benefit from emerging technologies. In the worst case, if energy policy is harming new technologies, young professionals get frustrated and take to the streets to demonstrate and voice their concerns, which may lead to violence and, unfortunately, loss of life.

Since energy policy is such a vast area, not all institutions worldwide can be a one-stop shop for all the topics that energy policies cover. The US Energy Information Administration (EIA) does an excellent job of catering to all the topics covered in an energy policy setting. But US energy utilities provide the data to EIA regularly. Eurostat provides data on EU countries, including electricity prices and renewable energy statistics [4].

Both state and federal energy regulators in the US also collect much data. Often this data is used to influence energy policy discussions. National labs also play a major role in the US energy policy space. National Renewable Energy Laboratory (NREL) provides an annual technology baseline (ATB), for instance, for public cost data for any new technology choice [5]. Although it is not uniform throughout the world, in many countries, government assistance is provided to help introduce new technologies.

Exercise 6.1 In your country, what are the major institutions that set energy policy? Pick one to focus on. What sets this institution apart? Who are the major leaders of this institution? Is this a political or nonpolitical institution? What is the educational background of the majority of staff who work at this institution?

6.2 Strategic Action Field Framework for Policy

To explain the role of institutions in energy policy, we apply the research findings from "The Strategic Action Field Framework for Policy Implementation Research," written by Professors Stephanie Moulton and Jodi R. Sandfort [6].

The strategic action field (SAF) framework paper identifies four forms of authority to implement policies. We believe in the concept of "authority," which is defined in the SAF paper as, "The significance of any particular authority source depends on how it is interpreted within a particular context" to explain the role of institutions in setting energy policy [6]. And we argue that we are providing context throughout this book based on our experience.

The four forms of authority are:

- Political authority: To put it simply, political leaders have political authority. That authority manifests itself in appointing federal and state regulators who are in tune with their political philosophy. While we believe in most instances that energy policy regulators at either the federal level or the state level are apolitical, the reader should know how politics and elections (since elections have consequences) influence energy policy.
- Economic authority: We believe most actors in energy policy have an economic authority, which is defined in our view as the ability to take action that results in commerce. By that broad definition of economic authority, the reader should take note within the US context, all the institutions we discuss have economic authority, except the EIA, which is an energy statistical agency of the US government.
- Professional authority: The SAF paper calls this "norms associated with professional expertise", hence we called this professional authority (see Chapter 5).

- Collective belief and value authority: The SAF paper calls this "beliefs and values shared by individuals and groups" [6], hence we call this "collective belief and value authority." Within the context of energy policy and institutions, we see this collective belief and value authority as the organizational culture [7]. And each energy policy institution has its own culture.

Does politics play a role in setting policy? This question is the fundamental question for readers. If readers think politics does not play a role in policy, they are mistaken.[4]

How politics plays a role is not straightforward until readers read between the lines and make friends who can speak to them candidly on their company positions. Some readers associate networking with politics; they think if their friend is schmoozing with an executive that friend is playing politics. This assumption is not always true. That friend might have someone to send a resume for a job down the road. Networking enables people to find a job. Landing on a good high-profile project is all about who you know – not about what you know!

So, how does politics play a role in policy? When President Trump was elected as the head of the executive branch of the US government, he appointed Commissioner Neil Chatterjee as head of the FERC. Why? According to the popular press, Commissioner Chatterjee hailed from Kentucky, which is the home state of Senate Majority Leader Mitch McConnell. If we believe press articles, perhaps, the connection with Mitch McConnell helped Commissioner Chatterjee's career path. It is possible that President Trump needed something from the United States Senate and hence went with the majority leader's choice.

Table 6.1 shows how to apply the SAF framework to our energy policy institutions. The table also indicates why we have picked these broad categories of energy policy institutions: to provide context on how institutions implement energy policies using the four forms of authority.

6.3 What Are the Major Institutions in US Energy Policy?

As an example for all countries, we first start with the US energy scene. We picked the following governmental institutions listed in Table 6.2 in US policy because we can provide context to the role these institutions play in US energy policy.

---

4 According to our external reviewer, the reader should know that the roots of the two words is the same: *politia* in Latin, *politeia* in Greek. So it's not even a question. Politics is the act of defining policies.

Table 6.1 Broad categories of energy policy institutions with their authorities to implement policies.

Broad Categories of Energy Policy Institutions	Political Authority	Economic Authority	Professional Authority	Collective Belief and Value Authority
Governments and Energy Ministers	Yes, elections have consequences	Yes, new governments mean new top officials at institutions with economic authority	Yes, when governments are also responsible for standards and certifications	Yes, governments and government institutions have their own cultures and value systems
Courts (e.g. Supreme Court)	Yes, in countries like the US, where courts are the judicial branch of government	No, courts do not have the economic authority	Within the legal community, courts are the ultimate professional authority Outside the legal community, they don't have professional authority	Yes, courts have their value system
Market monitors	No, they don't have any political authority unless market monitoring is a government agency function	Yes, market monitors have direct economic authority over market participants by their opinion on any market rule	Yes, market monitors have professional authority on organized markets	Yes, the individual market monitor might have an established norm with the market operator they oversee
Associations and Alliances (A&A) (e.g. Energy Storage Association)	A&A does not have direct political authority, but it influences politicians	A&A has direct economic authority by its members' combined revenue	Within A&A, there are professional authorities, but by itself A&A doesn't have professional authority – this is why it hires experts	Yes, within each A&A – there are committees and subcommittees And they have their norms and belief systems

Table 6.1 (Continued)

Broad Categories of Energy Policy Institutions	Political Authority	Economic Authority	Professional Authority	Collective Belief and Value Authority
Third-party actors (e.g. foundations, NGOs)	No, third-party actors don't have political authority	Yes, foundations/NGOs have economic authority in countries or regions with fragile economies	Yes, they do have professional authority (e.g. Engineers without Borders)	Yes, third-party actors have their norms and belief systems
Private companies (e.g. Google)	Yes, private companies, especially big companies, have political authority due to the number of jobs they create and lobbying activities	Yes, private companies have economic authority, especially big companies that create hundreds of jobs in a particular city/stage/region	No, private companies don't have professional authority unless they are the sole funding authority for new technology	Yes, we believe each business enterprise has its own culture

6.3.1 US Congress

In the United States, the US Congress has the political authority to pass laws that result in economic authority due to generated commerce. After the US Congress passes laws and the president of the United States (POTUS) signs, the DOE gets to work. POTUS can also sign executive orders (EOs) that come within the DOE scope.

In the 117th Congress (2021–2022), there are 70 energy bills from both the House of Representatives and the US Senate. All the Senate bills must follow this process, as illustrated in Figure 6.2.

Because there are too many energy bills introduced by the House, here is a table showing the 16 bills introduced by US senators in the Energy and Natural Resources Committee.[5] Note, of course, that not all of these bills become law, at least not in the form that they are introduced as bills for consideration by Congress.

Figure 6.2 How bills become laws if a bill is initiated in the US Senate.

---

5 Search for "energy" on https://www.congress.gov/advanced-search/legislation (accessed 11 December 2021).

Table 6.2 US governmental policy institutions with their authorities to implement policies (for brevity's sake, not all are shown here).

US Energy Policy Institutions	Political Authority	Economic Authority	Professional Authority	Collective Belief and Value Authority
US Congress	Yes, US Congress passes laws such as the Energy Policy Act	Yes, US Congress has the economic authority because legislation creates commerce	Among the Congressional community, members have professional authority figures based on their expertise and experience and political affiliations	Yes, public service is the common belief and value system that binds the Congressional community
Energy Information Administration (EIA)	No, EIA does not have political authority	No, EIA does not have economic authority	Yes, EIA has professional authority because it is the topmost energy data-oriented institution in the US	Yes, we believe EIA, like any other federal agency, has its values and belief systems
Federal Energy Regulatory Commission (FERC)	No, FERC is an apolitical organization	Yes, FERC has the economic authority in the industries it regulates	FERC is the top energy regulator in the US, hence it has professional authority over federal energy policymakers	FERC commissioners and staff all have collective beliefs and value systems, much like any institution
State legislature (Senate and House)	Yes, a state legislature has the political authority over energy policy in that state	Yes, a state legislature has the economic authority because legislation creates commerce in that state	Among the state legislative community, members have professional authority figures based on their expertise, experience, and political affiliations	Yes, public service is the common belief and value system that binds the legislative community

Table 6.2 (Continued)

US Energy Policy Institutions	Political Authority	Economic Authority	Professional Authority	Collective Belief and Value Authority
Public utility commissions (PUCs)	No, state PUCs are not political organizations even though there are states in which PUC commissioners are directly elected by people	Within a state, PUCs have the economic authority over industries they regulate	As a top energy regulator within a state, PUC commissioners and staff have professional authority in the state energy policy	State PUC commissioners and staff all have collective beliefs and value systems, much like any institution
North American Electric Reliability Corporation (NERC)	No	Yes, but not directly, because a reliability standard and the threat of a compliance fine had both direct and indirect consequences on commerce	NERC is the top electric reliability regulator in North America, hence it has professional authority alongside federal energy policymakers located in the US, Canada, and Mexico	Yes, the reliability of the bulk electric system is the common belief and value system that binds the NERC community
Federal Bureau of Ocean Energy Management (BOEM)	Yes, as an agency of the US government BOEM has the political authority in states that are near oceans and other water bodies that BOEM regulates	Yes, BOEM has the economic authority when it comes to oceans and offshore wind	We believe BOEM has professional authority over offshore wind energy policies	We believe BOEM has collective beliefs and a value system, much like any institution

Table 6.3 Summary table of all energy bills initiated by US senators in 117th Congress.

Bill title	What is the scope of the bill?	Detail summary
Interregional Transmission Planning Improvement Act of 2021	To require FERC to initiate a rulemaking to reform the interregional transmission planning process, and for other purposes	
Storing CO_2 and Lowering Emissions Act (SCALE Act)	To require the Secretary of Energy to establish programs for carbon dioxide capture, transport, utilization, and storage, and other purposes	
Wastewater Efficiency and Treatment Act of 2021	To amend the Energy Policy Act of 2005 to establish a program to provide grants and loan guarantees to improve the energy efficiency of publicly owned wastewater treatment facilities, and for other purposes	
Disaster Safe Power Grid Act of 2021	To require the Secretary of Energy to establish a grant program to improve the resiliency of the power grid to natural disasters and reduce the risk of wildfires caused by power lines, and for other purposes	
Renew America's Schools Act of 2021	To require the Secretary of Energy to provide grants for energy efficiency improvements and renewable energy improvements at public school facilities, and other purposes	
Fair Returns for Public Lands Act of 2021	To amend the Mineral Leasing Act to increase certain royalty rates, minimum bid amounts, and rental rates, and for other purposes	

Table 6.3 (Continued)

Bill title	What is the scope of the bill?	Detail summary
Open Back Better Act of 2021	To provide additional funds for federal and state facility energy resiliency programs, and other purposes	This bill directs the DOE to provide grants to federal and state agencies and tribal organizations to implement building projects that increase resiliency, energy efficiency, renewable energy, and grid integration. It also provides grants for projects that may have combined heat and power and energy storage as project components
		States must use at least 40% of grant funds to implement projects in environmental justice communities or low-income communities that have been adversely impacted by the COVID-19 (i.e. coronavirus disease 2019) pandemic
Electric Vehicles for Underserved Communities Act of 2021	To increase deployment of EV charging infrastructure in low-income communities and communities of color, and for other purposes	This bill requires the DOE to support the deployment of EV charging infrastructure in underserved or disadvantaged communities
		Specifically, the DOE must establish an EV charging equity program. Under the program, the DOE must provide technical assistance and award grants to increase the deployment and accessibility of EV charging infrastructure in such communities
		If practicable, the DOE must ensure that relevant programs promote EV charging infrastructure, support clean and multimodal transportation, provide improved air quality and emissions reductions, and prioritize the needs of such communities

Table 6.3 (Continued)

Bill title	What is the scope of the bill?	Detail summary
Clean School Bus Act of 2021	To establish the Clean School Bus Grant Program, and for other purposes	This bill directs the DOE to establish the Clean School Bus Grant Program in the Office of Energy Efficiency and Renewable Energy
		Under the program, the DOE must award grants for the replacement of existing diesel school buses with electric buses
Renewable Fuel Infrastructure Investment and Market Expansion Act of 2021	To amend the Farm Security and Rural Investment Act of 2002 to provide grants for the deployment of renewable fuel infrastructure, to finalize proposed rules relating to requirements for E15 fuel dispenser labeling and underground storage tank compatibility, and for other purposes	This bill establishes programs and requirements to expand access to renewable fuel
		Specifically, the US Department of Agriculture must establish a program to award grants for the deployment of renewable fuel infrastructure as specified by this bill
		In addition, the Environmental Protection Agency (EPA) must finalize a 2021 proposed rule titled E 15 Fuel Dispenser Labeling and Compatibility With Underground Storage Tanks. When finalizing the rule, the EPA must eliminate the labeling requirements for fuel pumps that dispense E 15 fuel (i.e. gasoline that contains 15% ethanol)
Nonprofit Energy Efficiency Act	To require the Secretary of Energy to establish an energy efficiency materials pilot program	
Keystone XL Pipeline Construction and Jobs Preservation Act	To authorize the Keystone XL Pipeline	This bill authorizes the TransCanada Keystone Pipeline to construct, connect, operate, and maintain the pipeline facilities in Phillips County, Montana, for the import of oil from Canada to the United States

Table 6.3 (Continued)

Bill title	What is the scope of the bill?	Detail summary
POWER Act of 2021	To prohibit the president from issuing moratoria on leasing and permitting energy and minerals on certain federal land, and for other purposes	
Conservation Funding Protection Act	To amend the Outer Continental Shelf Lands Act to require annual lease sales in the Gulf of Mexico region of the outer Continental Shelf, and for other purposes	This bill requires the Department of the Interior to hold at least two region-wide oil and gas lease sales per year in the Gulf of Mexico Each lease sale must include areas in the Central Gulf of Mexico Planning Area and the Western Gulf of Mexico Planning Area In addition, the bill establishes deadlines for completing environmental reviews of the lease sales
West Coast Ocean Protection Act of 2021	To amend the Outer Continental Shelf Lands Act to permanently prohibit the conduct of offshore drilling on the outer Continental Shelf off the coast of California, Oregon, and Washington	This bill prohibits the Department of the Interior from issuing a lease for the exploration, development, or production of oil or natural gas in any area of the Outer Continental Shelf off the coast of California, Oregon, or Washington
Florida Shores Protection and Fairness Act	To include the State of Florida in the Gulf of Mexico Outer Continental Shelf revenue sharing program, to extend the moratorium on oil and gas leasing in certain areas of the Gulf of Mexico, and for other purposes	This bill revises the Gulf of Mexico Outer Continental Shelf revenue sharing program Specifically, it expands the program to include Florida. Currently, only Alabama, Louisiana, Mississippi, and Texas are included in the program that shares the revenues from oil and gas leasing on the Gulf of Mexico outer Continental Shelf Additionally, the bill extends to 30 June 2032, the moratorium on oil and gas leasing in certain areas of the Gulf of Mexico

As Table 6.3 shows, there are bills on a wide variety of energy topics. Some include banning offshore drilling on the West Coast of the United States, whereas another bill includes quite the opposite: amend the current outer continental shelf lands act to promote offshore drilling.

There are bills on EVs for underserved communities as well as for schools. Like the Interregional Transmission Planning Improvement Act of 2021, some of the bills have been introduced in multiple sessions of the US Congress. As Table 6.3 indicates, some bills require the FERC to act on rules, whereas others require the Secretary of Energy to establish grants or programs. In summary, from the table, it is evident that senators and House members take the energy needs of the US. seriously. This form of legislation shows that the US Congress has both the political and the economic authority to implement energy policies in the US.

There are six specific laws passed in the US that provide the foundational elements for energy markets and advanced energy policies. They are:

- Energy Policy Act of 1992 (EPAct 1992)
- Energy Policy Act of 2005 (EPAct 2005)
- Energy Independence and Security Act of 2007 (EISA)
- Federal Power Act (FPA)
- Natural Gas Act
- Interstate Commerce Act.

The last three laws/Acts (Power, Natural Gas, and Interstate Commerce) grant FERC the authority to regulate interstate transmission, natural gas, and oil.

Exercise 6.2   Your friend is a policy adviser to the most powerful senator in your country. This policy adviser's role is to focus on energy, environmental, and carbon topics. And policy advisers must be familiar with energy tax incentives.

When a US senator brings a large energy policy package to the Senate floor, the policy adviser's role is critical. They have to coordinate with other senators' policy advisers. Additionally, they prepare committee hearing briefing materials and other memos; staff the senator at hearings, meetings, and other events; and represent the senator with constituents, federal agencies, and foreign officials. You want to support your friend and ultimately land a job similar to theirs in the near future. How would you help prepare a three-year strategy for the senator and their policy adviser on putting together an energy policy on a topic of your choosing?

Please note: if you are located in Europe, do the same exercise for Germany or France (including the EU aspects).

6.3.2 Department of Energy (DOE)

To provide institutional context on the important role the DOE plays in US energy policy, the DOE is the implementer of energy laws. Hence, in its way, the DOE has political, economic, and professional authority compared to the US Congress. The energy secretary has the final say in all matters related to energy in the US.

The DOE is a big organization. As Figure 6.3 indicates, the DOE manages:

- Nuclear energy security.
- Science and energy.
- EIA and other programs such as the Advanced Research Projects Agency-Energy (ARPA-E).

The DOE is headed by the Secretary and a Deputy Secretary. The FERC has a dotted line reporting responsibility to the Energy Secretary. Each major office is managed by an Under Secretary. One level below the Under Secretary is Assistant Secretaries who manage specific offices such as Nuclear Energy, Fossil Energy, Energy Efficiency and Renewable Energy (EERE), and Office of Electricity (OE).

Federal power companies such as the Bonneville Power Administration (BPA), the Southwestern Power Administration (SWPA), the Southeastern Power Administration (SEPA), and the Western Area Power Administration (WAPA) are managed by the OE.

DOE also manages 17 national labs spread across the country, which are not shown in Figure 6.3 [9].

6.3.3 Federal Energy Regulatory Commission (FERC) and Independent System Operator (ISO)

FERC is a major energy-policy-setting institution for the US. In the SAF framework, FERC has primary economic authority in setting energy policy with some political authority and mostly professional authority with energy lawyers.

FERC orders are how FERC sets the direction of energy policy. The president appoints FERC chairman. So, whenever there is a major party change in the White House, it is most likely that the FERC chairman changes. FERC commissioners are appointed by the president and approved by the Senate.

FERC, as an agency under the federal government, takes its authority from the US Congress. In addition to electric policy, FERC is also responsible for licensing hydropower facilities and natural gas regulations but not natural gas production.

As an institution, FERC has major offices that are individually responsible for executing the main goals of FERC. These include:

- Office of Public Participation (OPP)
- Office of Energy Market Regulation
- Office of Energy Policy and Regulation
- Office of Electric Reliability
- Other major FERC offices.

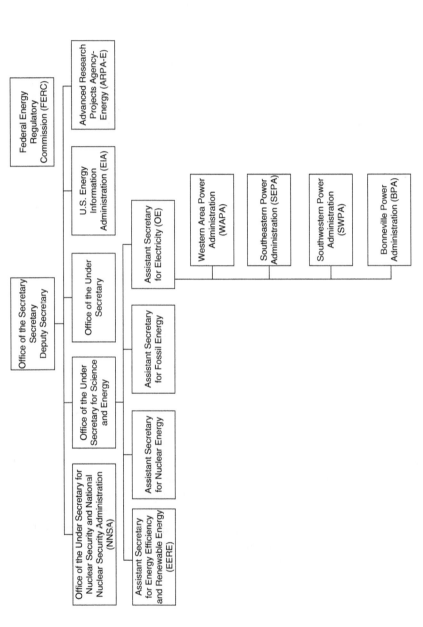

Figure 6.3 DOE Org Chart, from DOE. DOE Org Chart. Source: Adapted from [8].

History, major roles, and what is outside FERC's scope are available on the FERC website [10]. FERC interacts with other federal agencies, such as the Bureau of Ocean Energy Management (BOEM). Landmark FERC orders related to energy markets discussed in Chapters 4 and 9 include:

- FERC Order 1000 on competitive transmission planning.
- FERC Order 890 on regional planning.
- FERC Order 841 on electric storage resources.
- FERC Order 2222 on DER aggregations.

Before a FERC order is issued, FERC issues a notice of proposed rulemaking (NOPR). An NOPR on transmission line ratings, such as dynamic line rating (DLR), was issued in January 2021 [11]. A FERC order is pending. FERC has the economic authority because a FERC order mandates transmission providers such as ISOs to open up energy markets to newer technologies such as energy storage.

FERC regulators have successful careers before and after their commissioner stint. FERC Commissioner Pat Wood (Standard Market Design), Commissioner Jon Wellinghoff (Demand Response Order), Commissioner Philip D. Moeller, and Commissioner Cheryl LaFleur are some of the recent alumni of the FERC commissioner pool.

FERC commissioners dissent in some FERC orders. Readers are better off by reading both consenting and dissenting opinions of FERC commissioners on key topics. For example, Commissioner Richard Glick has dissented on an important PJM FERC filing called the minimum offer pricing rule (MOPR).

Most FERC commissioners are lawyers. Policy-focused readers should make it a habit to read and follow key announcements from FERC. Some FERC orders have ended up in the courts, like the demand response at the Supreme Court of the United States (SCOTUS) and FERC Order 841 at the DC Circuit Court of Appeals.

Because ISOs and Regional Transmission Organizations (RTOs) derive their regulatory authority from FERC, we don't discuss ISO/RTOs in this chapter. The role of the wholesale energy markets and organized markets is discussed extensively in Chapters 4 and 9.

Exercise 6.3 As an international energy regulator, you are interested in knowledge sharing and learning more about your US counterpart – the Federal Energy Regulatory Commission. You have a meeting with the FERC staff from the Office of External Affairs – State, International, and Public Affairs division. You want to learn about how FERC collects inputs from various stakeholders such as states, associations, and other industry participants. Since renewable hydrogen is a key topic of interest in your country, you want to focus on that topic during your discussions. What questions do you want to ask for this high-level regulator-to-regulator dialogue on the hydrogen economy?

6.3.4 Independent Market Monitors

Readers should know the important institutional role market monitors play in US energy markets. Market monitors have professional authority, not political or direct economic authority, because they don't establish energy laws or implement them. They monitor the markets for market manipulation.

The role of market monitors has increased in importance after the Enron crisis in the early 2000s. At some ISO/RTOs, market monitors are independent and administrated by an outside consultant reporting directly to a board of directors (BOD). In such instances, they are called independent market monitors (IMMs). Potomac Economics typically comes to mind as the IMM for ERCOT, ISO-NE, MISO, and NYISO.[6] A company called Monitoring Analytics is the IMM for PJM (the Pennsylvania–New Jersey–Maryland Interconnection). Monitoring Analytics indicates the company was spun off PJM's internal market monitoring department [12].

At some ISOs like Southwest Power Pool (SPP), there is an internal department called the Market Monitoring Unit (MMU) that serves a similar function. At the California ISO, there are both internal and external market monitors. The internal California ISO function is called the Department of Market Monitoring (DMM). California ISO also has the Market Surveillance Committee.

Like the State of the Union address at the national level, IMM issues a state of the market (SOM) address, typically in June, with market data from last year. SOM reports are full of market data with detailed appendices containing all the summary of key market metrics in a chapter on transmission congestion and financial transmission rights (FTR) markets (for example) such as the real-time value of transmission congestion, day-ahead congestion costs, and FTR funding, FTR auction revenues and obligations, and congestion on external constraints. A MISO SOM has detailed chapters on market trends and load, resource adequacy, market performance in both day-ahead and real-time markets, external transactions, competitive assessment, and demand response programs.

The ISO relationship with market monitors is not designed to be cozy. This is certainly the case with PJM's Monitoring Analytics. PJM's market monitor does not agree with most market improvements proposed by PJM.

Exercise 6.4 Jay is excited to find out that Potomac Economics (IMM for ERCOT, ISO-NE, MISO, and NYISO) and Monitoring Analytics (IMM for PJM) are willing to sit down with him for an interview.

Jay wants to learn how these IMMs were set up, who pays them, how they function in multiple markets, how much staff work for them, what it takes to establish

---

6 ERCOT: Electric Reliability Council of Texas; ISO-NE: New England's ISO; MISO: Midcontinent Independent System Operator; NYISO: New York Independent System Operator

an office, and what impact they had on energy policies. He also wants to go a bit deeper and understand how market monitors collect the data from ISO/RTOs, what problems they have run into, and what penalty mechanisms exist when data is not provided to the market monitor. In essence, Jay wants to understand how the market monitor built trust as an institution. Jay wants to apply all this information to his country/region. Can you help Jay document this in a report?

6.3.5 Energy Information Administration (EIA)

The EIA website says, "The U.S. Energy Information Administration (EIA) collects, analyzes, and disseminates independent and impartial energy information to promote sound policymaking, efficient markets, and public understanding of energy and its interaction with the economy and the environment" [13]. This sums up EIA's institutional role in the energy policy succinctly.

EIA does not have political or economic authority, but EIA has professional authority because it is the topmost energy-data-oriented institution in the US. More information regarding EIA's history and data quality guidelines, including how EIA assures data accuracy, can be found on the EIA website [14].

EIA's achievements in 2020 summarize the important role EIA plays in the US and international energy policy arenas [15]:

- EIA's monthly short-term energy outlook regularly addresses uncertainty in energy markets, and responded to economic changes related to COVID-19.
- In 2020, EIA began including a weekly estimate of US crude oil storage capacity utilization to address uncertainty in market conditions.
- EIA's energy disruptions map combines storm- and fire-related geographic data from the National Hurricane Center and the National Weather Service with EIA map layers to provide status updates on energy-related infrastructure.
- EIA explored the assumed future costs of renewable power generation technologies in the *Annual Energy Outlook 2020*, which feeds many study assumptions on the future cost of electricity.
- EIA provided a detailed analysis of Africa, Asia, and India in its international energy outlook in 2020.
- EIA expanded information on usage factors for utility-scale storage generators in its *Electric Power Monthly*, another popular report in the industry. EIA also expanded its analysis of battery storage and the US electric grid.
- EIA released preliminary energy consumption data for commercial buildings and the manufacturing sector, which is helpful to know where electricity is consumed in large amounts.

- EIA compiled new state-level biodiesel production and consumption estimates in its State Energy Data System.[7]

In summary, EIA is a statistical agency that influences energy policy in the US and elsewhere with its strong focus on data and reports.

6.3.6 North American Electric Reliability Corporation (NERC)

NERC is the electric reliability authority in North America. Before the September 2003 northeast blackout, NERC stood for the North American Electric Reliability Council. But after the blackout highlighted the need for mandatory compliance standards related to tree-trimming, the council became a corporation with a mandate to levy fines up to $1 million per day. NERC reports to FERC.

NERC does not have political authority, but it has indirect economic authority because a reliability standard and the threat of a compliance fine have both direct and indirect consequences on commerce. NERC has a tremendous amount of respect in the electric utility industry leading to professional authority on matters such as operator training, reliability standards, and lessons learned when events like blackouts happen, to name a few.

6.3.7 Federal Bureau of Ocean Energy Management (BOEM)

With the increase of interest in offshore wind (OSW) shown in US energy policy, the relevance of BOEM has increased. As an agency of the US government, BOEM has the political authority in states that are near oceans and other water bodies that BOEM regulates. BOEM also has economic authority when it comes to oceans and OSW because of the potential for job creation in regions where BOEM allows OSW.

BOEM's mandate to develop renewable energy comes from President Obama's term when the Department of Interior (in which BOEM is an agency) and FERC signed an agreement on their roles and responsibilities [16]. BOEM plays a major role in US energy policy, specifically on OSW, because of the federal government commitment to "establish a target to deploy 30 gigawatts (30 000 MW) of offshore wind by 2030, creating nearly 80 000 jobs" [17].

BOEM has offices all along the US coast, starting with the Alaska Regional Office in Anchorage, Alaska, the Pacific Regional Office in Camarillo, CA, and the Gulf of Mexico Regional Office in New Orleans, LA [18]. A good example of the role BOEM plays in determining wind energy areas in federal waters for the state of North Carolina is shown in Figure 6.4.

---

7 The State Energy Data System (SEDS) is the source of the US Energy Information Administration's (EIA) comprehensive state energy statistics. EIA's goal in maintaining SEDS is to create historical time series of energy production, consumption, prices, and expenditures by state that are defined as consistently as possible over time and across sectors for analysis and forecasting purposes. For more information, visit: https://www.eia.gov/state/seds/.

2014 BOEM announced three Wind Energy Areas (WEAs) after stakeholder engagement and deconfliction process

2017: Avangrid Renewables LLC won BOEM's competitive lease sales for Kitty Hawk WEA

March 2021 Governor Cooper asked BOEM to prioritize leasing NC WEAs and identify new WEAs before 7/1/2022

2021–2022 BOEM leasing process, coordinated with NC stakeholders

~2026 First of three 800 MW tranches is operational, remaining operational in 2028 and 2030

Figure 6.4 North Carolina offshore wind areas.

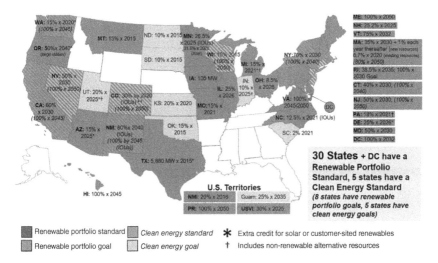

Figure 6.5 Renewable and clean energy standards map.

6.3.8 State Legislatures (Senate and House)

Similar to the US Congress at the federal level, the state legislative process in each individual US state has the political and economic authority to implement energy policies. States are free to set their statewide energy policy like the RPS in the past, and now, 100% carbon-free goal by 2050. As shown in Figure 6.5, many states have taken the RPS further by adopting clean energy goals.

6.3.9 Public Utility Commissions (PUCs)

In a traditional sense, the major actors in policy discussions start with the regulators at the state and federal levels. In some US states, commissioners are appointed by the state's governor. In other states, the commissioners are elected. Hence state public utility commissions (PUCs) (in some cases called public service commissions or corporate/commerce commissions) have economic authority similar to FERC at the federal level. And in states where utility commissioners are elected directly by people, those commissioners have political authority also.

Whether a utility commissioner is appointed or elected, we expect appointed commissioners to behave rationally based on their expertise, and elected commissioners to decide energy policy cases with an eye toward the next election.

In some states, there are advisers to the commissioners who either have a personal relationship with the boss or have the required expertise in the commissioner's focus area. As an illustration, Steven Olea was a policy adviser to the 2020 Arizona Corporation Commission (ACC) chair, Bob Burns. Mr. Olea was hired by Commissioner Burns, as they have a working relationship.

It is also interesting to see that state utility commissioners come from consulting or the industry, and then after their term ends they go back to consulting. They also come from citizen and environmental advocacy groups.

This consulting background of utility commissioners is certainly true of Commissioner Judith Judson, who came from Customized Energy Solutions, a consulting organization, before her term on the Massachusetts Department of Energy Resources (DOER). Commissioner Judson went to work for Ameresco after her term ended with MA DOER. Some commissioners at state come from their staff. This commission staff to commissioner promotion was certainly the case with Commissioner Sally Talberg at the Michigan Public State Commission (MPSC).

Another popular career track for state commissioners is employment at ISO/RTOs at executive positions. Before Dr. David Boyd was Vice President, Government, and Regulatory Affairs at MISO for five years, he was a Minnesota Public Utilities Commission (PUC) commissioner [19]. Dr. Boyd currently works as a consultant [20]. Asim Haque has a similar career trajectory at PJM. Mr. Haque joined PJM as Executive Director, Strategic Policy and External Affairs, in March 2019, after his commissioner and chairman role at the Ohio PUC. Mr. Haque is currently Vice President of State and Member Services at PJM [21].

So, major actors in energy policy from state utility commissions have successful careers in the industry. This success is also true in the federal energy policy arena.

6.3.10 National Association of Utility Regulatory Commissioners (NARUC)

Even though NARUC is an association, it is included here at this stage in the chapter because NARUC has the professional authority among state PUCs and FERC to an extent. By virtue of being an association of state utility commissioners,

NARUC does not have an economic authority because it is not a trade association, unlike the Solar Energy Industries Association (SEIA).

And NARUC does not have political authority, because NARUC does not establish energy laws in the US. But NARUC passes resolutions, which provide information on where collectively US state energy regulators are on any given topic.

6.3.11 The Role of Local City Governments

Some cities in the US are organizing themselves to combat climate change. A big city such as Los Angeles in California or New York City in New York comes to mind with its clean energy goals. We don't hear from small cities in a state like Minnesota or Iowa when they announce carbon-free goals. But many cities in the US are directly taking on energy efficiency, resiliency, or microgrid aspects related to climate change. There is an instance of a coalition of communities and cities in PJM RTO.

Because cities and communities are inside an RTO, the PJM cities and communities coalition (PJMCCC) is organized around decarbonization at the wholesale energy market level [22]. Communities and cities realize that there is a benefit of regional planning and interconnection of renewable resources in organized wholesale markets. As regional markets integrate renewable resources, the cities want to ensure that fossil-based generation leaves the grid faster and better for the entire community.

We can expect more community organizations around climate change in the next decade. Organized wholesale markets can no longer hide behind an energy policy that allows them to claim to be technology agnostic, which in turn can lead to more fossil-based generation. Organized markets must deal with the social and economic justice aspects of the electricity grid. This social justice aspect is a reason why FERC opened an OPP early in 2021 [23]. FERC OPP blurs the line between retail and wholesale markets for electricity and energy. OPP and PJM-CCC show that city governments' role in energy policy would only increase in the future, given their political and economic authority. This represents participative democracy. However, multiple jurisdictions dealing with technical matters with a long-term financial value necessitates considerable negotiation and compromise between the various institutional parties.

Exercise 6.5 You want to bring a positive change to your local city government on the topic of sustainability. You want your city to embrace solar power, to have a microgrid around city government buildings, for most of its properties to have EV charging and solar carports, and to set a goal to be 100% carbon-free by 2050. How would you put together a plan to achieve this goal to engage your city?

Professors Stephanie Moulton and Jodi R. Standfort jointly published a paper called "The Strategic Action Field Framework for Policy Implementation Research" [6]. This paper provides a framework called SAF. In this SAF framework, four forms of authority are identified to implement policies. They are: (i) political, (ii) economic, (iii) professional, and (iv) collective beliefs and value authority. Can you prepare an illustrative case for the SAF framework that puts together a plan to engage your city on meeting the 100% carbon-free goal by 2050? Identify the political, economic, professional, and collective beliefs and values authority in your city. Explore other frameworks that help you bring about positive change, as requested in this exercise.

6.3.12 Energy Advocates Role in US Policy

For most energy industry topics, there are several advocacy groups involved in influencing policy that are different from energy associations and alliances. Energy advocates influence both political and economic authorities.

In most cases, these foundations and advisory groups have a single focus. Understanding what their specific focus is and who are their major funders goes a long way toward understanding the unique role advocacy groups and energy foundations add to policy discussions.

In some cases, learning about the focus of the organizations is not straightforward. Additionally, several organizations in this space can pool their funding and go after a specific outcome in a docket or case before state or federal commissioners.

For example, the Environmental Law & Policy Center (ELPC) out of Chicago spearheaded opposition to a MISO transmission line in Wisconsin. Other clean energy organizations (CEOs) such as Fresh Energy, Clean Grid Alliance, and the Minnesota Center for Environmental Advocacy (MCEA) were in favor of the transmission line. Other notable energy advocates with a national presence in the US include:

- The Energy Foundation
- Vote Solar
- Natural Resources Defense Council (NRDC)
- ELPC
- Union of Concerned Scientists (UCS).

Exercise 6.6 You are interested in a job at the NRDC Sustainable FERC project. You read this chapter to understand what role institutions play in setting energy policy. On top of that, you have read whatever you can find about PJM RTO, including NRDC's PJM explainer [24].

The job that you are interested in covers the MISO region. Can you put together a MISO explainer similar to the PJM document? Your assignment in preparation for this NRDC job interview must address questions such as, "How are decisions made at MISO?", "How does a change at MISO get started?" and "Who has the final say at MISO?"

6.3.13 Stakeholder Working Groups' Role in Setting US Energy Policy

People with a stake in the outcome are called stakeholders. Stakeholders influence political, economic, and professional authorities.

Engineers with a stake in the outcome of the Loss of Load Expectation Working Group (LOLE WG) are stakeholders for that task. The management of these engineers with a stake in the resource adequacy policy at the ISO are stakeholders in the Resource Adequacy Sub Committee (RASC). The leaders of the management in executive positions are overseeing their organization's stake in the ISO at the advisory committee level, which typically has a direct line of sight to the BOD. That is how stakeholders set policy, and in an ideal world, there is alignment in the organization's position across all these three groups: working group, subcommittee, and committee.

Not all policy discussions happen in an organized setting like an ISO or RTO. Some happen with industry participants outside ISOs and utilities with advocacy and support groups whose main role is to bring diverse participants together. These groups have funding from large energy foundations. Meridian Institute is an example that comes to mind for convening a workshop on "replacing peaking units with storage" policy aspects.

Working groups are where the actual engineering work is done and presented. The structure is like a peer review carried out every month and sometimes twice a month. There is less push and pull around policy and more discussion around engineering assumptions, base models, and modeling results. Listening to a well-crafted presentation is like reading a study report. Because of the engineering underpinnings, working groups tend to be straightforward.

Accordingly, commissioners or leaders of most organizations do not attend or participate in a working group. Their engineers do. Readers are better off listening to or sitting in on subcommittee or committee meetings. But on the flipside, high-level policy aspects are presented at these committee meetings, and not enough engineering details are covered. Someone with a basic understanding of engineering concepts might struggle in a committee meeting. But on the other hand, a working group is right up their alley because the engineers at the working group level discuss assumptions such as the heat rate, the hurdle rate, the fixed and variable costs of generators, their assumptions of wind and solar in the area, and how much capacity factor they assumed in this particular calculation.

At the committee level, because the organization has representatives at the working group level, they are concerned less about engineering assumptions and more about their bottom line of whether the policy has financial implications for them compared to last year.

Some working groups need FERC critical energy/electric infrastructure information (CEII) clearance to pore over detailed transmission planning models. In this case, only engineers at relevant organizations can participate. These are closed meetings. Sometimes these closed meetings provoke consternation from the state commission staffers.

Exercise 6.7   Jay is visiting his friends in Pennsylvania. In addition to visiting historic sites, he wants to do some work. He found that the Commonwealth of Pennsylvania's PUC has opened a docket M-2020-3 022 877 related to the role of electric storage on the utility distribution system.

Once on the Pennsylvania PUC website docket search, Jay finds many organizations have filed comments on the subject matter, including electric distribution companies such as First Energy, PPL Electric Utilities, and Duquesne Light; associations such as the Energy Storage Association (ESA), the Retail Energy Supply Association (RESA), and SEIA; alliances such as the Advanced Energy Management Alliance (AEMA) and the Pennsylvania Energy Consumers Alliance (PECA); and numerous other organizations. Can you help Jay summarize the comments submitted by February 2021, specifically noting which organizations have advocated for the establishment of a working group and why?

6.3.14   Associations and Alliances Role in Policy

State and federal commissioners set policies based on the record in front of them. That record is set by receiving comments from the public and specific associations. For example, for a specific net metering policy concerning solar, it is expected that the state commission would receive comments from the SEIA. In some cases, the US national SEIA might hire a consultant to articulate SEIA's position in that state docket on net metering. All these reasons point to associations such as SEIA having economic authority.

If SEIA advocates on behalf of solar at the national level, the Edison Electric Institute (EEI) advocates for electric utilities. SEIA and EEI were on the opposite sides of FERC Order 841 on energy storage. SEIA is on the FERC side advocating for energy storage, and EEI is on the side of utilities who believe FERC overstepped its authority on distribution-connected storage. NARUC and EEI are on the same side of this energy storage order.

Another example where associations played a prominent role was when PJM came out with a market policy that long-duration storage was needed to implement the intent behind FERC Order 841. ESA and NRDC hired Astrape

Consulting and came out with a study that showed four-hour storage was enough for PJM's capacity needs [25].

Regulators have their association, NARUC. Annual and mid-year membership meetings are held in nice locations like Phoenix and Orlando during the winter. NARUC meetings are usually open to nonregulatory industry participants as well, even though the majority are sitting or past state and federal commissioners.

Consumer advocates have their association, the National State Utility Consumer Advocates (NASUCA). Their meetings are held around the NARUC meetings.

Exercise 6.8 In your country, you want to establish a program to evaluate advanced technologies in the energy industry. You want to model it around providing insights into this new technology for the engineers at the grid operator.

What are the initial program design aspects you should consider? What is going to be your role, and what would be the role of the technology provider? Take the help of PJM's Advanced Technology Pilot Program (ATTP) [26].

6.3.15 Summary of US Institutions

To summarize the major institutions in US energy policy and their primary authorities according to the SAF framework, please see Table 6.4.

Exercise 6.9 You have a job-related informational interview with the legal and policy adviser to the Illinois Commerce Commission (ICC), the state commission of Illinois trusted with energy policy regulation. You know and have read that the current chair of the ICC is focused on the interconnection of DERs and data access for aggregators.

Illinois is in both RTOs: MISO and PJM. As a result, the ICC is a member of both the Organization of MISO States (OMS) and the Organization of PJM States, Inc (OPSI). What questions would you be prepared to ask the legal and policy adviser to the chairman? What Illinois state rules and policies would you advocate being changed at the state level and RTOs for the smoother interconnection of DERs? What data access would you provide to aggregators of Illinois electric utilities? This job is a high-visibility role for you if employed at the ICC because you are the point person for both DER interconnection and data access policies at both PJM and MISO.

6.4 What Are the Major Institutions in International Energy Policy?

After setting the stage with the example of US institutions and their respective authorities in the SAF framework, we now turn our attention to institutions

Table 6.4 Summary table for US energy policy institutions and their primary SAF framework authority.

US Institution	Political	Economic	Professional
US Congress	√		
DOE	√		
FERC		√	
IMM			√
EIA			√
NERC			√
BOEM	√		
State Legislature	√		
State PUC		√	
NARUC			√
Local city governments	√		
Energy advocates	√		
Stakeholder Working Groups			√
Associations and Alliances		√	

involved in international energy policy. We also give examples from countries like Nepal and Congo (to showcase actors in fragile economies) and Australia (to showcase a leader in global renewable energy transition).

To provide the reader the necessary context from ground-level understanding, we surveyed professionals working in the international energy policy scene.

6.4.1 Examples from Strong Economies

6.4.1.1 European Union (EU)

As an institution, the EU has political, economic, and professional authority in Europe. In the words of our survey respondent, "With the recent Green Deal, and recent years increasing focus on green and inclusive transition, the EU has done a decent (moving to good) job in setting an agenda."

Just like US regulatory institutions, in the EU, the commissioners are changing while the staff is maintained. In general, our survey respondent has "noticed that physicists have a tremendous ability to understand and navigate in energy and energy policy."

6.4.1.2   European Commission (EC)

In the energy policy area, the European Commission (EC) has both economic and professional authority relative to the EU's political authority. According to one of our survey respondents, "Energy policies tend to be drafted by the EC under the form of directives. Loads of associations are trying to influence the EC to support their views. Well-established and powerful associations such as Eurelectric tend to have a stronger impact, but it could backfire as well as they are the 'bad guys.' Member states are critical as some can block a policy depending on the voting rules."

Also, EC is effective in advancing an energy policy for the entire EU. Some argue that EC has been too effective in advancing policy changes if we take the example of incentives for homeowners to install photovoltaic (PV) systems in Europe.[8] Europe is realizing the cost of those PV incentives, and as a result the EC has changed the policy, which led to homeowners' complaints at the courts.

6.4.1.3   International Energy Agency (IEA)

In the words of a survey respondent, the IEA is "a key data-hub for global energy data" because the IEA is "increasingly active on the green transition" has transitioned from being an "oil-importing country oversight body." The IEA has economic and professional authority based on our interpretation of the survey response.

The IEA has been criticized for consistently underestimating renewable development. At the IEA, the current director Fatih Birol has a long track record and experience within energy policy. Our survey respondent said, "He knows what he's talking about." The IEA is not a political organization, so it does not have political authority.

6.4.1.4   World Energy Council (WEC)

The World Energy Council (WEC) is rarely seen in international news. According to our respondent, what makes them unique is "their concept of the energy trilemma, i.e. balancing social, environmental, and security-of-supply issues."

---

8 Christophe Druet survey, "I would say the EC has been (and still is) instrumental in taking steps toward a decarbonized system. It's very complex as every decision, policy … is a compromise to accommodate the interests and views of a lot of people. This is why it's so slow. I believe it wouldn't be more effective to take more drastic decisions as the counter-actions would be stronger. Policymaking is a bit like damping: you need to generate enough change to produce the effect you want and if you provoke way more change than anticipated, you'll have to drastically change your policy again or face damaging consequences. That's what's happening with the green certificates here. They launch a system that was way too interesting for those installing PV. PV deployed massively. A success according to ecologist. But the cost for society to finance the green certificate exploded. They changed the policy. Owners complained and went to court."

The WEC is not a political organization, hence it does not have political authority. It has professional authority.

6.4.1.5 European Network of Transmission System Operators for Electricity (ENTSO-E)

In the words of our survey respondents, "EU level grid policies are drafted by TSOs under the umbrella of ENTSO-E. There is a solid consultation process before any approval by the Agency of Regulators. This agency tends to take a stronger and stronger position in amending TSOs proposals." Hence it is worth noting ENTSO-E's economic and professional authority in advancing energy policy in this chapter.

ENTSO-E in Europe has consolidated six European organizations such as the Association of the Transmission System Operators of Ireland (ATSOI), the Baltic Transmission System Operators (BALTSO), the European Transmission System Operators (ETSO), NORDEL,[9] the Union for the Coordination of the Transmission of Electricity (UCTE), and the UK Transmission System Operators Association (UKTSOA). All these six organizations were wound up on 1 July

Table 6.5 Scope of predecessors to ENTSO-E.

Organization name	Scope
Association of the Transmission System Operators of Ireland (ATSOI)	For the coordinated activities between EirGrid (Irish transmission system operator) and System Operator Northern Ireland (SONI)
Baltic Transmission System Operators (BALTSO)	BALTSO ensures the cooperation of Estonian, Latvian, and Lithuanian TSOs
European Transmission System Operators (ETSO)	An international association with a direct membership of 32 independent TSOs from the 15 countries of the European Union plus Norway and Switzerland
NORDEL	NORDEL ensures cooperation between TSOs in Denmark, Finland, Iceland, Norway, and Sweden
Union for the Coordination of the Transmission of Electricity (UCTE)	Represented 29 TSOs of 24 countries in continental Europe
UK TSOs Association (UKTSOA)	Coordinated activities between the TSOs of the United Kingdom

---

9 Nordel was founded in 1963 as a body for cooperation between transmission system operators in Denmark, Finland, Iceland, Norway, and Sweden. Its objective was to create the preconditions for a further development of an effective and harmonized Nordic electricity market.

2009, after ENTSO-E was founded the year before [27]. Readers should note that Brexit has had a big negative impact as technical and operational cooperation tends to be restrained for political reasons.[10]

ENTSO-E took over the activities of six previous organizations in Europe, shown in Table 6.5. To understand ENTSO-E's responsibilities, it is worth noting the scope of the six organizations. Today ENTSO-E represents 42 TSOs from across 35 countries in Europe.[11] The legal mandate for ENTSO-E comes from the EU's Third Legislative Package, which aims to deregulate the natural gas and electricity markets.

6.4.1.6 Australia

Australia sees an unprecedented transition into a renewable future [28]. Australia is setting the global trend. Hence, we want the reader to understand the countrywide actors there. Several of the Australian energy institutions are involved in renewable energy [29]. They are:

- Australian Energy Regulator (AER): makes decisions that promote efficient investment in, and efficient operation and use of, energy services for the long-term interests of energy consumers. This is a government institution, hence has political and economic authority.
- Australian Renewable Energy Agency (ARENA): funds innovation and shares knowledge about renewables, accelerating Australia's shift to a renewable energy future. A government institution with professional authority.
- Climate Change Authority: provides expert advice on Australian Government climate change mitigation initiatives. Also, a government institution but has professional authority.
- Clean Energy Council: the peak body for the clean energy industry in Australia and is the leading one of all the private renewable energy organizations, hence has economic and professional authority.
- Australian Solar Council: the peak industry body for the solar industry in Australia, hence has economic authority in the solar industry.
- Australian Energy Storage Council: seeks to advance the uptake and development of energy storage solutions in Australia, hence has economic authority in the storage industry.
- SEIA: formed in 2007 in response to demand from within the industry, hence has economic authority in the solar industry.
- Australian Photovoltaic Institute (APVI): comprises companies, agencies, individuals, and academics interested in solar energy research, technology, and policies. Its objective is to support the increased development and use of PV

---

10 External reviewer comment.
11 As of 1 January 2022, this number is 41 because National Grid ESO will be leaving ENTSO-E. Source: Christophe Druet.

Table 6.6 Summary of multiple renewable energy-based institutions in Australia using the SAF framework.

Australian Renewable Institution	Government?	Political	Economic	Professional
Australian Energy Regulator (AER)	√	√		
Australian Renewable Energy Agency (ARENA)	√			√
Climate Change Authority	√			√
Clean Energy Council			√	
Australian Solar Council			√	
Australian Energy Storage Council			√	
Solar Energy Industries Association			√	
Australian Photovoltaic Institute (APVI)	√			√
Climate Council				√
Electric Vehicle Council			√	
Australian Solar Thermal Energy Association (AUSTELA)			√	

via research, analysis, and information. Also, a government institution with professional authority.

- The Climate Council: supplies independent, authoritative climate change information to the Australian public in the belief that Australians deserve to have independent information on the state of their climate, hence has professional authority.
- Electric Vehicle Council: this body (launched May 2017) represents members involved in producing, powering, and supporting EVs, hence has economic authority in the EV industry.
- Australian Solar Thermal Energy Association (AUSTELA): works with other renewable energy organizations to improve investment in solar thermal power generation in Australia, hence has economic authority in the solar industry.

Because of multiple renewable energy-based associations, councils, and agencies, including regulatory authorities in Australia, we summarize the institutions using the SAF framework in Table 6.6. The key takeaway in Australia is: not all government institutions have political authority. As we see, ARENA, the Climate Change Authority, and the APVI have professional authority in Australia.

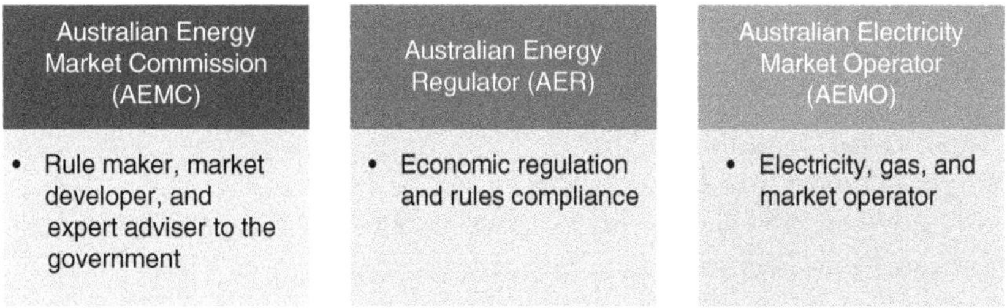

Australian Energy Market Commission (AEMC)	Australian Energy Regulator (AER)	Australian Electricity Market Operator (AEMO)
• Rule maker, market developer, and expert adviser to the government	• Economic regulation and rules compliance	• Electricity, gas, and market operator

Figure 6.6 Market regulatory institutions in Australia.

Additionally, there are a few institutions in Australia (shown in Figure 6.6) that are involved in regulating the markets [30]. So, these institutions first have economic authority and professional authority. These are:

- Australian Energy Market Agreement (AEMA): The AEMA sets out how energy policies are developed through the jurisdictions.
- Australian Energy Market Commission (AEMC): The AEMC is an independent statutory body with two key roles: making and amending rules for the National Electricity Market (NEM), elements of the natural gas market, and related retail markets, and providing strategic and operational advice to the COAG Energy Council.
- Australian Electricity Market Operator (AEMO): Day-to-day responsibility for operations of wholesale and retail markets (electricity and gas), e.g. operates wholesale market operator for NEM (eastern market) and WEM (Western market).
- Reliability Panel: Monitors, reviews, and reports on the safety, security, and reliability of the national electricity system, plus performs any other functions or powers under the NEL or National Electricity Rules (NER).
- Energy Security Board: This board is made up of senior members from AEMC, AEMO, and AER.

6.4.1.7 Energy Regulators Regional Association (ERRA)

Like NARUC in the US, Energy Regulators Regional Association (ERRA) is an association that brings together regulators from countries in Europe, Asia, Africa, the Middle East, and North and South America. Hence our understanding is that ERRA has professional authority.

ERRA was established with USAID support in the 1999–2008 time period [31]. NARUC is an associate member of ERRA. A complete list of ERRA members can be found at this location [32].

The influence of ERRA in member countries' energy regulation can best be summarized in the ERRA purpose statement: "to increase the exchange of information and experience among its members and to expand access to energy regulatory experience around the world" [33].

6.4.1.8 China

Energy institutions that affect the Chinese power sector are generally not known in the Western world.[12] However, much renewable transition is happening in China.

Decentralized electricity (e.g. microgrids, virtual power plants, demand response) is a trendy topic in China. However, it is not mature. The Chinese government's focus is on a centralized energy plan. This focus probably also has to do with the centralized government structure.

The renewables are in the west and north, and the load in the east and central, parts of the country. Hence China is focusing on high voltage DC (HVDC), ultra HVDC, and ultra HVAC. UHVDC and UHVAC are typically over 800 kV, and HVDC is a few hundred kV to less than 800 kV. The National Energy Administration drafts the policy and sets the national and industrial standards and requirements, and the grid and generating companies set the enterprise standards and technical requirements.

Before 2002, National Electric Power Company was the only centralized electric power company in China which covered the generation, transmission, distribution, and supply in most of China. In 2002, China implemented its first electric power reform, and the National Electric Power Company was separated into five generating companies and two grid companies. So now, in China, there are big five state-owned generating companies and a few smaller ones. The generating companies are competitive.

The two grid companies mainly manage transmission and distribution. Among them, the State Grid Corporation of China (SGCC) covers over 80% of China. It is also one of the largest grid companies in the world. Another state-owned grid company called China Southern Power Grid (CSPG) covers the five southern provinces of Guangdong, Guangxi, Yunnan, Guizhou, and Hainan. The grid companies don't compete because they cover separate geographies. There are also some small local electric power companies in some provinces owned by the local government. See Figure 6.7 for a high-level outline of the structure described in this section.

Also, after the 2002 reform and recent reorganization, the planning, design, and construction companies originally under the National Electric Power Company are separated and formed two big state-owned planning, design, and construction companies, i.e. China Energy Engineering Corporation Limited and Power China. On top of these companies, there are additional state-owned electric power equipment manufacturing companies in China, and they are: China National Electric Equipment Group Company, China Dongfang Electric Group Company, and

---

12 This entire China section would not have been possible without our friend Ming Ni's help.

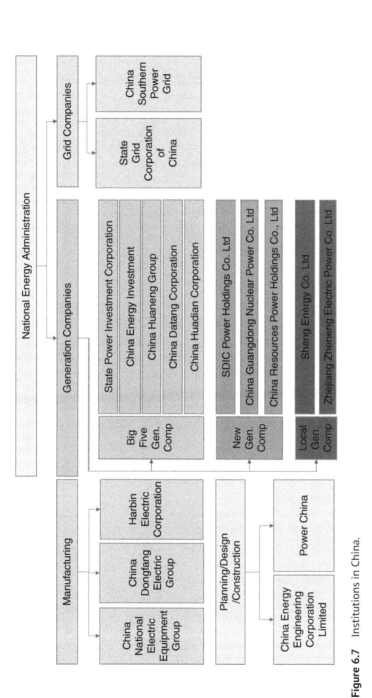

Figure 6.7 Institutions in China.

Harbin Electric Group Company. There are also some local government-owned or private electric power equipment manufacturing companies in China.

The generating and grid companies initiate generation, transmission, and distribution projects, but they must first be approved by the National Energy Administration or its provincial branches depending on their size or voltage level. The detailed planning, design, and construction of these projects are done by the planning, design, and construction companies. Though decentralized power systems are all the rage right now, they are not as established. The grid companies have the mandate to build the lines and provide power even in the remotest villages. Citizen groups are not as active in China, and individuals generally don't pay much attention to power system planning and operations.

Exercise 6.10 The energy institutions in China are typically run by the state. What kind of authority do they possess?

6.4.2 Examples from Fragile Economies

6.4.2.1 Nepal

Like the actors from strong economies who are leading the global renewable transition, several fragile economies have immense renewable potential. But due to various reasons (technical, economic, and sociopolitical), they are unable to exploit it. They are leading in their transition in their own way. For example, in Nepal, the Nepal Electricity Authority is leading a top-down transition. It is the central actor and a government monopoly (see Figure 6.8). Hence the NEA has political authority. Other key organizations are shown below with their main role [33]:

- Policy-level institutions: the following institutions have political authority:
  - o Ministry of Energy: responsible for power sector policy formulation, water resource development, oversight and regulation of the NEA, and private power development.
  - o National Planning Commission: responsible for the coordination and development of the government's five-year multisector investment program.
  - o Water and Energy Commission: responsible for policy advice to the government on technical, legal, environmental, financial, and institutional matters related to water resource planning and development.
  - o National Water Resources Development Council: responsible for government guidance on strategic issues and policy regarding integrated water resource development.
  - o Environment Protection Council: responsible for policy development and preparation of environmental regulations and environmental protection guidelines for environmental assessments, permits, licensing, inspection, and monitoring of environmental licenses.

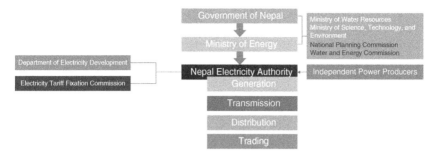

Figure 6.8 Major institutions in Nepal. The Nepal Electricity Authority is the government monopoly that is responsible for the generation, transmission, distribution, and trading of electricity.

- Regulatory-level institution:
  - Electricity Tariff Fixation Commission (ETFC): reviews and approves the tariff applications submitted by the NEA, hence has economic authority.
- Operational-level institutions: the following institutions have economic and professional authority:
  - NEA: responsible for electricity generation, transmission, and distribution throughout Nepal; energy exchange with India; and purchase of electricity from independent power producers (IPPs) as a single-buyer agency.
  - Butwal Power Company: a nonprofit organization under the United Mission of Nepal, responsible for undertaking rural electrification in Nepal.
  - IPPs: responsible for the development of private power plants and the generation of electricity.
- Implementation-level institution
  - Department of Electricity Development, Ministry of Energy: responsible for implementation and promotion of the government's private power policy, management of bidding process for IPPs, issuance of survey licenses, provision of guidance to private investors, and technical support to the ETFC, hence has economic authority and professional authority.

In summary, as shown in Table 6.7, Nepal's energy policy institutions have all four forms of authority as outlined in the SAF framework, if we assume each institution has its culture, values, and belief system.

6.4.2.2 Democratic Republic of the Congo

The Democratic Republic of the Congo (DR Congo) is another representative of several more fragile economies [34]. Like Nepal, the country's energy resources are in development. There are fewer energy institutions than in some other economies. However, the central actors here also exert political authority.

Table 6.7 Summary of Nepal's energy policy Institutions using the SAF framework.

Nepal Institution	Level	Political	Economic	Professional
Ministry of Energy	Policy	√		
National Planning Commission		√		
Water and Energy Commission		√		
National Water Resources Development Council		√		
Environment Protection Council		√		
Electricity Tariff Fixation Commission (ETFC)	Regulatory		√	
Nepal Electricity Authority (NEA)	Operational		√	
Butwal Power Company			√	
IPPs			√	
Department of Electricity Development, Ministry of Energy	Implementation		√	

DR Congo has a population of 85 million people. The electrical access rate is 19%, the access is only 1% in rural areas and 41% in urban areas, according to the World Bank [35]. There is a high hydroelectric resource potential, but it has only been partially developed. Transmission and distribution are minimal and often in poor repair.

Government energy institutions with political authority:

- Société nationale d'électricité (SNEL), national utility
- Autonomous regulatory agency of the government (ARE), in development
- ANSER rural electrification agency of the government, in development.

International energy, finance, and aid institutions with economic authority:

- United States Agency for International Development (USAID), Power Africa
- United Nations High Commission for Refugees (UNHCR)
- NGOs such as CARE and Alight.
- African Development Bank
- World Bank.

Private and semiprivate entities with political and economic authority:

- Sinohydro and China Exim Bank, new 120 MW hydroelectric dam.
- Two governments dams funded by Exim Bank of India.
- Various small solar and hydro IPPs.
- Direct sale/rental of small standalone solar household systems (SHS), African/international companies such as Bboxx.

For further details on energy issues in fragile economies, please see Chapter 10 of this book.

Exercise 6.11 You are asked to prepare a policy brief on "The Race to a Fossil Fuel-Free Transportation Sector: What the US can learn from Nordic countries to accelerate its transition." The new Deputy Secretary of Transportation at the US Department of Transportation wants to offer you an internship. You are pretty excited that the US is back in the Paris Climate Accord. You are also happy that General Motors (GM), a car manufacturer, has released a Super Bowl advertisement that challenges the US to do better on electric vehicles (EVs) than Norway. If you don't live in the US and are not interested in it, feel free to prepare such a brief for your country/region.

As an intern to the Deputy Secretary, you have to communicate with the US embassies of Denmark, Finland, Iceland, Norway, and Sweden. Your task is to manage multiple high-level ministers' schedules and top corporate leaders such as the prime minister of Norway, the Finnish Minister of Transport and Communications, the Icelandic Minister for the Environment and Natural Resources, and the Senior Vice-President of Global Public Policy, General Motors. What will you include in the draft policy brief to impress your Deputy Secretary and foreign leaders?

6.4.3 Examples from the Private Sector

Private enterprises, by virtue of their economic authority, often take the lead in steps toward decarbonization. Our international survey respondent comments point out that Bloomberg New Energy Finance (perhaps, specifically Michael

Liebreich) has been quite outspoken. Danish utility Ørsted. Big tech (Apple, Microsoft, and others) has its carbon targets. These developments show that, both in the US and internationally, private companies are leveraging their economic authority to take the lead in decarbonization.

In Australia, there are many examples of private companies taking the lead in decarbonization. The renewable energy certificate (REC) market generally is an example in many countries, including in Australia. It is mostly a voluntary market. But companies are innovating around it and implementing it, e.g. time-based RECs from Google.

Another example, applicable globally, is a voluntary move to go for zero emissions (could be tied to RECs or not). Siemens Smart Infrastructure recently announced that they want to go for zero emissions. They announced plans to become carbon neutral by 2030. And zero in their headquarters by 2023. Another example is the distribution network service providers (DNSPs) being transformed into distribution system operators (DSOs) in Australia. The government proposed rule changes and mandates for nonnetwork alternatives. The distribution providers are heavily invested in exploring options and conducting trials.

Exercise 6.12 Google Cloud is pioneering a new concept called the Time-based Energy Attribute Certificate (T-EAC) in the Midwest in the US and in Denmark [36].

This T-EAC is one way to verify clean energy at Google's data centers. What are other ways out there? What are the available RECs in the marketplace today? Are they only available for renewable energy? What is the definition of renewable energy in these tracking systems? Is there a role for government and other institutions in this certificate market? Explore the relationship between Google, a for-profit company, and Midwest Renewable Energy Tracking Systems (M-RETS), a nonprofit company based in Minnesota. Is the role of tracking certificates a mandate for government or nonprofit enterprises?

6.4.4 Summary of International Institutions

To summarize the major institutions in international energy policy and their primary authorities according to the SAF framework, please see Table 6.8.

6.5 The Role of Environmental Regulations, Goals, and Pledges in Setting Policy

A special form of institutional effect derives from worldwide climate change, low-carbon, and renewable-energy goals. These energy policies have significant influence over political and economic authority-based institutions.

Table 6.8 Summary table for international energy policy institutions and their primary SAF framework authority.

International Institution	Governmental	Economic	Professional
European Union (EU)	√		
European Commission (EC)		√	
International Energy Agency (IEA)			√
World Energy Council (WEC)			√
European Network of Transmission System Operators for Electricity (ENTSO-E)		√	
Nepal Electricity Authority (NEA)	√		
Australian Energy Regulator (AER)	√		
DR Congo – Société nationale d'électricité (SNEL)	√		
Energy Regulators Regional Association (ERRA)			√
National Energy Administration	√	√	√
Private enterprises (e.g. big technology companies)		√	

Over the last two decades, many entities and institutions have established standards and goals for clean energy. This is being done by institutions at the local level, the country level, and even by pledges and treaties at the international level. The standards and goals strongly influence how electrical generation, storage, transmission, and distribution are planned. The power of these standards varies substantially according to whether they are good faith pledges or laws and regulations. For illustrative purposes, the following are some examples:

- Many cities, states, provinces, and even countries have adopted standards and goals for percentages of clean energy. Some of these standards are tied to concessions or rules for utility operation. They sometimes go by the name of RPS: renewable portfolio standards.
- Individual residents and businesses have taken advantage of tax or rate subsidization programs provided through local governments and utilities to install their own renewable energy equipment. Many of these installations are connected to and can sell electricity back to the grid at favorable rates.
- As mentioned earlier in this chapter, corporations, nonprofits, and government buildings often set internal goals for their carbon neutrality and install or purchase renewable energy or renewable energy credits (RECs). These RECs

for renewable energy produced in a remote location are also closely related to carbon/CO_2 abatement trading, which is done on various markets in development around the world.

This is a complex topic and still very much in development. It is often the subject of political and popular debate as it has a great effect on the world's environment and the investment required for our modern energy systems. This book provides the practitioner with the basic tools to better understand the tradeoffs, outcomes, costs, and players in dealing with this critical topic.

Exercise 6.13   Owing to multiple drivers in your country, hundreds of renewable projects seek to interconnect the transmission system. The transmission operator is overwhelmed. These operators are looking for policy direction from you. If the transmission grid operator approves all of the renewable project interconnection requests, the grid's reliability may go down. If they don't approve these requests, the country cannot meet its RPS. Can you help propose some policy changes that take effect in one/two/three years?

How do you propose to bring all the stakeholders to the table? What are the technical solutions that are possible? How would you address the need for more load flow models to study this queue backlog? Assume the interconnection queue keeps increasing in the future. What is your plan for annual project intake?

6.6   The Role of the Courts

Yes, energy policies do end up in courts. But that is the last resort. Let us take some examples from the US. The first venue to resolve the differences is the energy regulator. If that does not work, usually, the aggrieved party takes the case to the state or regional regulator. If that doesn't work, the supreme court at the state or regional level is involved. If that doesn't work, then the federal appeals process is petitioned. Lastly, if the petitioner doesn't like lower court rulings, they are free to take the case even higher, to the SCOTUS. There are several instances where energy policies have ended up in the SCOTUS.

The following are examples of when policies ended up in the highest US court:[13]

---

13 In order to give a wider international flavor, we have not provided an in-depth analysis of all of the energy-related policies that have ended up in the Supreme Court of the United States. For example, according to Jessica Bell, the reader should note a zero emissions credits case (certificate denied) and a uranium mining case (Virginia won) in the past few years. The Supreme Court also just heard oral argument in the *PennEast* case, which considers whether a certificate holder under the Natural Gas Act can take state land or if the state can successfully raise a sovereign immunity defense. There are others on the Clean Air Act front. This website provides some of the additional cases, see Allco https://statepowerproject.org/connecticut (accessed 11 December 2021).

- FERC Order 745 on-demand response compensation ended up at SCOTUS. SCOTUS ruled in favor of FERC that FERC had the authority over setting prices for demand response programs participating in the wholesale energy markets [37]. For more on SCOTUS's 25 January 2016 decision regarding *FERC* vs. *Electric Power Supply Association et al.*, please visit the Schiff Hardin site [38].
- Demand response provider Voltus complained to FERC about the lack of market opportunities at MISO. A copy of Voltus's complaint in the Federal Register can be found online at the Federal Register [39]. At the time of writing, FERC has issued Order 2222-A, which does not extend the opt-out for DER aggregations that include demand response, but that an aggregator of just demand response would still be subject to the RERRA opt-out possibility [40]. There are pending requests for rehearing on this issue. It is also worth noting that in FERC Order 2222-B [41], FERC is seeking to investigate removing the opt-out as to demand response. Voltus had argued that state opt-out harms its demand response prospects at MISO.
- LS Power, an independent transmission company, petitioned SCOTUS to sue Minnesota public utility commissioners regarding Minnesota's Right of First Refusal (ROFR) law. LS Power felt that the Minnesota ROFR law was anticompetitive [42]. SCOTUS refused to hear this petition [43].

The following is an interesting case from Europe:

- This case concerns the definition of system operation regions. TSOs made a proposal, but the European Agency for the Cooperation of Energy Regulators (ACER) made a different decision. Then, ENTSO-E appealed to the board of appeal of ACER and obtained to get from ACER either another decision or a better explanation. ACER provided a better explanation, and ENTSO-E appealed again, but right before submitting the appeal, ACER withdrew the better explained decision.

Exercise 6.14 If you live in the US, write a blog posting of 1500–2000 words on Minnesota Power's petition for approval of the EnergyForward Resource package. The specific natural gas project called Nemadji Trail Energy Center (NTEC) is rated 525 MW, located in Wisconsin near the Minnesota border. This project is fascinating because Wisconsin state law does not allow Minnesotan companies to own power projects. Hence Minnesota Power's Wisconsin affiliate South Shore Energy LLC equally owns NTEC with Dairyland Power Cooperative (DPC), a Wisconsin generation and transmission electric cooperative.

Pay close attention to environmental reviews such as Minnesota Environmental Policy Act (MEPA), when a public utility commission assigns an administrative law judge (ALJ), why PUC rejected ALJ's recommendation, when the utility files for appeal, and who are all the stakeholders at the table including when both Minnesota Power and the Minnesota PUC are on the same side. What is the "but-for causation standard"? Also, understand the role of institutions such as the Minnesota Supreme Court.

If you live outside the US, write a blog posting of 1500–2000 words on the hottest electrical system court case, of your choice, in your region.

6.7 Visioning

Policy development in the future will bring many discussions with actors from diverse business models and interests. It is becoming increasingly complex, and also that the policies are more and more debated from a legal point of view rather than a technical/environmental point of view, hence the procedure or process being challenged instead of the objective pursued by the policy. This challenge is logical, but it slows down the process quite effectively.[14]

Depending on the circumstances, readers may be better off by focusing on a single area or a single institution at a time, for example, working at an organization that sits at the intersection of different policy debates.

One mistake a reader can make is thinking they have now learned everything about a rapidly changing topic. That may immediately become out of date if the industry moves in a different direction, for example a reader who is an expert on how independent system operators (ISOs) function and has studied ISO/RTOs in the US for quite a while. That knowledge immediately becomes a liability if they don't realize that DSOs are on the horizon.

The TSO knowledge set becomes an advantage if the student can pivot to distribution system discussions and contribute effectively. All of which depends on when the transmission to distribution shift happens, based on their industry network. Proactive students do not wait. They act based on the information available to them.

The following ideas enable a student to keep an eye on the industry for gainful employment or more research into specific policies:

- Write a position paper.
- Attend policy seminars.
- Gain internship experience.
- Network purposefully.
- Seek challenging projects.
- Listen to mentors.
- Tour plants, offices.
- Follow an association closely.
- Follow a specific policy at FERC or state commission proceeding.
- Understand the details.

---

14 Comment added based on Christophe Druet's review.

Case 6.1 *Convincing Your Peers You Have a Legitimate Complaint and How to File It*

Situation

For the US audience: Your assignment is to help your client file a Section 206 complaint at the FERC. You must research the Federal Power Act Section 206. One source is PJM [44].

For the non-US audience: Please find the appropriate top energy regulatory authority and understand the process of filing a complaint.

Analysis

Based on the nature of the client business and the RTO decision, your client might have a reasonable chance of reversing the RTO decision. Some of the questions or tracks to follow could include:

- Past Section 206 complaints at FERC, and where they landed.
- FERC needs to see evidence that your client has raised this issue at RTO stakeholder committee meetings. The meeting minutes should reflect your participation. The RTO committee should have voted on your client proposal – so, you need to gather all this evidence.
- If your client prevails at FERC, what would you do?
- What is the next step if your client does not prevail at FERC?
- What is the cost of a good FERC lawyer to represent your client?

Possible Solution

Filing a complaint at FERC is not a frivolous task. Your client must bring along the rest of the stakeholders who have the same grievance with the RTO and support your FERC position. All the above steps take time. Have you considered whether the benefit of a favorable order from FERC would outweigh the costs of litigation?

Case 6.2 *Convincing Political Leaders to Take Batteries and Battery Storage Technology Seriously*

Situation

Recently, South Korea took the energy storage industry and made it a national priority [45]. Can you prepare a similar business case for your state/regional political leader to act on the battery industry?

Analysis

Countries like South Korea do not want to depend on China for the latest battery technology, for national security purposes. Your country or region might have similar issues in the energy industry. Take a look at how energy-intensive some industries are in your region. Understand the source of your raw materials. Do they come from a friendly nation?

Possible Solution

One possible solution here is a transitional goal toward achieving a long-term objective. For example, from the South Korean president's speech, "Lithium-sulfur batteries will be commercialized by 2025, all-solid-state batteries by 2027 and lithium-metal batteries by 2028," you will note that it is specific to battery chemistry and has a timeline. Can you think of something similar for your state/region and work with your political leader to develop a national policy?

Case 6.3 *Creating a Local Energy Market with P2P Trading in a Fragile Economy* You work for a distribution company in a fragile economy – Nepal. It is a vertically integrated monopoly. Its mandate is to "generate, transmit, and distribute power by planning, constructing, operating, and maintaining all generation, transmission, and distribution facilities in [country's] power system" [46]. You want to go above and beyond on your regular work and showcase the latest available technology and help in the democratization of energy.

Your company is seeing an unprecedented rise in DERs, especially rooftop solar. After reading this book and other sources, you are convinced that P2P trading and local energy markets are the keys to the democratization of DER power in your region. It will empower the people and ensure that your wired upgrades (significant investments) can differ, and the potential "duck-curve" can be flattened. Further, the country already has net-metering rules in place.

Create a techno-economic brief for your managing director highlighting the need for P2P trading. What technologies are available? What legal/commercial changes are needed? What could be a phased approach to implement this (the costs can be contained)? Who holds the power of decision-making? What organizations are involved? Can you apply an SAF framework to analyze this? How can P2P trading and local energy market (LEM) empowerment help alleviate the climate change issue?

If you don't live in Nepal, please do this exercise for your region.

Case 6.4 *NetZero In Your Workplace* You work for a small/medium-sized enterprise that does not have corporate targets for lowering emissions. Your CEO believes that he needs to focus on his top-line and bottom-line targets and that having targets for lowering emissions is not-productive. Write a letter/email to your CEO recommending the benefits of accepting net-zero targets. Try to persuade him.

Hint: two ways to have quick impacts are (i) improving energy efficiency and (ii) installing renewable microgrids. Also use the SAF framework, or another framework of your choice, to identify the key decision-makers and map them.

References

1 IEA (2020). Electricity demand from the electric vehicle fleet by country and region, 2030. https://www.iea.org/data-and-statistics/charts/electricity-demand-from-the-electric-vehicle-fleet-by-country-and-region-2030 (accessed 11 December 2021).

2 New York Times (2021). Bitcoin uses more electricity than many countries: how is that possible? https://www.nytimes.com/interactive/2021/09/03/climate/bitcoin-carbon-footprint-electricity.html (accessed 11 December 2021).

3 Federal Energy Regulatory Commission (2021). FERC order terminating the proceeding. https://www.ferc.gov/media/e-3-ad18-7-000 (accessed 11 December 2021).

4 Eurostat (2021). Statistics explained. https://ec.europa.eu/eurostat/statistics-explained/index.php/Main_Page (accessed 11 December 2021).

5 National Renewable Energy Laboratory (2021). Annual technology baseline. https://atb.nrel.gov (accessed 11 December 2021).

6 Moulton, S. and Sandfort, J.R. (2016). The strategic action field framework for policy implementation research. *Policy Studies Journal* 45 (1): 144–169.

7 Schein, E.H. (2017). *Organizational Culture and Leadership*, 5e. Hoboken, NJ: Wiley.

8 Energy.gov (2021). Leadership org chart. https://www.energy.gov/sites/default/files/2021/01/f82/OrgChart_20210124_0.pdf (accessed on December 11, 2021)

9 Energy.Gov (2021). S2 E4: 17 labs in 17 minutes. https://www.energy.gov/podcasts/direct-current-energygov-podcast/s2-e4-17-labs-17-minutes (accessed 11 December 2021).

10 Federal Energy Regulatory Commission (2021). What is FERC? https://www.ferc.gov/about/what-ferc (accessed 11 December 2021).

11 Federal Register (2021). Managing transmission line ratings. https://www.federalregister.gov/documents/2021/01/21/2020-26107/managing-transmission-line-ratings (accessed 11 December 2021).

12 Monitoring Analytics (2021). Home. https://www.monitoringanalytics.com/home/index.shtml (accessed 11 December 2021).

13 EIA (2021). About. https://www.eia.gov/about (accessed 11 December 2021).

14 EIA (2021). Information quality guidelines. https://www.eia.gov/about/information_quality_guidelines.php (accessed 11 December 2021).

15 EIA (2021). Accomplishments. https://www.eia.gov/about/accomplishments/2020 (accessed 11 December 2021).

16 Bureau of Ocean Energy Management (2021). President Obama, Secretary Salazar Announce Framework for Renewable Energy Development on the US Outer Continental Shelf. https://www.boem.gov/sites/default/files/boem-newsroom/Press-Releases/2009/press0422.pdf (accessed 11 December 2021).

17 Bureau of Ocean Energy Management (2021). BOEM advances offshore wind in major US East Coast energy market. https://www.boem.gov/boem-advances-offshore-wind-major-us-east-coast-energy-market (accessed 11 December 2021).

18 Bureau of Ocean Energy Management (2021). Operating status alert. https://www.boem.gov/operating-status (accessed 11 December 2021).

19 Midcontinent Independent System Operator (2019). ISO's David Boyd set to exit post in 2019. https://www.misoenergy.org/about/media-center/misos-david-boyd-set-to-exit-post-in-2019 (accessed 11 December 2021).

20 AESL Consulting (2021). Dr. David C. Boyd. https://www.aeslconsulting.com/david. (accessed 11 December 2021).

21 PJM Inside Lines (2021). PJM Interconnection announces executive appointments. https://insidelines.pjm.com/pjm-interconnection-announces-executive-appointments (accessed 11 December 2021).

22 World Resources Institute (2021). PJM cities and communities coalition. https://www.wri.org/initiatives/pjm-cities-and-communities-coalition (accessed 11 December 2021).

23 Federal Energy Regulatory Commission (2021). FERC announces the Office of Public Participation Virtual Listening Sessions & Public Comment Period. https://www.ferc.gov/news-events/news/ferc-announces-office-public-participation-virtual-listening-sessions-public (accessed 11 December 2021).

24 Natural Resources Defense Council (2021). Understanding energy: PJM explained. https://www.nrdc.org/sites/default/files/media-uploads/pjm_explainer_-_sustainableferc.pdf (accessed 22 February 2022).

25 Astrape Consulting (2021). Publications: Astrape capacity value of energy storage in PJM. http://www.astrape.com/publications (accessed 11 December 2021).

26 PJM (2021). Advanced technologies. https://www.pjm.com/about-pjm/advanced-technologies (accessed 22 February 2022).

27 ENTSO-E (2021). Former organizations. https://www.entsoe.eu/news-events/former-associations (accessed 11 December 2021).

28 Verdonck, R. (2021). To see a true energy transition look to Australian rooftops. https://www.bloomberg.com/news/articles/2021-10-27/to-see-a-true-energy-transition-look-to-australian-rooftops (accessed 11 December 2021).

29 Energy Matters (2021). Australian renewable energy organisations and agencies. https://www.energymatters.com.au/misc/australian-renewable-energy-organisations-agencies (accessed 11 December 2021).

30 Australian Energy Market Commission (2021). National energy governance. https://www.aemc.gov.au/regulation/national-governance (accessed 11 December 2021).

31 Energy Regulators Regional Association (2021). What is ERRA? https://erranet.org/about-us/what-is-erra (accessed 11 December 2021).

32 Energy Regulators Regional Association (2021). ERRA member list. https://erranet.org/about-us/members (accessed 11 December 2021).

33 Asian Development Bank (2021). Nepal energy sector assessment, strategy, and road map. https://www.adb.org/sites/default/files/publication/356466/nepal-energy-assessment-road-map.pdf (accessed 11 December 2021).

34 USAID (2021). Democratic Republic of the Congo: Power Africa fact sheet. https://www.usaid.gov/powerafrica/democratic-republic-congo (accessed 11 December 2021).

35 International Trade Administration (2022). Democratic Republic of the Congo: country commercial guide. https://www.trade.gov/country-commercial-guides/democratic-republic-congo-energy (accessed 6 March 2022).

36 Google Cloud (2022). A timely new approach to certifying clean energy. https://cloud.google.com/blog/topics/sustainability/t-eacs-offer-new-approach-to-certifying-clean-energy (accessed 22 February 2022).

37 Federal Energy Regulatory Commission (2021). Demand response. https://www.ferc.gov/industries-data/electric/power-sales-and-markets/demand-response (accessed 11 December 2021).

38 Schiff Hardin (2021). Supreme Court issues ruling on FERC order no. 745 . https://www.energyenvironmentallawadviser.com/2016/02/supreme-court-issues-ruling-on-ferc-order-no-745 (accessed 11 December 2021).

39 Federal Register (2021). Voltus, Inc. v. Midcontinent Independent System Operator, Inc.: notice of complaint. https://www.federalregister.gov/documents/2020/10/30/2020-24016/voltus-inc-v-midcontinent-independent-system-operator-inc-notice-of-complaint (accessed 11 December 2021).

40 Federal Energy Regulatory Commission (2021). FERC addresses demand response opt-out for certain DER aggregations. https://www.ferc.gov/news-events/news/ferc-addresses-demand-response-opt-out-certain-der-aggregations (accessed 11 December 2021).

41 Federal Energy Regulatory Commission (2021). FERC sets demand response opt-out for further consideration. https://cms.ferc.gov/news-events/news/ferc-sets-demand-response-opt-out-further-consideration (accessed 11 December 2021).

42 LS Power (2021). LS Power petitions Supreme Court to overturn anticompetition electric transmission laws that hurt consumers. https://www.lspower.com/ls-power-petitions-supreme-court-to-overturn-anti-competition-electric-transmission-laws-that-hurt-consumers (accessed 11 December 2021).

43 Supreme Court of the United States (2021). Search results. https://www.supremecourt.gov/search.aspx?filename=/docket/DocketFiles/html/Public/20-641.html (accessed 11 December 2021).

44 PJM (2022). Federal law guides changes in PJM governing documents. https://www.pjm.com/~/media/about-pjm/newsroom/fact-sheets/federal-power-act-sections-205-and-206.ashx (accessed 22 February 2022).

45 Office of the President (2022). Remarks by President Moon Jae-in at K-Battery Development Strategy Presentation. http://english1.president.go.kr/BriefingSpeeches/Speeches/1030 (accessed 22 February 2022).

46 Nepal Electricity Authority (2020). Gender equality and social inclusion strategy and operational guidelines 2020. https://energypedia.info/images/1/11/NEA_GESI_Strategy_and_operational_guidelines.pdf#:~:text=The%20Nepal%20Electricity%20Authority%20%28NEA%29%2C%20established%20in%201985%2C,transmission%20and%20distribution%20facilities%20in%20Nepal%27s%20power%20system (accessed 22 February 2022).

7

How Does the Power System Work?

7.1 Introduction

We all know about the control centers that are showcased during satellite launches. However, we rarely understand that the electric utility or system operator has a similar control center [1]. Electricity control centers form the backbone of modern power system operation. The power flowing through the wires is monitored and controlled from these centers (generally redundant centers, see Section 7.6). Control center operators monitor and maintain it.

The problem for the electricity control centers is further complicated because power cannot be stored, and the energy generated must be consumed almost at the same time. Hence there should always be a balance between generation and consumption. Let us explore this complexity.

When you charge a mobile phone, the phone's power must be generated simultaneously. Since energy cannot be stored (commercially at grid scale for more extended periods), how does a power plant operator know when to crank up a generator for you to be able to charge your mobile phone? Who conveys this message?

Further, some power plants take days to start (or stop). How does the system operator ensure that whatever is generated is instantly consumed? How does the system operator ensure that all the new constructions (generation and load) are considered? Who plans whether the generators should be renewables? Nuclear? Others? How do you decide to build new generators? Where do you make them? When? How about new loads?

There are so many other questions that must be considered while you remain aloof from all of this, and all you want is to charge your mobile phone.

After reading this chapter, this person (e.g. graduate student, engineer, nonengineer, decision-maker, or members of the informed public) will be:

Modern Electricity Systems: Engineering, Operations, and Policy to address Human and Environmental Needs, First Edition. Vivek Bhandari, Rao Konidena and William Poppert.
© 2022 John Wiley & Sons Ltd. Published 2022 by John Wiley & Sons Ltd.
Companion website: www.wiley.com/go/bhandari/modernelectricitysystems

- Acquainted with the technologies used in the power system, how these technologies interact, and how they form the grid of the future (spread throughout the chapter).
- Acquainted with how the electrical energy system works (Section 7.5).
- Acquainted with the general principles that guide this operation (Section 7.6).
- Able to use this knowledge to make and appreciate power system planning and operations (Section 7.8).

A reliable power supply is the backbone of our modern society. Hence, the electric control centers have more at stake than those used to monitor or control satellite launches. After reading this chapter, if not anything else, you will appreciate this fact.

7.2 Guiding Principles for a Power System

Energy access is a significant social indicator. According to the World Bank, over 89% of the global population had electricity in 2018. But is the supply reliable? Is it affordable? Is the supply secure? Is it clean? In other words, "access to electricity" should include reliability, security, affordability, and clean factors. Just having electrical connections is not enough, and the customer should be able to consume it productively.

For example, several parts of rural Nepal are powered by microhydro plants. The nominal frequency in Nepal is 50 Hertz (Hz). But several of the microhydro plants operate tens of Hz below the nominal. Electricity at such low frequencies is unreliable and of poor quality, damaging the connected equipment and reducing its overall life. Its urban supply operates at around 50 Hz. But it is plagued with frequent power cuts.

Another example is small, medium, and large factories slashing production or shutting down in Nepal due to unreliable supply (frequent power cuts). One of the leading newspapers there writes, "Factory owners say that power cuts have not only pushed up their cost of production but also caused losses to daily wage laborers" [2].

The details around proper access to electricity are discussed in Chapter 10.

This chapter introduces this topic related to the systems deployed for the power system's adequate functioning.

Exercise 7.1 The Australian government is working on reducing the price of electricity, delivering it with reliability, and increasing more and more renewables. For example, they have invested over 30 billion AUD in a portion of their electricity market. Like Australia, several countries have policies and investment plans to promote reliable, affordable, secure, and clean electricity. Jay, the consultant, needs your help to prepare a policy brief on this topic. What aspects of the guiding principles of the power systems of your country does he need to know about? Does your country also follow the directions above? Why (not)? What are the salient features of these principles?

7.3 Schematic of the Modern Energy System

Table 7.1 illustrates the different generation sources and their attributes.

Exercise 7.2 What other newer forms of generations are being tried in your country? How do you think the generation mixture for 2050 should look for your country? Why?

7.4 Governing Bodies and Actors

Governing bodies make the policies to ensure that the key objectives to provide reliable, secure, and affordable power are met. To understand or challenge policies, or to formulate newer policies related to the electrical systems, an adequate understanding of power systems operations and planning is essential (see Chapter 6 for details).

Exercise 7.3 Can you draw a concept-board type (freehand or using drawing tools like word or Visio or a similar platform) drawing of the different organizations that govern the electrical system operations for your country/region? Is it similar to your most prominent neighbor? Why (not)?

7.5 Power System Operations Management

Power system operations can be divided into multiple parts: generation, transmission, distribution, end-use/prosumers (consumers who produce as well as consume energy), and markets. Several areas of these parts are operationally and physically different, yet others are the same. Hence, the management techniques sometimes are similar and, at other times, different. In this chapters, we look at each one of these topics.

For example, transmission lines are balanced; they are polyphase, have adequate measurements (using SCADA/EMS meters)[1], and are physically visible. Distribution lines, on the other hand, could be unbalanced. They could be single or multiple phases, they may or may not have telemetry (or SCADA measurements), and

---

1 SCADA: supervisory control and data acquisition; EMS: energy management system.

Table 7.1 Generation sources used in the modern electrical systems and their techno-commercial and environmental attributes

	PV	Onshore Wind	Offshore Wind	Hydro Power	Nuclear	Coal	Natural Gas	Combined Heat and Power	Diesel Generator	Biomass	Thermal Storage	Electrical Storage	Hydrogen	Electric Vehicles (G2V and V2G)	Virtual Power Plant	Microgrid
Renewable Supply	Y	Y	Y	Y	Y*	N	N	N	N	Y	Y	Y*	Y	Y*	Y*	Y*
Supply Energy	Y	Y	Y	Y	Y	Y	Y	Y	Y	Y	Y	Y	Y	Y	Y	Y
Supply Capacity[1]	N	N	N	Y	N	Y	Y	Y	Y	Y	Y	Y	Y	N	Y	Y
Supply Flexibility	N	N	N	Y	N	Y*	Y*	Y	Y	Y	Y	Y*	Y	Y*	Y	Y
Burns Fuel	N	N	N	N	Y	Y	Y	Y	Y	Y	N	N	N	N	Y*	Y*
Capital Costs	$$	$$	$$$	$$$	$$$	$$$	$$$	$$	$$	$$$	$$	$$$	$$$	$$	$$	$$
O&M Costs	$	$	$	$	$$$	$$	$$	$$	$$	$	$	$	$	$	$	$$
Salvage Costs	$	$	$	$	$$$	$$	$$	$$	$$	$	$	$	$	$	$$	$$
Risk to Environment	L	L	L	L*	M	H	H	H	H	L	L	L*	L	L	L	L
Risk to investment	L	L	L	L	M	M	M	M	L	M	M	M	H	L	L	L
Conventional or New	Ne	Ne	Ne	C	C	C	C	C	C	Ne	Ne	Ne	Ne*	Ne	Ne	Ne*
Requires access to robust T&D	N*	Y	Y	Y	Y	Y	Y	Y	Y	M	Y	Y	Y	Y	Y	N*
Carbon Release	L	L	L	L*	L*	H	M	M	H	M	L	L*	L	L	L	L*

Legend

Y Yes, N No, * Depends; $ Cheap, $$ Moderately Expensive, $$$ Expensive; L Low, M Medium, H High, * Depends; Ne New, C Conventional, * Depends

<hr>

[1]The intermittent generation sources like wind/solar can provide energy but not capacity or flexibility primarily because of unpredictable and intermittent nature.

they could be underground (hence invisible). So, though they both are principally poles and wires, the operations and planning techniques could fundamentally be different.

In the following subsections, we will first look at general principles that apply to each of these systems and then later look at these systems in some detail.

Exercise 7.4 Jay is required to support a request for proposal (RFP) process. He advises your country's most significant transmission network system provider (TNSP) to replace their transmission management system (TMS). What functions does your TNSP run? What systems should he seek from a TMS provider vendor at a high level? Current prominent TMS vendors are Siemens, ABB, General Electric, and Emerson. Please also refer to their web pages.

7.5.1 Energy Management Systems (EMS)

EMS generally comprise generation management, transmission management, distribution management, market management, and end-user management. However, in some parts of the world, EMS include generation and transmission management but not others. Irrespective of the naming conventions, we need to understand what each contains.

7.5.1.1 Generation Management System (GMS)

Meridian Energy in New Zealand uses a SCADA-GMS to manage their fleet of hydrogenerators and control the water flow in their canal. Without this system, they will not be able to generate electricity while following the instructions from Transpower (local TSO). They will also not be able to optimize and control water flow in their canals. A typical architecture of a Generation Management System is shown in Figure 7.1.

Generating utilities and independent companies like Meridian use GMS to manage their generation fleets. It consists of primary, secondary, and tertiary control systems, including SCADA systems (see Chapter 1 on layers of control and their needs).

Primary control is provided by the generation controllers, e.g. inverter control that comes with a solar photovoltaic (PV) system or power plant controller of a hydrogen plant. They are speedy and take prompt action when they see any disturbance. For example, as soon as a primary controller of a PV system sees a frequency event, it would act to control the inverters to take immediate action. This information/instruction set is stored in a local device.

Other such controllers include underfrequency relays and controls to support dynamic and transient stability. For instance, if there is a frequency event (sudden drop in frequency) at the point of interconnection of a solar and battery farm in New South Wales in Australia, the primary controller of this system will

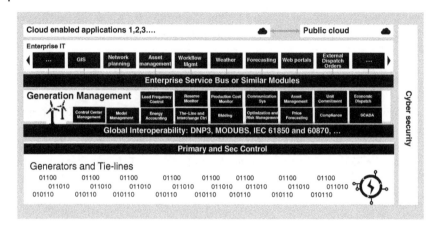

Figure 7.1 Generation management system (typical architecture).

take immediate action to remediate the problem. The frequency relays could be triggered if this local controller failed to act. The solar and battery farm could deteriorate the grid stability and, when combined with other events, could even cause a blackout.

But having just a primary controller is not enough. On top of the primary control, there is a secondary control. It provides functions like automatic generation control (AGC) (see Section 7.7.10) that act within a few seconds. AGC follows the load and balances it with generation. Doing so looks at the system frequency and tries to bring it back to normal before the system is impacted. In the example above, the reaction from the primary controller might not be enough. Australian Energy Market Operator's (AEMO) needs could still see that the generation and load are not following one another. AEMO runs AGC every few seconds and makes sure that the generation and load are balanced. If there is a frequency event (at a system level or national electricity market level), AEMO's AGC will quickly bring the frequency back to normal (50 Hz), as shown in Figure 7.2. Without AGC, the system shall remain abnormal.

On top of AGC, there is a tertiary monitoring and control layer. It includes functions like asset management (see Section 7.7.9), and SCADA (see Section 7.7.7). It also includes optimization of assets and bidding where the market participants like generators come up with their offers to the market. The market management system takes these offers, optimizes them, and provides the dispatch-order. For the example above, in an ideal situation, the action of the primary controller and AGC would be enough to balance the grid and make it stable. However, the lack of rotating mass in the AEMO's grid has opened avenues for exploring synthetic inertia. Take an example of the plot of system inertia for one of AEMO's regions. As illustrated in Figure 7.3, minimum monthly inertia is consistently below the minimum operating limit from 2015 onwards.

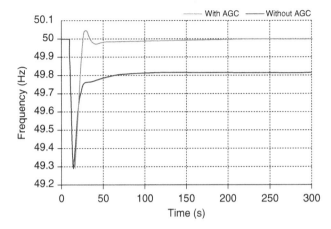

Figure 7.2 Frequency response with and without AGC. Source: [3]/IEEE.

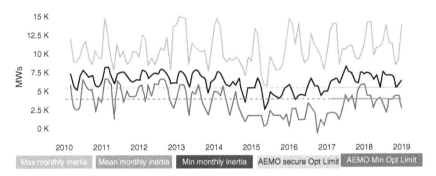

Figure 7.3 AEMO inertia example. Source: [4]/WATTCLARITY BY GLOBAL-ROAM PTY. LTD.

This is primarily happening because of the rise of distributed energy sources (DERs), intermittent generation, and traditional inverter-based controllers. AEMO is consistently exploring ways to compensate for such a lack of system inertia. It may do so by opening markets for synthetic inertia. Suppose AEMO runs a market for synthetic inertia and plans around this event. It would be a tertiary monitoring and control layer. However, the primary controllers and high frequency switching inverters will most likely provide synthetic inertia. Figure 7.4 illustrates different control schemes deployed in a power system over various time steps.

7.5.1.2 Transmission Management System (TMS)

Amprion GmbH and 50Hertz Transmission GmbH in Germany are two transmission companies. They need a TMS to ensure that the transmission lines are utilized fully while not burning them by creating thermal or other violations. Without a

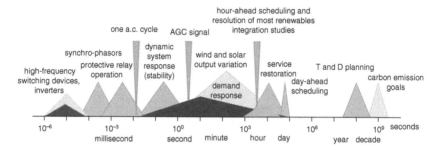

Figure 7.4 Different generation control schemes utilized to control power systems at different time scales (10^{-6} to 10^{-3} primary control, 10^{-3} to 10^{-1} secondary control, rest tertiary control and planning). v / Pacific Northwest National Laboratory (PNNL) / Public Domain.

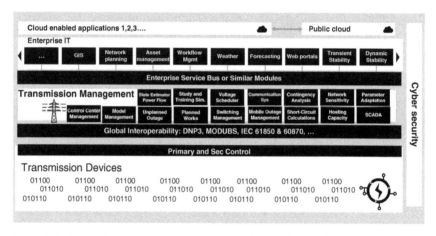

Figure 7.5 Transmission management system (typical architecture).

TMS, they cannot monitor the line, run analytics, plan, or control the equipment. Mismanagement of such lines could result in a blackout.

The transmission management consists of three components: (i) transmission network applications (TNA) (see Sections 7.7.1.1, 7.7.2.1, 7.7.3, 7.7.4.1, 7.7.5, and 7.7.6.1); (ii) outage management (OM) (see Section 7.7.8); and (iii) transmission SCADA (see Section 7.7.7). A typical architecture of a TMS is given in Figure 7.5.

TNA is the control center application that a transmission network operator operates. Such tools are used to manage the operations of a transmission network reliably. The TNA tools monitor, analyze, and optimize the grid. It provides both real-time monitoring/control and quick offline study capabilities.

A typical TNA suit consists of state estimator (SE), dispatcher (or operator) power flow, contingency analysis (CA), parameter adaptation, fault calculation, network sensitivity analysis, voltage stability analysis, and optimal power flow (OPF) for steady-state operational stability (minutes to hours scale).

These days, with the growing complexity of the grid, transmission systems work close to their steady-state stability limits. Hence, a typical steady-state analysis is not enough. The operator needs situational awareness for dynamic stability (a sub-second to a second scale). These tools typically include transient and small-signal stability and help understand the dynamics within a transmission system.

OM is the set of procedures and functions to manage planned work and unplanned transmission outages. Transmission SCADA is the platform for receiving telemetry and other messages, processing them, monitoring, and sending the controls out.

7.5.1.3   Distribution Management System (DMS)

AusNet Services in Victoria, among other things, is a distribution company. They need to monitor and control the distribution network (at least from the substation to the distribution transformer and the associated customers). If there is an outage, they need to inform their customers and dispatch the crew. If there is a phone call to report an outage, they need a system to take the phone call and plan accordingly. They need to balance the unprecedented rise of DERs and run analytics. Without a distribution management system (DMS) in AusNet, the Victorians would not have reliable power in their houses, industries, and other compounds.

The DMS primarily consists of three components: (i) distribution network applications (DNA) (see Sections 7.7.1–7.7.6), (ii) OM (see Section 7.7.8), and (iii) distribution SCADA (see Section 7.7.7). A typical architecture of a DMS is given in Figure 7.6.

DNA is the control center application operated by a distribution system operator (or a distribution network operator). Such a set of applications provide automatic verification of completeness/quality of network data, calculation of the current state of the network, and detection of erroneous measurements, and can conduct "what if" and advanced active network management (ANM) optimization, for example, to reduce network thermal and voltage overloads.

A typical suite of distribution management platforms consists of distribution SE and power flow, generation, and load forecast for a look-ahead scheduling of the load, fault management (fault location, fault isolation, service restoration), volt-var-watt control, feeder reconfiguration, and short-circuit calculations (SCCs).

OM is the set of procedures and functions to manage planned work and unplanned outages. Distribution SCADA is the platform for receiving telemetry and other messages, processing them, monitoring, and sending the controls out.

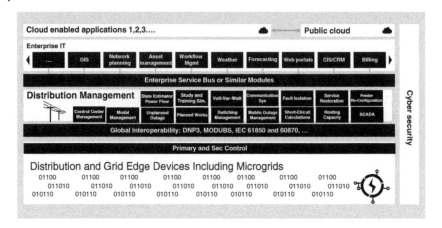

Figure 7.6 Distribution management system (typical architecture).

Exercise 7.5 Can you help Jay again? You need to help him expand his understanding of how the distribution utilities work on an outage. He lives remotely and has a traditional meter. His distribution transformer only serves a few rural houses. A tree has fallen in front of his home, and his house has a power outage. It has already been two hours. Jay thought, in the world of IoT, the utility company should be immediately aware. Why hasn't the utility company sent repair personnel? Could it be that the utility company is unaware of this incident?

Hint: even in developed countries, the distribution utilities generally operate a trouble call management system. The availability of SCADA and other forms of telemetry in a distribution system is still rare. It is, however, changing with the emergence of the smart meters.

Distributed Energy Resource Management System (DERMS)

With more and more DERs coming onto the distribution grid and newer rules for managing the grid appearing (e.g. see FERC Order 2222 on DER aggregation in Chapter 9), grid complexity cannot be solved merely by the use of a DMS.

A distributed energy resource management system (DERMS) caters to planning for increasing DERs and their operations, including rooftop solar, behind-the-meter generation, electric vehicles, etc., to deliver vital grid functions. In short, it is a generation management and additional function available for DERs.

Figure 7.7 shows a typical architecture of a DERMS. It includes DNA and SCADA (shown as distribution grid operations), planning tools (offered as grid planning and simulations), building management (shown as grid edge

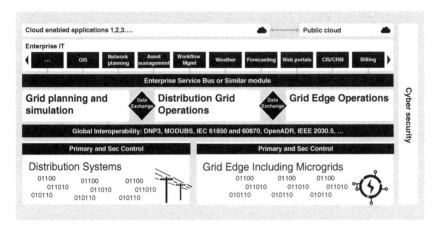

Figure 7.7 DERMS with other adjacent systems (architecture).

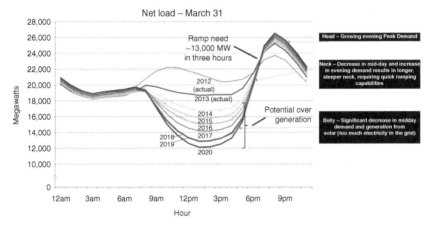

Figure 7.8 Duck curve. Source: [5]/California Independent System Operator Corporation.

operations), and protection and control (shown as primary and secondary control) for the efficient management of distribution scale DERs.

Exercise 7.6 Figure 7.8 shows a "duck-curve" from the California independent system operator (ISO). It illustrates a plot of net-load vs. time of day. For a selected day in spring, owing to large output from solar, the net load is lower during the day. During the evenings, there are high ramp-up requirements. Traditional generators may or may not be equipped to perform such a quick ramp. How can we potentially mitigate such a challenge? Can (how can) DERMS and peer-to-peer (P2P) trading help in alleviating the needs of high ramping?

Head-end Systems (HES)

CitiPower, Powercor, and United Energy are distribution companies in Victoria in Australia. As a distribution company, on top of maintaining the poles and wires, they are also meter data providers (MDPs). As a MDP they are responsible for orchestrating smart meter data for example for billing and settlement. They deploy meters from various vendors. And these vendors collect data from the smart meter into a head-end system (HES) and pass it to other systems for further processing. Without an HES, the meter data would not be able to be posted for a meter data management system (MDMS), or other systems, for timely billing. Hence, the whole process of buying and selling power every five minutes in the AEMO's market would be disrupted.

HES is a set of hardware and software systems that receives the data from the meters brought to the utility company through the metering infrastructure. This system connects to the meters. Hence it requires communication, including communication with programmable logic control (PLC), general packet radio service (GPRS), 3G/4G/5G, and other Internet protocol (IP) based communication methods. HES typically includes functions like alarming, low voltage monitoring, control (connect/disconnect, load control, and firmware updates), prepayment options, reporting, etc. Some of the HES also could contain data validation techniques. A typical architecture of a HES is given in Figure 7.9.

Meter Data Management System (MDMS) and Revenue Protection System (RPS)

HES passes the data collected from the meters into a MDMS. It is as critical as a HES. Without a proper MDMS, CitiPower, Powercor, and United Energy cannot

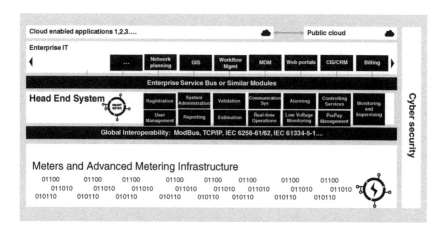

Figure 7.9 HES with other adjacent systems (architecture).

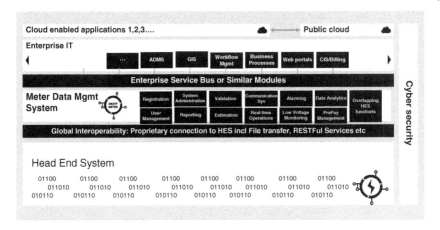

Figure 7.10 MDM with other adjacent systems (architecture).

meet their obligations for a timeline and accurate delivery of the five-minute smart-meter data.

MDMS is a software system that connects to the HES system downstream and customer information system (CIS) system upstream. MDM will typically import the data, perform validation, clean it, and process it. It includes applications for these functions and may also include smart-meter deployment and management systems, meter network assessment, some overlapping functions with HES, smart-meter provisioning, workforce management, asset management, prepaid functions, reporting, and data analytics, etc. Figure 7.10 shows a typical MDM architecture.

Customer Information Systems (CIS)

CitiPower, Powercor, and United Energy in Australia only manage the meter data; they do not do billing. The retailers in Victoria in Australia do billing and customer management. So, the data from MDMS is passed to the billing system of the various retailers in Victoria, who will then send the bills to the residents of Victoria. Without CIS, the retailers cannot manage the customer relationship, including the bills.

CIS and billing systems are typically used once the data from the MDM and HES is processed. CIS is an end-to-end customer relationship management (CRM) suite that allows end-to-end customer billing and business process management. This software enables the utilities/retailers to manage payments, deposits, customer accounts, taxes, and meters. As illustrated in Figure 7.11, it also includes functions that allow reporting, historical billing access, customer portal access, data analytics, service order management, etc.

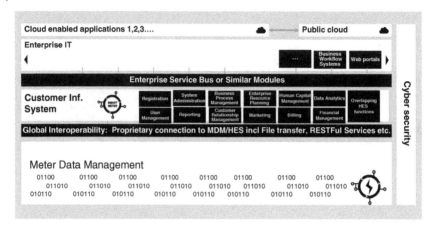

Figure 7.11 CIS with other adjacent systems (architecture).

Virtual Power Plant and Demand Response System

In the early 2010s, a highly flexible biogas plant was planned in Germany. It was unique with aims to become flexible and help stabilize the grid while increasing its revenue. They installed three physical combined heat and power (CHP) units with a combined installed capacity of around 4 MW. They created a virtual plant that aggregated the physical CHP units. They then started offering energy and ancillary services from this virtual plant into the spot market. They could provide services to the grid and improve their revenue without impacting their biogas plant operations. This concept is called a virtual power plant (VPP).

VPPs are not new. They were also used for grouping large generators during an AGC (see Section 7.5.1.1) or wholesale market operation (see Section 7.7.11). Like the German example above, the physical generating units were grouped into virtual logical groups. These groups could be several generators in a plant or an offshore wind farm, or an entire independent power producer. AGC and markets control the logical group, while the local controllers control the physical units. This is a predecessor to a modern-day VPP.

Nowadays, VPPs are re-emerging primarily because of the Internet of Things (IoT) and the unprecedented rise of DERs. On the one hand, the IoT is making individual control and monitoring possible. On the other hand, primarily due to the large volume of data, it is not technically feasible for the distribution operators to model each IoT device (e.g. each thermostat in your house could be modeled as a load, and a battery in your EV could be modeled as a generator or a load).

Hence, VPPs are needed as an intermediate stage. VPPs would aggregate these loads and generators controlled by IoT devices and present them to the system operator as an aggregated load or generation.

VPPs allow for smartly offering aggregated energy and ancillary services in the market (where available). If the market operator/dispatcher requires the service (energy and ancillary service), the VPP gets awarded. The VPP then does smart disaggregation to implement the award. VPPs also manage the customer account and help with registration, program management, and settlements. They implement the logic in real time to ensure that the bulk requirements of the system operator for energy and ancillary services are met. They help to make better use of DERs and other loads.

Demand response (DR) is like a VPP with a few caveats. Let us take an example. Vector, a utility company in New Zealand, saw a rise in decentralized solar, batteries, and other DERs. This caused spikes and drops of load/generation and bidirectional flows at its distribution feeders, causing stress to the distribution network. It could choose to make a significant capital investment in upgrading and replacing existing transformers and wires to accommodate the changing grid's needs. It could also choose a "nonwire alternative" by optimizing loads and DERs, and better channel generation to areas of high demand hence reducing the network stress. In doing so, among other things, it implemented a DR program. Without this program, Vector would have invested heavily in poles, transformers, and wires that would ultimately be stranded.

DR is like a VPP. It includes the reduction in demand (VPP also includes control ramp-down or ramp-up of a generator) that occurs at a connection point because of a specific request to the customer from a party like a market operator or a dispatcher. Second, there is no real-time feedback in DR. DR is a fire-and-forget type control. Hence it is estimated by comparing actual consumption (e.g. after meter reads are obtained) against a predicted baseline.

Peer-to-peer Trading and Flexibilities System

In the early 2020s, the whole world, including Australia, was shut down due to a global pandemic. Access to bars and restaurants was severely impacted. And the brewers were sending several million gallons of beers down the drain. Hence, new business models were emerging to avoid this situation. Wouldn't it be cool to get beer delivered in exchange for your access to solar power? Asahi Group's Carlton & United Breweries started implementing a program to buy access to solar power from households and pay them in beer cans. The patrons would get chilled beer delivered. The brewery would sell its beer and meet its target to use 100% renewable by 2025. Further, the grid operator would also benefit without being deeply involved.

This project is an exciting example that illustrates trading between peers (households and the brewery), one of the many promising ways to manage the unprecedented rise of renewables. A typical architecture of a P2P system is shown in Figure 7.12.

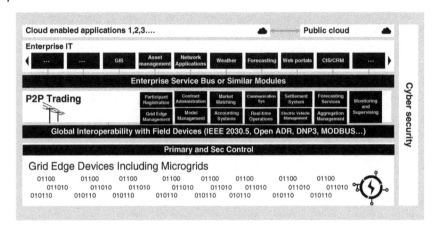

Figure 7.12 P2P management (typical architecture).

As more DERs emerge, one of the ways to manage them is to allow DER suppliers (generators) to trade their generation directly to the nearby consumers across the electricity network. There are several forms of P2P trading systems that are available today. Some allow the buy/sell of energy and services for money; others even allow the buy/sell of energy and services for other commodities.

For example, in the beer industry example or a distribution operator buying flexibility services (ability to ramp up, ramp down, curtail, or generate quickly) from its end customers, or two neighbors purchasing and selling their rooftop solar among themselves in the local energy market (see Chapter 1) are all examples of P2P trading.

7.5.2 Market Management System

Energy Imbalance Market (EIM) was launched in 2014 with the California Independent System Operator. Recently, the benefits of EIM surpassed $1.28 billion. Systems like EIMs are a part of wholesale market system. A typical architecture of a wholesale market system is shown in Figure 7.13. California ISO uses sophisticated software, including a wholesale market management system (MMS), to orchestrate and operate EIM while finding and delivering low-cost energy and improving reliability and reducing emissions. Without MMS, such benefits could not have been realized.

The MMS is required for operating a market. It includes functions like registering the market participants (e.g. generators, loads, transmission, distribution network operators, and VPPs), running dispatches, unit commitments, markets (i.e. real-time, intraday, or multiday ahead), billing and settlement, forecasting applications for load and demand, a communication system for communicating

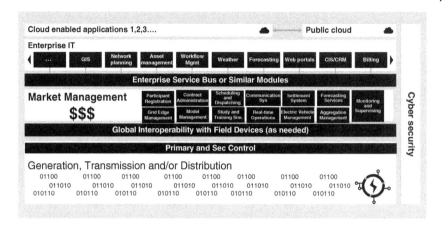

Figure 7.13 Wholesale market management (typical architecture).

with external systems, and network applications for ensuring adequate operations of the power system.

7.6 High-level Architecture and Redundancies

The systems mentioned in Section 7.5 are critical to the operation of any economy. Developing countries could quickly come to their knees if such systems were impacted. Without such systems, a fragile economy will not make its electricity affordable, reliable, or secure.

Here is a typical architecture of the systems mentioned in Section 7.5. The real-time (e.g. SCADA and communication systems) and non-real-time (user interface, historian, and network applications) servers are placed in physical or virtual servers. They are segregated into development, test, training, preproduction, and production environments or systems. A typical hardware architecture is shown in Figure 7.14.

Some of these server functions might be merged, or they might be named differently in some geographies. However, a production system is generally the name of servers and applications that do real-time monitoring, analytics, and control. Other servers serve supporting functions like testing, training, and development. Owing to the critical nature of such systems, there is typically redundancy in the communication lines that come from the remote terminal units (RTUs) or external applications. Each of the servers is redundant, the environments are redundant, and even the control center is redundant. At some places, there are even three control centers. For example, Midcontinent Independent System Operator (MISO) in

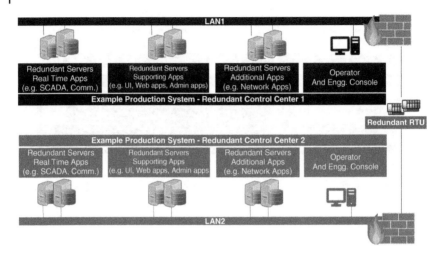

Figure 7.14 Reference hardware architecture of an energy management system. The servers and operator consoles may be physical or virtual.

the United States (US) has one control center in Indiana, a second in Minnesota, and a third in Arkansas.

7.6.1 Cybersecurity Architecture: Critical Infrastructure

Owing to the critical nature of some of these systems, they are typically carefully isolated from the Internet into a so-called electronic security perimeter (ESP). Where needed, the reach to the Internet is carefully orchestrated.

For example, a distribution operator would have a call-taking system that the call center operator accesses to record outages. It would then have an OM system for handling the life of an outage (from when it is registered to when it is fixed and archived). It would also have a SCADA system used for doing the actual controls. These three functions – call taking, OM, and SCADA – will be put in servers in a less secured zone to the most secured zones, respectively.

All the data in and out of these systems (in transit and at rest) is encrypted. Strict procedures are followed on who can access the data, the applications, and even sometimes who can develop such systems. Such regulations are typically set by global associations that are governmental or quasigovernmental. For example, in North America, NERC Critical Infrastructure Protection (NERC CIP) standards are a set of requirements designed to secure the assets required for operating North America's bulk electric system. Similarly, in Europe, the General Data Protection Regulations (GDPR) and Bundesverband der Energie- und Wasserwirtschaft

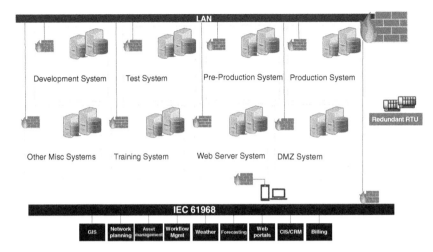

Figure 7.15 Reference high-level cybersecurity architecture.

(BDEW) whitepaper-based regulations are typically followed. Figure 7.15 shows a typical cybersecurity architectural layout that enhances operational cybersecurity of the systems by firewalling and zoning.

7.6.2 Change Management of Software Changes

Software changes to these environments are carefully tested before deploying for field monitoring and control. Let us take an example. Red Eléctrica de España (REE) wants to implement a change in its EMS system. It wants to implement software that curtails solar, causing network instability. REE will ask its vendor to do new development (assuming the existing EMS does not have this feature). The vendor will develop the software using security cleared engineers to work on the REE project.

Once the development is done, REE will perform a factory acceptance test (FAT) in the "development system" on the vendor's premises. Once the vendor is satisfied with the development work (REE is typically welcome to witness the test), they sign the test books and then transfer the change to the development system on REE's premises. REE's engineers will then do other sets of on-site tests in the development system or test system. They will then slowly promote the changes into preproduction and then finally into production, aka there is rigorous testing before any minor change is transferred into production. If the tests fail in any one of these steps, the changes are rolled back, and the process restarts from fixing and testing at with the vendor's development systems. Figure 7.16 shows this process.

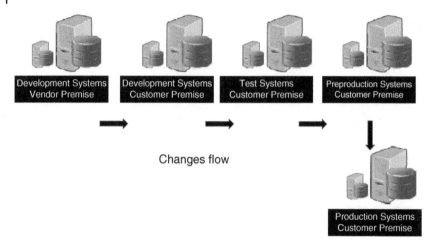

Figure 7.16 Reference change management workflow among different servers.

7.7 Advanced Concepts of Power and Control for Electrical Energy Systems

The world of electric utilities and system operators is changing rapidly [6]. Hence, advanced and flexible solutions need to be deployed to manage generation, transmission, and distribution systems (while minimizing the costs).

7.7.1 Power Flow

The reliable operation of power systems is vital for the proper functioning of any country or region. Power flow determines the voltages and active and reactive power at various places in the electrical network. This knowledge is essential to continuously evaluate the performance of the power system such that necessary remedial control actions can be taken if needed. For example, AEMO, in its projected assessment of system adequacy (PASA) process, runs a look-ahead power flow, among other things, to determine the state of the network based on the forecasted conditions. This information is further used to run additional analyses to develop control measures in an emergency.

7.7.1.1 Transmission Power Flow (TPF)

Transmission power flow (TPF) is typically (i) dispatcher or operator power flow or (ii) OPF. TPF is used to calculate the network status when the load and generation conditions vary in a transmission system. TPF calculates the magnitudes and angles of voltages for all the nodes/buses and active/reactive power for the relevant buses. It forms the basis of understanding the status of the network to detect potential violations.

Advanced applications (unless they are rule-based) run based on the power flow results. For example, security check switching ("What happens if I open or close a breaker?") utilizes a power flow results solution to detect and identify violations.

Similarly, CA also uses TPF to identify which of the monitored contingencies (predefined sets of "what ifs") would create a voltage or a thermal violation.

OPF is like dispatcher power flow, just that OPF is used to enhance the operations of the transmission system. Example objectives used by OPF are: (i) minimization of active/reactive power production costs, (ii) rescheduling of active/reactive power for minimizing violations, (iii) minimization of transmission losses, and (iv) a combination of these.

OPF is highly time-consuming and requires an army of engineers. Hence, OPF is often discussed in academia but rarely used for power system operations.

7.7.1.2 Distribution Power Flow (DPF)

The distribution power flow (DPF) is slightly different from the TPF. DPF is used to calculate the network status when the load and generation conditions vary in a symmetrical, balanced (e.g. typical European network) or asymmetrical, unbalanced distribution system (e.g. the specific US network), in a radial (e.g. standard rural network) or a meshed network (e.g. typical city network), or in a single-phase (e.g. line connecting a household) or three-phase element (e.g. a line connecting an industrial load). The transmission power flow calculates the magnitudes and angles of voltages for all the nodes/buses, active/reactive power for the relevant buses. It forms the basis of understanding the status of the network to detect potential violations.

Advanced applications (unless they are rule-based) run based on the power flow results. For example, security check switching (what happens if I open or close a breaker analysis) utilizes power flow results to detect and identify violations. Similarly, ANM (like centralized volt/VAR control) also uses power flow to determine what control actions are to be taken to get voltage or reactive power support without overloading the network.

OPF is an emerging concept in the distribution network. It has high applicability in future distribution systems. However, currently, it only remains a subject of research.

7.7.2 State Estimation

Power system operations depend on the available data, the analytics run on this, and the actions taken to overcome the issues. However, not every point in the network is measured. There are missing states. SE, based on the available set of redundant measurements, determines the steady-state condition of the given power system in real time. Hence, SE is vital to a power system operation. One of the major blackouts in 2003 that affected over 50 million people in the US and Canada could have potentially been avoided if MISO's SE and CA software had not been faulty. The software problems and other issues led to the unavailability of real-time system information that ultimately led to blackouts. This highlights the importance of SE.

7.7.2.1 Transmission State Estimator (TSE)

In transmission systems, the purpose of the SE is to provide a complex solution (missing states) based on the available measurements. The transmission state estimator (TSE) solution is based on real-time measurements (generally redundant), load, generation, and user-entered conditions. The world can be divided into an observable and unobservable network by TSE. TSE can estimate quantities (e.g. voltage magnitude and angles) that best match the measurements for the observable network. For an unobservable network, TSE uses the loads, generation, scheduled values, and any scattered measurements to estimate the solution (e.g. voltage magnitude and angles). The entire solution (for observable and unobservable networks) is done simultaneously. TSE and TPF work together to calculate voltages and branch flows. TSE/TPF can be used with future conditions for conducting look-ahead type functionalities and helps detect insufficient data.

7.7.2.2 Distribution State Estimator

In distribution systems, compared to transmission, there are very few measurements. Therefore, the function of the SE is slightly different. Thus, a distribution state estimator (DSE) adapts loads to match the weighted measurements and detects anomalies. It typically starts from their static load curves or load/generation forecast and adjusts it to fit the incoming height, assigned weights, and the network topology. DSE and DPF work together to calculate nodal voltages and branch flows. DSE can also consider the future conditions and create short-term load/generation schedules that could be further used for look-ahead type functionalities.

Exercise 7.7   Please research the mathematics behind the transmission and distribution of state estimation and present it to your friends. Please make it concise (3–5 pages of PowerPoint-type presentation).

7.7.3 Contingency Analysis (CA)

AEMO, in its PASA process, runs around 2800 contingencies for every SE/power flow run. In other words, it has 2700 active "what if" scenarios that are run on top of a baseline power flow solution to come up with a plan. For example, there is an HVDC link between Tasmania and Victoria. This link is used to draw power from generation in Tasmania to load in Victoria. What happens if this HVDC fails? AEMO would have modeled this potential failure as one or more of its 2700 monitored contingencies. It is impractical to monitor the loss of each of the power system components. Hence, as soon as any unfortunate event happens, AEMO adds additional contingency to its active list. Therefore, CA is used to determine the steady-state security of the power system under predefined events. Without adequate contingency planning, we face several blackouts and brownouts.

Contingencies are defined as predefined "what if" events. Such events are grouped as credible and others. CA alerts the operators with a list of thermal or voltage violations should a contingency event happen. It also provides the operator with a list of controls to alleviate the overloads.

Generally, with CA, there are N-n criteria. If $n = 1$, the requirements are defined as follows. For a single ($n = 1$) credible contingency (e.g. loss of prominent DC link, loss of single transformer or generator or capacitor/reactor/static VAR compensator), the power system should be capable of operating within a particular minimum performance (e.g. voltage deviation between $\pm 10\%$, equipment loading up to 150% of the nominal capacity, etc.).

Exercise 7.8 Do you know if any transmission utility uses N-n criteria in their operations? Why (not)? What measures (if any) are used in your country?

7.7.4 Fault Management

The power system is not static. It changes during operations (e.g. switching of breakers or changing generation or loads, etc.) and during planning (e.g. where to add a new generator, etc.). Such operations and planning activities require evaluating what equipment to select, what ratings the relays should be programmed at, and how it should be managed if there is a fault. Hence there is a need for a fault management function, another vital function of any power system.

If one chooses the wrong setting in the relay, it could cause unnecessary tripping or no tripping. This could ultimately damage the equipment in the power system and even cause blackouts. For example, the 2013 Super Bowl blackout could have been avoided if newly installed relay settings were correct [7].

7.7.4.1 Transmission Fault Management (TFM)

In a transmission system, transmission fault management (TFM) computes the fault currents and compares them against the nominal ratings of the devices. Since transmission systems contain adequate measurements, the relay/protection system does the fault isolation and restoration part; and there is little to no need for centralized fault location/isolation and restoration function. TFM computes short-circuit currents for symmetrical and unsymmetrical faults: three-phase faults (L-L-L), three-phase faults with the ground (L-L-L-G), two-phase faults (L-L), two-phase faults with the ground (L-L-G), and single-phase faults with the ground (L-G).

7.7.4.2 Distribution Fault Management

It is more challenging to manage fault in a distribution system than in a transmission system because it could be balanced, unbalanced, radial, meshed, single-phase, or three-phase. However, fault management is needed in both T&D worlds.

The fault management applications are used to locate a fault, isolate the faulty section, and restore the service. Fault location finds permanent faults[2] (e.g. outage faults like those due to short-circuits) and creates outage or non-outage faults like ground faults that might not cause a network outage.

Fault location is performed by reading the measurements from the relays, looking at the network topology, and utilizing manual information (especially in the distribution world where you might not have sufficient and reliable telemetry). For example, in a line section, two relays indicate that the fault is between them; fault location takes this information and presents it to the control system operator.

The transmission system will likely contain overhead lines with enough measurements and is well equipped with protection relays. Hence, it allows the operator to take quick action based on this information. If it is a distribution fault, the lines could be underground, and they may (not) have reliable telemetry. So additional algorithms like impedance or current-based search to improve accuracy are required. Because the system operator needs to decide whether to dig below Bob's office, or Jay's, in the distribution world, straightforward searches and fault management techniques fail due to such complications.

Once the fault is located, we need to isolate it. The local isolation might have already happened if the line had been equipped with adequate protection and control systems. Otherwise, the central system runs a fault isolation function.

The isolation function determines a set of switching operations that will isolate an area of the network so that it does not affect other areas of the network. Centralized fault isolation can also be run on top of the local isolation to improve the isolation further. This functionality can also be used to optimize switching during planned outages.

Fault isolation considers switches per phase to develop a technique that causes minimum outage or minimum switch operations while ensuring network security.

---

2 Transient faults and temporary faults are generally skipped by these applications and are taken care by the local protection devices.

The final stage in fault management is to restore the service. Service restoration in transmission systems is easy because, typically, only a few switches would have operated to isolate the fault. The fault needs to be cleared and switching from above should be undone.

There could be a multistage process in distribution: partial restoration before the fault is fully cleared and full restoration once the fault is cleared. In either of these, there are multiple ways to restore the services.

So, the service restoration function restores the service to deenergized sections while ensuring that the power system is secure. In doing so, the system operator can surgically say what not to restore yet (as the fault may not have been fully cleared), injection sources that can be used to energize the section, and even ask the system operator to restore every area to the prefault stage.

Finally, the fault management function in a distribution system also contains SCC. SCC is like TFM. It calculates the current results of a fault or an incorrect connection. It is primarily used to determine a circuit breaker's maximum/minimum current rating and during fault management (for calculating fault currents). It can be used for three-phase faults (L-L-L), three-phase faults with the ground (L-L-L-G), two-phase faults (L-L), two-phase faults with the ground (L-L-G), and single-phase faults with the ground (L-G).

7.7.5 Volt-VAR-watt Control

Active monitoring of voltage, active and reactive power is vital for power system stability. In the operations world, both in Transmission and Distribution, especially with the increase in DERs and their intermittency, volt-VAR-watt optimization is required. This technique is also sometimes part of ANM. It determines the control actions for transformers, voltage controllers, capacitors, batteries, generators, and flexible contracts to improve network operations while maintaining its security.

The technique is run to minimize system violations (T&D), minimize operational costs (T&D), optimize toward a target power at an injection source (mainly transmission), minimize power losses (T&D), and minimize power consumption (T&D). At times it is rule-based.

For example, one of the largest TSO in Spain uses a function that translates to "solar curtailment" to define triggers and a sequence of actions to curtail solar to secure their transmission network.

Similarly, another utility in the UK runs a system-wide volt-VAR-watt control to minimize violations in their distribution network. In system-wide (non-rules-based) schemes, network constraints and power flow equations are considered.

Exercise 7.9 What is open and closed-loop volt-VAR-watt control? In countries like Australia, an extension of volt-VAR-watt control and hosting capacity is being considered. It is commonly referred to as a dynamic operating envelop. What is the difference between a traditional volt-VAR-watt control and a dynamic operating envelop? Please explore.

7.7.6 Optimal Network Reconfiguration

Power needs to be routed from generation through transmission and distribution lines to the load centers. If there is an issue (e.g. loss of line, outage, etc.), the power must be rerouted through another set of transmission and distribution lines (if available). Hence, the network would need periodic reconfiguration.

7.7.6.1 Transmission Capacity Assessment

Power transfers create operational stability issues. Hence the margins to operating limits need to be calculated. Transmission capacity assessment (TCA) provides such information to the operator. Margins are calculated based on the maximum shift in generation for the corresponding increase in load. It can be used to monitor the voltage stability margins (where at the network could a voltage collapse?), how much power can be transferred, and how much additional capacity is available for transferring the power based on the power flow and contingency conditions – and using this information to reconfigure the network to make it stable.

7.7.6.2 Distribution Optimal Feeder Reconfiguration

Optimal feeder reconfiguration is used to determine switching plan options for distribution network reconfiguration to reduce distribution network overloads or optimize the distribution network to minimize violations, losses, or load balancing among transformers. It is generally triggered after DSE identifies a breach or activates a distribution operator manually.

7.7.7 Supervisory Control and Data Acquisition (SCADA)

In 2013, the supervisory control and data acquisition (SCADA) system of a small dam in New York was accessed by Iranian hackers. In 2015, the Ukrainian power grid had a significant impact causing an outage for nearly a quarter of a million Ukrainians. The SCADA equipment was inoperable, and the restoration of the system had to be done manually. There are so many other attacks happening in this control infrastructure. Why? Because it is critical to the operation of the power system.

SCADA is the core/backbone of transmission, generation, and distribution operations. It typically includes three functions: telemetry, monitoring and alarming, and controlling. In a SCADA system, the telemetry from the RTUs originate using industry-standard protocols like DNP3, MODBUS, IEC 101/103/104, IEC 61850, etc., or open standards like OpenADR, IEEE 2030.5, or via meter reads using multispeak type protocols. The front-end systems collect this information from the field devices that ultimately feed it to a SCADA-type system. Such a system receives the telemetry and processes the data for monitoring and further use.

For example, a SCADA system would have logics to handle digital data (on–off messages), analogs (power or voltage measurements), and payloads from external systems (like bids and awards). It would also allow local control operations, the generation of an alarm as needed, the application of limits, and the substitution of the values, and would allow these to be aggregated. SCADA is not only used for receiving telemetry and monitoring. It is also the one-stop place for sending the control out to the field devices.

Since SCADA is already equipped to send and receive signals, SCADA functions are extended to create a sequence of events for islanding and black start/resync in a microgrid setting. Similarly, in an EMS/ADMS setting, SCADA functions are designed to develop automated switching.

Exercise 7.10   What are the islanding functions of a microgrid? What is a black start? Please explore these terms.

7.7.8   Outage Management

Utilities are judged by how reliably and securely they provide power to their customers. Despite all their best efforts to avoid power outages, e.g. by continuous and proactive monitoring and control, outages happen. For example, a utility company cannot prevent a storm, bush fires, or a power outage in general. They can only minimize the impact. They need a mechanism to monitor and manage the outages. This is done by OM.

OM is equally applicable in both the transmission and distribution world. OM is a collection of functions, tools, and techniques used by the system operator to detect, locate, isolate, correct, and restore the unplanned outages; and plan switching/upgrade works in the system.

It also calculates and monitors the key performance indicators (KPIs), e.g. System Average Interruption Duration Index (SAIDI) in the distribution network, transmission lines unavailability duration per year in the transmission network, and reporting/dashboarding.

The transmission system is well telemetered. Hence, managing unplanned outages and planned works requires different tools than those necessary for distribution.

In a distribution system, the lines may or may not be telemetered, i.e. the only way for your utility company to know a tree has fallen onto a power line is by somebody making a phone call. Hence, distribution OM has an integral function for trouble call management. Both systems use OM.

7.7.8.1 Unplanned Outage

OM is well equipped to handle unplanned events. It includes fault detection and outage record creation, service restoration, and archiving outages. OM provides support to the operator to detect a fault, create an outage record, collect data from SCADA, collect data from trouble call management (prediction based on calls), AMI meters, manual updates, or network applications. It then calculates the estimated time to restore and complete this outage. This is generally done by adapting historical values over time.

Service restoration requires the following steps: partial or complete repair using manual knowhow or automatic process (e.g. fault management), crew assignment and tracking, switching step creation, study, approval, execution, and finally archiving all the events for further auditing.

7.7.8.2 Planned Outage/Work

Planned work is typically maintenance activities in a network. OM allows a system operator to agree upon a plan to isolate, repair, or change a part of the network and provides end-to-end services to manage the approvals, track the process, conduct studies, execute the switching, and archive the steps.

7.7.9 Asset Management

As generation, transmission, and distribution companies own assets, it is essential to create an inventory of them, map them, monitor their health, and plan for their maintenance and growth.

Hence, the asset management programs apply to the generation, transmission, distribution, and end-use assets. The objectives for asset management can be categorized as short-term (day-to-day operational), mid-term (maintenance), or long-term (strategic planning) objectives. The asset management systems allow for the mapping of assets and their inventory, predicting the health of assets (where to focus), and how to grow (where to build DERs).

7.7.10 Automatic Generation Control (AGC)

Not so long ago, before the use of AGC, a designated generating unit would take responsibility for balancing the difference between load and generation to maintain the system's frequency. All the other units would be controlled using their primary controller (e.g. droop-based control) that would allow them to contribute toward the load based proportionally to their ratings. Such a scheme is still used in a few parts of the world, e.g. microgrids, power systems, or fragile economies. However, AGC was used as the power system became more complex and interconnected.

AGC is used in generation management (see Section 7.5.1.1) or microgrids. Microgrids are smaller versions of the bulk electric power system. They contain some of the functions of the bulk electric power system. AGC is generally not applicable on the distribution grid (as few large controllable generators). DERMS is supposed to provide AGC functions in a distribution system. AGC is a secondary control utilized by the control authority, which, every few seconds, is responsible for making sure production matches the sum of consumption and losses.

AGC has several functions: load frequency control (LFC), which looks at the frequency of the system to change load or generation; reserves monitoring, which ensures that there is an adequate amount of reserve capacity available from the participating generators; unit commitment; economic dispatch, which ensures that the cheapest resources are deployed in real-time to meet the load (generally done in the absence of a market); and production cost monitoring, which ensures that the generators are operating at scheduled target costs.

7.7.10.1 Load Frequency Control

LFC controls the system frequency by balancing system generation with the load. If there is a mismatch between load and generation (including flows in the interchanges), the system frequency changes. LFC picks this change and then allocates and implements (through SCADA) the total desired generation among the generators to match the current load.

7.7.10.2 Unit Commitment (UC) and Economic Dispatch (ED)

Unit commitment (UC) and economic dispatch (ED) determine the most economical method to meet the system load with the available generation. The ED process assumes that the generators are already running and computes the economically sustained generation for each unit to achieve the best possible cost position for the overall network. The unit commitment process, among other things, also considers starting and stopping units. And security constrained UC or ED considers the power flow constraints while calculating commitments or dispatch.

For example, a microgrid at Blue Lake Rancheria in the US runs hourly UC with a look-ahead feature to develop the least cost generation to meet the forecasted demand.

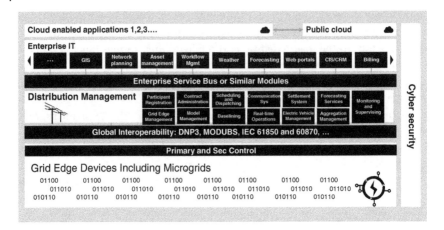

Figure 7.17 Demand response and virtual power plant (typical architecture).

7.7.10.3 Reserve Calculations

Each balancing authority (BA) – authorities that are responsible for making sure load is met with the secure, cheap, and reliable generation – is responsible for allocating reserves from the generators to cater for fluctuations and contingencies in real-time operating conditions. Depending on the response time to convert capacity into delivery energy, different classes of reserves are defined for each BA. The differences in the response time allow the reserve with the fastest response time to be utilized before those with slower response times.

For example, in the National Electricity Market of Australia, the market and out-of-market reserves are maintained. In the market, reserve represents capacities that are bid into the market but are not yet dispatched (in each five-minute dispatch) or frequency control ancillary service (FCAS) to keep the system frequency within limits for a credible contingency event (seconds to tens of minutes). Out-of-market reserve represents reserves procured to provide a specific purpose, e.g. as a last resort to meet demand.

7.7.11 Market Operations

Electricity is a commodity capable of being sold, bought, and traded. Markets and power exchanges enable this process. Bids are put to buy, and offers are called to sell using financial or obligation swaps. Market operations may or may not include AGC. It has functions like registration, dispatch, scheduling, settlement, etc. (see Figure 7.17 for further details). Please refer to Chapters 1 and 9 to learn more about the electricity markets. Market operations utilize the market management systems as described in Section 7.5.2.

7.7.12 Model Management and Digital Twin

Model management is another vital function of a power system. The more accurate and consistent the model is, the better results one can expect. For example, if an ISO has a planning model in PSSE® (a planning tool from Siemens), an operational study model in the form of PSSE and the bids/offers and forecasts for its assets in its internal operational planning tools, and real-time operating SCADA model in a vendor control-center software, SE model in a fourth software, and geographic information system (GIS) maps in the fifth software, imagine how easy/hard will be to plan and operate this ISO? Alas, this is happening at several ISOs today in 2021.

The model of their power systems is kept at different places, inconsistently. There is no single source of truth. Hence, a central repository to put all the static data related to the power system (e.g. line ratings, drawings, one-lines, network parameters, etc.) and a means to exchange this data with external systems easily (e.g. from planning to operations, or from ISO to another utility) is vital in today's world. Digital Twin is such a method to maintain data in one central repository.

Storing at a single place is not enough. This data should be interoperable with other systems. One of the methods that allow the exchange of such data is called the common information model (CIM). The CIM defines a set of schemas for describing information collected for network and systems management. It can be used to store static information (e.g. line ratings, line parameters, etc.) and dynamic information (e.g. power flow solution, outage information, etc.).

Exercise 7.11 What is a digital twin for power system planning and operations? Why is it important?

7.7.13 Dynamic Line Rating

Ratings of power system equipment are generally classified as short-term, mid-term, or long-term rating. These ratings refer to the capacity of the equipment to carry an additional load over the rated limit for a given period. The equipment might take more considerable overloads for a shorter period and lesser overloads for a more extended period. Dynamic line rating (DLR) seeks to identify the rating of the transmission lines over time to help alleviate grid congestion and thus promote renewables.

During CA, the grid operators operate the lines within their specified limits. When the transmission lines hit the rating, there is congestion. DLR suggests that such ratings of the lines (transmission or distribution) vary over time, depending on changing environmental conditions like temperature, wind, solar radiation, etc. Take, for example, a TSO in your country that runs CA. Traditionally, the operators look for overloads against a static limit (long, medium, or short term). Passing more current through these lines heats them. However, the extent of line heating depends on whether it is a cold day or a hot day. Hence, these limits might be different depending on the weather. Like the weather, it may also depend on the insulation type, physical properties of the conductor, and other real-time values. Thus, there is a need for a line rating that is dynamic (calculated in real time). This is called DLR.

DLR has been available since before the 1990s. However, only recently have they started to gain attention because the renewables in several places were curtailed, citing congestion – when there was no congestion. Hence, primarily to promote renewables, any EMS software today can calculate DLRs. It is a part of typical network applications or SCADA.

7.7.14 Other Basic Control Center Functions

Power grids are operated from control centers. The power system assets (including the control centers) are considered "critical infrastructure" in many parts of the world. Therefore, the operations need to be cybersecure and reliable.

High-availability architectures are maintained to ensure reliability and redundancy. Redundant communication of the measurements through the RTUs is the first layer of redundancy. This communication generally arrives at two control centers.

The control centers may be hot-hot (both the control centers can actively receive/process the data and send out selected controls) or hot-standby (only one control center receives/processes the data and sends the control out. The standby control center is synced with the hot control center).

Hot-hot and hot-standby configurations are generally used in North America. There can be several variations to this. In Europe, they operate hot-standby with regional control centers. In other continents, there is no backup control center.

There is a hot-hot or hot-standby in another region and a data center cold periodic backup (the data is backed up every 15 days or so in a different environment). Receiving data, processing it, storing it, and maintaining high availability (e.g. 99.9999%) are some of the basic functions of a control center. Additionally, cybersecurity functions are essential in these control centers. Cybersecurity includes end-to-end communication encryption, data encryption in rest and transit, data access, application safe listing, following ISO certifications, hardening of the products, regular patch management, periodic penetration testing, and more. Please refer to Section 7.6 for the hardware and cyber architecture of a control center.

Exercise 7.12 What is NERC CIP? What is BDEW Whitepaper? How about ISO27001/2? Please create a summary (three slides for each one). Are these standards followed in your country? If not, what is followed? Which of these local standards trace back to NERC CIP, BDEW, or ISO?

7.8 Power System Planning

Power system planning generally includes generation, transmission, and distribution planning. Generation planning traditionally focuses on planning for peak load. The process starts with load forecasting, including extreme cases, followed by reliability evaluation to understand when additional generation is needed. In the end, economic evaluations are done to determine optimal capacity expansions. However, with the advent of renewables, it should now shift to using net load (instead of peak) and coming up with requirements for a reliable, secure, and flexible power system.

Transmission planning is done to optimize the use of the generation portfolio. It ensures that the infrastructure can economically deliver generation to the load and continue during normal conditions and contingencies.

It includes a more accurate modeling of variability created by intermittent generation sources like solar and wind. Similarly, the distribution planning is now incorporating capabilities to include the variable generation sources. Energy planning includes processes for forecasting (gathering data, building profiles, creating scenarios, allocating forecast, etc.), analysis (understanding the capacity, protection, reliability, automation, and contingency needs, etc.), and solution identification (balancing risk, analyzing costs/benefits, deliberation within and outside the planning organization, etc.). From a time horizon perspective, the planning process is classified in Section 7.8.1–7.8.4.

7.8.1 Long-term Planning

Long-term planning means having a horizon of 15–25 years. These plans help verify the technical feasibility of long-term generation selection. It is more strategic and essential for a region to have this plan to conduct strategic implementation and investigate future infrastructure development. For example, the Brazilian government identifies this plan as an "essential condition for the country to position itself competitively [meets economic, social, and environmental aspirations]." Such plans are also used as a vehicle to spark debates and start the deliberation process with society.

7.8.2 Medium-term Planning

In a larger sense, medium-term plans are prefeasibility studies of the proposed alternatives. Medium-term planning generally looks at a horizon of 10–15 years.

It analyzes the power system and energy requirements outlined by the long-term plan. For example, in medium-term planning in Brazil, the cost–benefit of capacity increase must be considered and its social and environmental impact assessed; the impact of hydro resources (being a predominantly hydro-based generation) in power system operations is also required.

7.8.3 Short-term Planning

Such planning is carried out every year (or every other year) for 5–10 years. It looks at specific projects/connections identified in the medium-term plan. Such a plan could also be used to respond to unforeseen changes of generation/load/transmission that could not have been otherwise predicted earlier.

7.8.4 Operational Planning

Operational planning is generally done days or weeks ahead. It is done to identify "what-ifs" if a contingency happens. This is very popular with system operators. For example, in AEMO-NEM, this process is called short-term process and resource adequacy. The process looks (weeks) ahead to identify contingency events and remedial actions for those.

Until recently, it was easy to figure out how to reliably operate the transmission grid by looking at the peak load (generally summer or winter days) and ensuring that the fossil-fuel generators would turn on when needed. Replacing fossil-fuel-based generators with renewables requires replacing energy and capacity and other grid-related requirements (e.g. avoiding shortfalls). The renewables should have enough adequacy to meet all the grid-related needs. For example, the concept of resource adequacy suggests that there should be sufficient generation facilities within the system to satisfy the consumer load or system operational constraints. The contribution of any resource for ensuring this adequacy is typically referred to as a capacity credit.

In short, effective load carrying capability (ELCC) is a probabilistic method to estimate the capacity credit. In other words, ELCC is the ability of a resource to provide energy when the grid is likely to experience shortfalls. Take an example of a 200 MW solar farm. If it has an ELCC of 30%, it means that this farm, keeping every other variable constant, could supply 60 MW toward avoiding shortfalls.

Just replacing fossil-fuel-based generation with renewable generation is not enough. Renewables should also provide capacity and other grid requirements. Hence, better estimation of metrics like ELCC is key to ensuring a 100% renewable future transition.

On top of this, not only are the large-scale renewables emerging. The power is flowing bidirectionally. Heat, gas, and water systems are also coupled with the electrical system. DERs are growing on the distribution side, the wholesale markets with newer products are emerging on the transmission side, and the environmental needs are more pressing than ever. Hence, accurate planning of an electrical system becomes more vital. Unfortunately, owing to technologies and sociopolitical silos, current planning processes are happening in silos. To plan for a better future sector, integrated planning processes with transparency toward models and methods should be followed.

Exercise 7.13 What is sector coupling? What is happening in your region? Are heat and electricity sectors planned separately? Is the electric transportation planned separately?

7.9 Visioning

Now that you understand several systems, principles, and actors. How does a power system work? What happens when you flip a light bulb? Or turn on/off a TV? Or a start/stop an extensive industrial process? Let us take some examples.

In the first case, imagine that you are in Melbourne in Australia. AEMO (depending on the size and periodicity of your load pattern) must have already forecasted your consumption and scheduled the generators to meet the load during its operational forecast. It will also prepare for "what-ifs" in case a contingency happens.

The awarded generators will be spun to provide you with the electrons in real time. The generator owner will use a GMS to control its generation to follow the real-time market dispatch orders from AEMO. The electrons will travel through transmission and distribution network operators, and they will be paid for using their network services.

The network operators use transmission and distribution management systems to route these electrons to your place. Your retailer will be the energy reseller and will sell it to you. The retailer will settle the prices with you at the end of the month and with AEMO/distribution network service provider (DNSP), etc., at a separate time step. The retailer will use a CIS system for billing and CRM.

The DNSP, on the other hand, will submit the reading from your smart meters to AEMO at every five-minute interval (from October 2021) using a combination of HES and MDM. All this discussion assumes that you consumed what AEMO had initially forecasted.

What if you consume more than the forecasted amount, or if you generate from your rooftop solar (which was invisible to AEMO)? In such cases, which are the new normal, AEMO will deploy the reserves or take other necessary measures to compensate for the necessary change. All of this is guided by national policies and laws. Interestingly, all of this happens before you notice anything. You flipped your switch, and the light was on.

Let us take another example. You are a sawmill that wants to operate in the Dolpa district of Nepal. In Dolpa, there is no national grid. In this case, there is no day-ahead planning or forecasting.

If the local generator has enough capacity to serve your load, the electrons from this generator will arrive through the T&D infrastructure before you notice any difference. If the generator is incapable, it will trip or provide you with unreliable electricity. Nevertheless, interestingly all of this happens before you notice anything. You flipped your switch, and the light was on. But the power system operations are much more complicated. They must be planned and operated meticulously. The poor electron had to follow the laws of physics, legalities, policies, and the social aspects of the society before it reached your household or your industry.

As you can imagine, understanding the intricacies in such detail is not easy for everyone. Let alone for the common public, even for electrical engineers, other than what is learned through the textbooks in mathematics, power systems, and matrix algebra, the actual day-to-day working of the power system and electricity markets is generally a mystery. This chapter explained the working dynamics of power systems from their planning to day-to-day operations. Now you should understand what happens when you flip a switch.

Case 7.1 *Mining Company and DERs*   You work for a mining company. You have a grid connection but, as a backup, have five diesel generators and two transmission lines. You also have an extensive solar and battery system. Your loads (demand) are mining loads. Your company wants to make a microgrid out of the available assets. It has a corporate target to optimize consumption from solar. You need to explore the market to check what is available. You know the typical functions in a generation, transmission, and DMS. Please prepare (five pages maximum) an expression of interest document seeking your desired functions for the microgrid in a mine.

Hint: you may want functions like SCADA, unit commitment, islanding, black start, transmission network application, etc. You may also want to think about P2P trading type functions.

Case 7.2 *Mining Company Functional Tradeoff*   For Case 7.1, the indicative prices were $4–$5 million just for the control system. Why? Would you drop the functions to meet the budget? Or would you go to your management to seek additional funding? For either case, prepare a two-page justification report.

Case 7.3 *Too Many Inverters and Converters Issues and Solution*   Traditionally, the power system had large rotating machines that produced electricity, and there were loads like heaters, lightbulbs, and industrial appliances. Both loads and generators are becoming more inverter/converter based with time.

For example, renewable generators, fridges, electric vehicles, etc., all use this technology. Inverters do not typically possess the characteristics of large rotating generators. They possess challenges like loss of dynamic and transient stability of the system (loss of inertia that would have allowed the ride through minor disturbances), frequency, and voltage-VAR regulation (abrupt and rapid switching used in the inverters could impact the frequency and voltage stability of the grid). Some methods to solve these issues are synthetic inertia and grid forming inverters. What are these? Explore this further. How big is the issue of a lack of system inertia for your utility grid? What are the mitigating measures? Using fewer than 20 slides, summarize the findings for your peers.

Case 7.4 *Additional Challenges and Solutions for Inverters and Converters*
Increasing use of an inverter-based system would also carry additional challenges like

- Black start: It is the ability to restore the system from an outage. How should an inverter provide enough active/reactive power to support the motors and other loads and stably operate the lines?
- Communication and synchronization: How do you ensure accurate communication between thousands of inverters for adequately synchronizing them?
- Cybersecurity: When there is communication and third-party ownership of the system, could there be issues around cybersecurity?

What are other topics currently being considered? Are these topics of potential concern for your utility grid provider? Based on the information collected in Exercise 7.3 and this exercise, prepare an opinion piece for your local newspaper.

References

1 Statnett (2022). How the power system works. https://www.statnett.no/en/about-statnett/get-to-know-statnett-better/how-the-power-system-works/#:~:text=The%20power%20system%20is%20a,time%20as%20it%20is%20generated (accessed 22 February 2022).

2 Dhungana, M. (2021). Manufacturing encounters energy problems just as industries were recovering from pandemic blow. https://tkpo.st/3dKZa6lhttps://kathmandupost.com/money/2021/04/07/manufacturing-encounters-energy-problems-just-as-industries-were-recovering-from-pandemic-blow (accessed 22 February 2022).

3 Gupta, I. (2018). A case study on load frequency control with automatic generation control on a two area network using. MiPower Proceedings of the National Power Systems Conference (NPSC), 14–16 December, NIT Tiruchirappalli.

Indiahttps://www.iitk.ac.in/npsc/Papers/NPSC2018/1570475534.pdf (accessed 22 February 2022).

4 McArdle, P. (2020). AEMO's notice of inertia shortfall in South Australia. https://wattclarity.com.au/articles/2020/08/aemos-notice-of-inertia-shortfall-in-south-australia/ (accessed 22 February 2022).

5 Burnett, M. (2016). Energy storage and the California "duck curve." http://large .stanford.edu/courses/2015/ph240/burnett2/ (accessed 22 February 2022).

6 Conejo, A.J. and Baringo, L. (2018). *Power System Operations*. Springer International.

7 Scriber, B. (2013). Super Bowl blackout: was it caused by relay device, or human error? https://www.nationalgeographic.com/science/article/130208-super-bowl-blackout-cause (accessed 22 February 2022).

8

How Are Changes to Power Generation Operation and Control Relevant Today?

8.1 Introduction

Artificial intelligence (AI) is more profound than fire, electricity, or the Internet, says Google chief executive officer (CEO). Some version of this statement was run by major news outlets like CNBC, Business Insider, etc. in 2021. We disagree with this statement. Without electricity, there is no AI. In fact, without electricity, there is no Google, or Amazon, or college professor who has the luxury to teach online classes, or a doctor to adequately take care of their patient, or a pharmacist who wants to develop a new medicine, or a biologist who wants to develop a new vaccine. Without electricity, you (our reader) and we (the authors) will have difficulty living our everyday lives. But we don't talk much about electricity. All the electricity talks and works happen in the background. In Chapter 7, we explain the details of the different systems that operate today's power system.

Further, in previous chapters, we introduce the fundaments of electricity, it's engineering, and the economic, policy, and operational aspects of a power system. But managing electricity in today's world is more complicated than ever. Hence this deserves a separate discussion and explanation. For this discussion, this chapter expounds on two important reports one from the NREL [1] and the other from the IEA [2].

First, there is more variability and uncertainty in demand and supply. Low-carbon technologies, especially wind and solar, introduce huge variability compared to their traditional counterparts. For example, more than one-third of global electricity comes from low-carbon technologies. Wind and solar are at 5.3 and 2.7%, respectively.

Second, there is sector coupling. For example, combined heat and power (CHP) technologies, including district heating/cooling, couple heat, and electricity. Electric vehicles (EVs) couple transportation and electricity. Sector coupling means a change in one sector would impact another. The Texas power crisis of 2021 is a

Modern Electricity Systems: Engineering, Operations, and Policy to address Human and Environmental Needs, First Edition. Vivek Bhandari, Rao Konidena and William Poppert.
© 2022 John Wiley & Sons Ltd. Published 2022 by John Wiley & Sons Ltd.
Companion website: www.wiley.com/go/bhandari/modernelectricitysystems

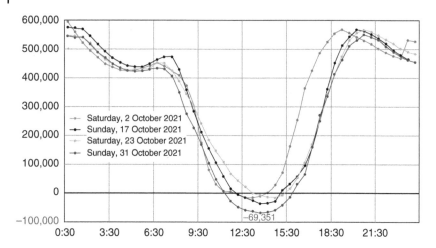

Figure 8.1 SAPN net KW load from ElectaNet in October 2021. Such a figure is referred to as a "duck-curve" in Chapter 7. Source: [3]/Reneweconomy Pty Ltd.

prime example. As the temperatures dramatically dropped in Texas, the demand for gas surpassed its predicted amount and outpaced power generation. Since the same natural gas was used in both sectors, an increase in demand in the heating sector impacted the power sector.

Third, the electricity grid was initially supposed to be unidirectional. The grid was designed to buy from traditional generators, route the electricity through transmission and distribution, and be retailed and consumed in houses, industries, or commercial spaces. Now, the grid has become multidirectional. The residential, industrial, and commercial places install rooftop solar and batteries and feedback to the grid. For example, the IKEA Group (a furniture producer) chose to go 100% renewable in the late 2010s. They wanted to help in the shift toward a low-carbon future. They wanted to become energy independent and make money by not consuming from the grid and selling the excess back. While this is interesting, the distribution system previously feeding IKEA's stores was not designed to treat IKEA as an energy-independent company. So, the grid operator would have their opportunities and challenges associated with this transition. Forget IKEA. You and I, retail customers, can also feed our solar electrons back to the grid and even make money depending on where we live. Figure 8.1 is a graph from South Australia Power Network (SAPN) from October 2021, where the electrons flowed from the distribution toward the transmission for several days in a single month.

The unidirectional grid installed decades ago would have never envisioned such a fast-paced multidirectional power flow. Such a rapid transition is both an opportunity and a challenge.

In summary, generation, operation, and control as we know them have changed due to economic incentives on flexibility, resiliency, and reliability, and goals toward a cleaner future. A single large baseload unit on the electric grid does not provide automatic generation and control (AGC). Several distributed generation sources offer this. The generation source is not always alternating current (AC). Some generators generate AC, whereas others generate direct current (DC). Hence, power electronics converters and inverters have increased for converting DC to AC, and vice versa. Such inverter- and converter-based resources are the norms of today and the near future. We explore this complex scenario in this chapter and discuss how the changes to electrical engineering are relevant today.

We also discuss the global transformations that are currently happening in the power systems along with the opportunities/challenges that they bring. At the end of this chapter, the reader should use this knowledge to make an informed decision to make (or support making) policy, regulatory, and financial interventions to help better plan and manage the power system. The reader will appreciate the "distributed" framing of traditional power generation, operation, and control. The reader will also understand how the increasing penetration of distributed energy resources (DERs) in the distribution grid calls for newer actors like aggregators and how they are likely to shape the future of power systems. The complexity of the ongoing transformations should give an appreciation to the reader that AI can potentially be used to support power system transformation. But it is not more important than the electricity. There is no AI without electricity.

8.2 Current Events and Future Changes in Power Systems

Power systems are undergoing a fundamental change primarily because of the following trends.

8.2.1 The Costs of Renewables Are Declining

We all want the future power system to be cleaner and use renewable generation sources. The costs of cleaner/renewable generation sources have a vital role in this future. To move to a grid with more renewables, the price of the renewables relative to that of fossil fuel should be lower. Figure 8.2, derived from OurWorldinData, shows the prices for renewables and fossil fuels without subsidies. Focusing just on solar and wind, you will see that the costs of these renewables are now significantly cheaper than what they were before. The electricity from utility-scale solar photovoltaic (PV) cost \$359/MWh in 2009, and in 2019, it was ~\$40/MWh. The price has declined by ~89% in just one decade. This is massive.

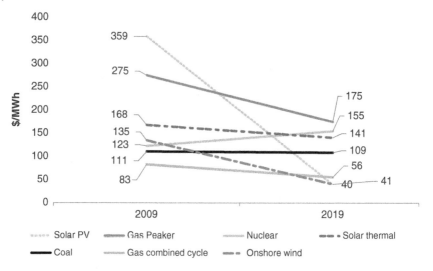

Figure 8.2 Levelized costs ($/MWh) of generation 2009 vs. 2019.

Moreover, the relative price of solar compared to fossil fuel has now reversed (see Figure 8.2). Let's put it into perspective. This is equivalent to saying that if in 2009 you were willing to invest ~$35 000 for a car of your choice, and assuming this car would see the exact price decline as solar, you would pay $4000 for the same car, with better features, in 2019.

Why did this happen? Why didn't the expensive solar technology die in its early phase? This is because solar found (i) niche use cases to nurture and improve the technology, (ii) the price of the technology rapidly decrease, and (iii) the need for a clean technology rapidly increase. Solar in the 1960s and 1970s, though expensive, was used in places like satellites, lighthouses, remote railroad crossings, and even for refrigeration of vaccines [4], and later it was used in rural electrification. Similarly, the price of solar modules has declined by ~100% since the 1970s primarily because there had been more demand, room, and interest for improvement. In other words, solar technology showed rapid learning and economies of scale due to increasing demand and room for improvement. Hence, levelized costs ($/MWh) of solar, primarily a function of development costs (as there are no fuel costs), also rapidly declined. Wind, like solar, has also shown significant rapid learning and has benefitted from the economies of scale due to increasing demand (also see the law of Chapter 2, Section 2.3.8). Renewable demand has increased because of the global focus on renewables. This happened due to climate change concerns. For example, among other things, people have become more concerned about climate change. They knew the electricity sector was one of the main contributors. Hence, they wanted to go green and opt for renewable energy. This increased the demand for solar and impacted its costs.

However, other sectors, like coal, remained stagnant because there is no room for improving the coal power plants. The levelized costs are significantly affected by the coal prices (compared to renewables with zero costs). On the other hand, gas declined because the gas prices were significantly lower, but the technology did not have much room for improvement. Finally, nuclear, unfortunately, shows a price increase primarily because of increased regulation due to safety concerns. And as we are not building new nuclear plants that often, there is no rapid learning and benefits of economies of scale. Having said that, the future of small modular reactors could be looking up. Perhaps they are a decade away.

Renewables that are variable and unpredictable are on the rise. Coal and nuclear are more predictable but are more expensive. How should renewables be handled? What can be done about their variability and unpredictability? Can we do a better job in forecasting? How has this impacted the grid?

These questions, on one hand, have led to an evolution of independent system operators (ISOs) and balancing authorities (BAs) who need larger coordination areas to drive down renewable variability. In many cases, the ISOs and the BAs are the same entities. To manage large geographic areas, we need entities with mature balancing functions, hence ISOs also became BAs in the United States (US); see an example of how Midcontinent Independent System Operator (MISO) became a BA in Chapter 9.

Let's come back to our topic on renewable variability and ISOs. Let's take an example of an ISO/BA: California ISO, which is an ISO on the West Coast of the US. California ISO operates the US grid from the south of Canada to northern Mexico. It helps increase the integration of renewables and reduce the need for curtailment of renewable resources. California ISO does this by offering excess energy across the West by imposing a fee on carbon and allowing dispatch using the least-cost resource (generally renewables) in the region.

On the other hand, the declining costs of renewables and the narrative around resiliency have significantly increased the penetration of customer-sited renewables systems. These two countervailing trends (greater cohesion using system operators and promotion of decentralization) changed the power system's dynamics (physical and political). We don't yet know where the equilibrium is. Is it in the cohesive, centralized system of operator-led markets or the decentralized, rooftop systems that give energy independence to the end-user? We don't yet fully understand this, and the answer depends on who you ask.

Exercise 8.1 How does the generation mix for your country look? What is the primary fuel source? Has it changed over the last decade? How do you expect it to change in the upcoming two decades? How would you contribute to this change?

8.2.2 Sectors Are Increasingly Coupled

Sector coupling means replacing the traditional approach of considering electricity, transportation, heating/cooling, industrial processes, traditional consumption, etc., separately in favor of a holistic approach. Sector coupling could help make a better case for renewables and produce an overall better outcome for society.

As the population increases, the need for electrical energy increases. It means that we will need to build new power plants and transmission and distribution infrastructure. We will also need to reinforce and improve the existing coupled sectors while promoting this energy's sustainable and efficient use.

Power systems around the world, both renewable or fossil fuel based, among other things, compete for water and land that are already scarce. In other words, building a new power plant or upgrading one will impact the water and land sectors. What does this mean? Should we not build them? No, we could build them, but we need to do it wisely. We need to adapt better accounting techniques to understand the sector coupling better. Lifecycle analysis and material flow analysis can teach us to do some of this.

Exercise 8.2 What is lifecycle analysis? What is material flow analysis? Please explore these terms.

Parallel evolutions are happening at the distribution end. For example, there is an increasing trend in distributed rooftop solar installations. More and more commercial and industrial spaces want to install microgrids. Many entities want to install superefficient appliances, and some are also opting for thermal storage. The electrification of transportation has opened new avenues like the grid-to-vehicle (G2V) that promotes the usage of electricity from the grid to charge vehicles efficiently. There is another future-looking concept: vehicle-to-grid (V2G). V2G promotes electricity usage to or from parked electric vehicles to provide grid services like energy or ancillary services. So, do we treat the electrical sector separately from transportation? How about the industrial sector? How about the natural resource sector?

Exercise 8.3 The use of electricity in all energy-related processes, whether transport, heating, or manufacturing, would revolutionize the energy world. What are other sectors coupled in your country (in your region)? Now, you must help Jay. He has taken a consulting assignment for a German think-tank. In Germany, the primary source of renewable power is wind and solar. They are intermittent, variable, and uncertain. They may not be available to meet demand, and storing electricity using a commercially viable battery is still a significant issue. Jay needs to prepare a report on sector coupling. Could interconnection of sectors help to make a better case for renewables? Why (not)? How (not)?

Exercise 8.4   How would your answer change if you had to apply sector coupling to a fragile economy? Please refer to Chapter 10 on a discussion about fragile economies and energy poverty.

8.2.3   Energy Security, Reliability, and Resiliency Goals Are Important

We define the guiding principles for a power system in Chapter 7. There we talk about reliability, security, affordability, and clean factors. We will elaborate a bit more here. Additionally, we will also talk about resiliency.

The International Energy Agency (IEA) defines energy security as the uninterrupted availability of energy sources at an affordable price. It is highly contextual. For example, some countries have abundant solar and wind close to the demand centers (e.g. India and Mexico); others have it far from the demand centers (e.g. China and the US). So, renewables may or may not be considered an uninterrupted economic source. Further, in some countries, imports of coal, oil, and natural gas are raising energy security concerns (e.g. Germany and most of the European Union), whereas, in others, abundant supply and use of natural gas offer increased opportunity for low-cost utilization (e.g. US) but have also created questions around long-term pathways for a carbon-free future. So, should the EU go for local renewables and the US go for natural gas? The answer is not that simple. The economic potential of renewables and their uninterrupted availability to mitigate security concerns vary from country to country, region to region.

Unlike energy security, grid resiliency can minimize the consequences of extreme weather, reliability and quality issues, or malicious cyber or physical attacks on the grid. Since both extreme weather events and cyber or physical attacks on the grid are increasingly likely, being able to bring parts or the whole of the grid back online is critical. This is traditionally offered by using "redundancies for reliability" at communication lines, servers, and even control centers, creating backups and isolating/fire-walling the critical infrastructure from the rest of the public network. Newer methods look for better mechanisms to restore critical parts of the grid quickly and independently. Such needs may differ from one location to another. For example, bushfires are one of the primary threats to Australia's grid. Whereas in Nepal, power reliability and quality are the primary reasons to seek a resilient grid.

Grid resiliency also requires the hardening of the grid. Grid hardening is a set of actions that are required to make grid infrastructures stronger, using wired and non-wired alternatives, to plan and better protect end-users during extreme events. It involves having proactive plans that require redundancies, including documentation, storm preparation, risk mitigation plans, and additional budgeting.

Exercise 8.5   How should Jay consider the concept of sector coupling with the needs of security, reliability, and resiliency in Germany? How would it change if he was consulting for your country?

Exercise 8.6   Is your utility provider considering grid modernization efforts? To better prepare for extreme events (mainly due to climate change and change in weather patterns), does the effort include a budget for grid hardening? Why (not)?

8.2.4   Innovations and the IoT Are Opening New Doors

The traditional grid was unidirectional. It had sensors primarily for improving reliability. The utility provider typically installed these sensors, and the industry used operations technology (OT). An OT monitors, controls, and manages industry-type real-time processes, assets, and industrial types of equipment. The information technology (IT) boom in the 1990s permeated the electrical power industry. IT and OT once considered separate domains that employed their unique systems and processes have now converged in the electrical power industry. IT industry also brought a new player: the IoT. The IoT describes the network of physical objects/things embedded with sensors, software, and technologies that connect and exchange data among themselves, other devices, and other systems over the Internet.

The electric power system that protected itself from cyberattacks by using the principle of obscurity has moved into using more open protocols. This system is now cautiously connected to the Internet, and remote users can log in. Sensors, including IoT devices, create better transparency and a ubiquitous flow of operational and market data. Such data can help improve the network's reliability, help relieve congestion in real-time, reveal investment needs, help improve efficient switching patterns, and more.

The convergence of such OT and IT techniques in the power industry encourages innovation in data engineering and opens avenues utilizing AI, better cyber considerations, and data protection considerations. These trends are opening newer doors for a better future of the power system.

Exercise 8.7   What is the principle of "security by obscurity"? Please explore this concept. Are there cybersecurity regulations that apply to the electrical power system of your country? Or do you follow industry best practices like ISO27001/2? Do you need such standards? Why (not)?

8.2.5   Customers Are Becoming More Aware

Power system dynamics are changing, and so is customers engagement. Customer expectations are shifting online, and mobile experiences are becoming the new norm. The customer (traditional consumer) is evolving to become an interactive prosumer (a consumer of who also produces energy). First, the climate and renewable dialogue is becoming a mainstream customer mandate. More and more customers are asking their utilities/retailers to provide electricity from a greener source. For example, AGL, a generating company and a retailer in Australia, highlights how it has been innovating with customers at its center. They installed an innovative 245 kW solar carpark with electric vehicle charging stations at Pernod Ricard's two Barossa Valley wineries in South Australia to help their customers achieve the goal of sourcing 100% of their electricity from renewables [5]. Second, the investment trend is affecting customer behavior. For example, customers are installing more energy-efficient appliances, distributed generation, and IoT devices and buying EVs. Third, technological innovations are helping the customer to shift their demand and participate in the market.

Such changing customer awareness has proven to be slightly challenging to the traditional utility business model but has proven fruitful for better power system management.

8.2.6 New Actors Like the Aggregators Are Emerging

Energy management systems can loosely be broken down into generation management and transmission management for the transmission scale power systems and distribution management for the distribution scale power system (see Chapter 7 for further details on such systems). Note that the need to manage generation and markets has only traditionally existed at the transmission level (as the large generators only existed and were only connected to the transmission grid). However, in today's power system, generation like solar, home batteries, EVs, and geothermal are growing at the distribution scale. The power system was initially designed to take from large-scale utility generators. They are capable of providing energy as provided by their centralized counterparts. They also can provide new services due to their distributed nature and proximity to the loads. Hence to manage such generation that the traditional generation management does not manage, systems like distributed energy management systems, concepts like demand response, virtual power plant (VPP), and localized energy markets (LEMs), and actors like aggregators are emerging.

Resiliency needs, reliability (including hardening and security) needs, IoT innovations, prosumers, and the emergence of aggregators are the new forces (see Section 8.5.2) to define the grid of the twenty-first century, and to these can be added the consideration of geopolitics and the existing energy mix and energy infrastructure.

These trends are likely to persist, and so is the change in the future of the power system. We should change our current business models to adapt to these changes. It requires deliberate and proactive collaboration among stakeholders (see Chapters 5 and 6 for details on how policies are made/enacted). The traditional business model to buy and sell energy needs to be changed. Power system flexibilities must be considered. Actors like aggregators need to be enabled.

Flexibility is the ability to manage the variability/uncertainty of supply and demand across all timescales, from instantaneous stability to long-term security of the supply. Flexibility is essential today and, in the future, primarily because of the variable generation sources at grid scale, like wind turbines, or the residential scale, like rooftop PV systems.

According to the IEA, different countries, regions, and states are at different stages of variable generation, as shown in Table 8.1. The more they enter higher phases, the more flexibility is required. For example, Australia's system operator just announced that it wants to take the instantaneous variable generation by the mid-2020s. It has done this in some parts of its grid. Doing it once (Phase 1–3) is a relatively easy task compared to continuously (Phase 5 and 6).

Exercise 8.8 What phase of variable renewable generation does your country use? How are flexibility-related topics handled by your system operator or your utility companies?

Flexibility is an essential characteristic of any power system component. However, we need to change our business models, enable policy reforms, make

Table 8.1 Different phases of integration of variable renewable generation.

Phases of variable generation integration	Definition	Example countries and regions
Phase 1	A variable generation has no noticeable impact	South Korea, Africa, Thailand
Phase 2	A variable generation has a minor to moderate impact	Mexico, India, Japan, China, US, parts of Australia, Belgium
Phase 3	Variable generation determines the operating pattern	Kyushu, Sweden, Texas, California, Italy, Greece, Inner Mongolia, UK, Uruguay, Germany, Spain, Portugal
Phase 4	The system could periodically meet total load using variable generating sources	Ireland, other parts of Australia, Denmark
Phase 5	A growing amount of variable generation creates surplus (days to weeks)	
Phase 6	Seasonal surplus or deficit due to variable generation	

new rules, and deploy a range of operational and investment-based interventions to use such flexibilities in the future entirely. According to the IEA [2], by 2023, 40+ countries will have their annual variable renewable generation within 10–20% of their annual generation. This section is adapted from the IEA report.

How can we adapt to these needs? A wealth of strategies is available. We need to focus and tailor our strategies around the diverse trends already highlighted in this chapter. However, who needs to be involved? How should it be done? What hardware and infrastructure should be used? And when should this happen?

A set of measures is identified for each broad category of institutions in Table 8.2. The measures could be energy strategies, legal frameworks, policies and programs, protocols, pricing mechanisms, etc. Let's look at some of these measures by taking examples from various countries around the globe.

An energy strategy is a plan of action created to manage supply, costs, consumption, and all aspects of energy across all areas of a country or a region. More and more countries are considering flexibility in their strategies. For example, China's 13th five-year plan (2016–2020) for the power sector has targeted nearly 220 GW of its thermal fleet to be retrofitted for flexibility and performance improvements.

A legal framework should be created that includes the core component of the legislation itself and its context (e.g. the institutional, administrative, political, social, and economic conditions or arrangements). Such holistic means make the legislation available, accessible, enforceable, and effective. The legislation also needs to be flexible, constantly relevant, and adaptable over time. It needs to be futureproof.

Exercise 8.9 Futureproofing legislation is not straightforward. Let us take a hypothetical situation. Your country's government and energy ministry passed an act to allow customer DERs to participate in the market. Jay is tasked with creating guiding principles for making the legislation futureproof. Can you help Jay? Identify 3–4 points that will potentially make this act futureproof.

Like legislation and regulatory frameworks, governments can create policies and programs to support flexible services – governments then further support such legislation by allocating budgets for conducting trials or proof of concepts. For example, in Germany, Siemens AG, Kempten University of Applied Sciences, AllgäuNetz GmbH & Co. KG, and Allgäuer Überlandwerk GmbH were awarded a trial by the German government (following its policies to support renewables and innovation) to develop technology to expand the role of the grid operator. The consortium developed a market-based intervention mechanism that capitalized on the flexibility of DERs and virtual power plants (VPPs).

Such mechanisms set by governments and energy ministries (if needed) are to be interpreted by the courts, and disputes (if any) need to be settled. For example, the Ontario (Canada) government introduced the Green Energy and

Table 8.2 Various actors, policy frameworks, and hardware infrastructures to deploy flexibility.

Broad categories of energy policy institutions and actors (who)?	Policy and market framework (How?)	Hardware and infrastructure (What?)	When?
Governments and energy ministries	Energy strategies	Power plants	Now
	Legal frameworks		
	Policies and programs	Electrical grid infrastructure – transmission and distribution	
Courts (e.g. Supreme Court)	The legal interpretation of the above		
Market Monitors (e.g. FERC in the US, AEMC in Australia)	Regulatory frameworks and decisions	Energy storage	
	Power system planning exercises		
	Electricity pricing	Grid edge devices, including DERs	
Associations and Alliances (e.g. Energy Storage Association)	Inform power market rules and codes, system operation protocols, and connection codes		
Third-party actors (e.g. foundations, NGOs)	Inform power market rules and codes, system operation protocols, and connection codes		
Private companies (e.g. Facebook, Siemens Energy)	Compliance with power market rules and codes		
	Compliance with system operation protocols		
	Compliance with connection codes		
Emerging actors (e.g. aggregators, VPPs, microgrid providers, localized markets)	Compliance with power wholesale market rules and codes		
	Formulation of local market rules, identification of stakeholders and processes		
	Compliance with system operation protocols		
	Compliance with connection codes		
System operators and utilities (e.g. AEMO in Australia)	Implement and enforce power market rules and codes		
	Implement and enforce system operation protocols		
	Implement and enforce connection codes		

Green Economy Act in 2009. It was successful in terms of increasing renewable procurement. However, over time, people started claiming health-related issues. And Canada was at the center of wind farm health-related court challenges, with 17 separate hearings for its 7.8 GW of wind energy capacity and a population of 35 million [6].

Exercise 8.10   Are there any current legal disputes related to an energy act or a regulatory framework that you know exists for your utility in your region?

Market monitors then work on allocating the costs and risks of providing such services and make decisions. For example, On 15 July 2021, the AEMC in Australia published a final determination and a final rule to introduce two new market ancillary services to ensure power system stability and security.

Though not explicit in our examples, in coming up with such decisions the market monitors could also include (or ask to include) flexibility topics in their power system planning exercises. They could also work on coming up with retail pricing structures to promote flexibility. For example, some jurisdictions allow consumers to participate in the wholesale market directly (e.g. FERC Order 2222 in the US) or to choose their retailer (e.g. Power of Choice in Victoria, Australia) or choose a demand response provider (e.g. AEMO in Australia allows customers to register as a wholesale demand response service provider [WDRSP] or engage a WDRSP to act in the market on their behalf).

Once all of this is done, the rules must be formulated, the protocols must be enacted, and connection codes must be established and enforced. Several people and institutions are involved in this process. While the utilities and the system operators set and work to enforce these, the foundations and NGOs watch and inform the process, and the private companies comply with it.

In doing this, the actors can use the new and (almost all) existing power system devices to deploy flexibility. For example, conventional older fossil-fired generation, especially gas, can be retrofitted to provide flexibility. Wind and solar generation, including battery storage, can be asked to provide quick ramp-up/down services. The customer can be provided with adequate price signals to change their consumption or generation into the grid.

Exercise 8.11   Create a similar table for your country or region. Who are the actors? What are they doing? What should they do? How are they addressing the flexibility needs? How can they improve?

8.3 Summary of Our Lessons, and Present and Future Applications

Understanding the current operations and planning for a better future require understanding the fundamentals. If you have read the chapters sequentially, you will have learned about:

- Foundational engineering concepts like active and reactive power, inertia, frequency, voltage, and current (Chapter 1).
- Essential financial concepts such as discounting and levelized cost, to effectively participate in policy matters (Chapter 2).
- Debates around tradeoffs by looking at some of the current crises to allow the reader to start appreciating that there is no right or wrong answer when it comes to energy policy (Chapter 3).
- The power of wholesale energy markets in reducing carbon allows the reader to see linkages between market constructs with an established value proposition and developing markets with foundational concepts such as the capacity benefit margin, allowing the introductory concepts of market fundaments and providing information on transmission project cost allocation (Chapter 4).
- Ways to detangle the regulatory processes that are time-consuming, process-driven exercises that set the record and involve the diverse interests of all stakeholders to make an impact on energy policy (Chapter 5).
- Demystify the roles of institutions and the people that drive policy changes in any country and link it to climate change and its role (Chapter 6).
- The actual day-to-day working of the power system and electricity markets is generally a mystery even to some electrical engineering graduates (Chapter 7).

The fundaments (Chapters 1–7) that we have learned are equally applicable and relevant for the future of the power system. This chapter introduces the concept of flexibilities and highlights the role of aggregators in the future of power systems. The concept of flexibility would mean that the future of market operations (where it exists) lies in balancing inverter- and converter-based resources. Whereas offshore wind, energy storage, other DERs, and innovative technologies are the future of grid investment and planning. The concept of aggregators would mean that there are new actors that can optimize the use of DERs by bundling DERs to engage as a single entity (like a VPP) with wholesale electricity markets or distribution markets or even with LEMs (as described in previous chapters). The future power system requires clear pathways, newer elements, and innovative approaches.

8.4 Pathways to Make Informed Decisions for the Future of the Power System

National Renewable Energy Laboratory (NREL) in the US has identified various pathways for transforming power systems [1]. In the subsequent subsections, we expand on some of those pathways and add our nuances to them.

8.4.1 Transformation into an Unleased Distribution System Operator (DSO)

The number of DERs is increasing, and most of it is happening in the distribution grid. However, market operations or generation management is typically the job of the transmission provider. The transmission grid does not have clear visibility of these DERs. One way to address this is to communicate the data from DERs to the transmission grid operator, thereby increasing their visibility and control. Another way is to improve the regulatory and policy frameworks to empower the existing actors of the distribution grid, aka the distribution system operators (DSOs). DSOs, depending on the region, already own the distribution network and maintain it. They, however, don't have the infrastructure or mechanisms to run a distribution scale market system or adequately control generation. Depending on the jurisdiction, they may (not) own the intelligent meters or the end-customer accounts. But in either case, DSOs are well poised to innovate and promote DERs. In this model, the DSOs can run distribution level markets or promote LEMs or even perform or promote peer-to-peer (P2P) trading. They provide financial and other signals to invite customers (e.g. residential, small and medium-sized enterprises, and commercial and industrial, etc.) and service provides (e.g. VPPs, aggregators, network service providers, distribution energy service companies, local market operators and P2P traders, etc.) to participate and ensure that the DERs are adequately managed.

Though DSOs are better poised than the TSOs to handle increasing DERs, it is not a simple process and requires a gradual shift. The DSOs are traditionally distribution utilities that focus on increasing their profitability by increasing the number of served customers and ensuring that their customers get electricity while maintaining their reliability indices. The new function, managing increasing DERs, would require decoupling their existing revenue mindset (they can still increase revenues but not just by selling electricity to more customers) to performance-based incentives for promoting DERs. It would also require them to make a capital investment in improving their advance distribution management systems (ADMS) and distributed energy resource management systems (DERMS) to facilitate liberalized distribution markets and the formulation of newer bodies like distribution market operators (DMOs) and special regulations to promote entities like aggregators.

If we want to unleash the DSO, DMOs would be critical agencies. A DMO would maintain open access between DERs, service entities, and transmission entities. Loosely speaking, they are the distribution version of the wholesale market operators (about whom we have talked at length in other chapters). They would adhere to the rules set by the distribution regulator, which most likely is a transformed version of the current regulator that regulates the distribution system. DMOs support the approved infrastructure upgrades, perform or oversee the management of the distribution network, administer programs to promote DERs, provide provisions of default service for the nonparticipating distribution customers, liaise with and clear bids from the participants like aggregators and distribution service companies, and interact with transmission operators such that they can provide electricity to or take electricity to maximize overall system benefits.

Several countries are looking at this pathway. Some examples include Australia, where the distribution network service providers (DNSPs) are mandated to evolve into an intelligent DSO. Another example are the utilities in Thailand. They are also looking into evolving as DSOs. So are the grid companies in China. In doing so, such utilities install demand response, VPP, or DERMS programs as an initial step of this evolution.

Exercise 8.12   Unleashing a DSO requires significant institutional coordination and a gradual shift from the current system. Is there a regulatory strategy for it in your region? If so, please summarize it. If not, how would you put together a regulatory strategy? Summarize the critical technical, policy, and financial challenges and opportunities and discuss them with your colleagues. (Hint: also refer to Chapter 4. We talk a little about regulatory strategy there.)

8.4.2   Encouraging (Re)innovation for Cleaner Restructuring

In the previous pathway, DSO was unleashed to create distribution markets and control DERs as it fits to their needs. In this pathway, the lessons from the global wholesale market reform are applied. The market is re-innovating. This is more palatable as the existing bodies remain, and they will only add additional functions or reshuffle their existing ones. This is more palatable as the existing bodies remain, and they will only add additional functions or reshuffle their existing ones. Traditional lessons learned in the wholesale and transmission side are: (i) open access to the grid, (ii) appropriate screening and grid connection criteria, (iii) independent planning and dispatch, and (iv) a competitive wholesale market are the key to today's flourishing power system. This pathway requires the addition of newer elements to encourage cleaner restructuring like DER-related incentives and trading schemes, better and integrated planning/forecasting, better coordinated dispatch of DERs, and additional platforms for prosumer engagement.

In this pathway, the existing market operators, adhering to newer rules from the wholesale regulators, would run the markets by allowing DERs to bid. DSOs would maintain open access between the DERs, service entities, and transmission entities. The existing system operator and transmission operator will maintain their functions. Together all the actors coordinate to maximize overall system benefits.

Exercise 8.13   Cleaner re-innovation and restructuring can happen quicker than "unleashing the DSO." Do you agree? Why (not)? If yes, please summarize. Is there a regulatory strategy for such restructuring and renovation in your region? Is there another pathway than the ones mentioned here? Please discuss this with your colleagues.

8.5 Newer Elements of the Power System

Whether the DSO is unleashed or where it happens by re-innovation and cleaner restructuring, critical elements of the power system are changing, and newer elements are introduced. The following section highlights some of them.

8.5.1 Mini and Microgrid and Their Roles in Top-down and Bottom-up Electrifications

Mini and microgrids are more petite self-contained (with their generation and load and mechanism to balance them) versions of a traditional utility grid. They are emerging both in stable and fragile economies. In strong economies, the reason for such a grid is primarily resiliency, hardening, and reliable access for remote regions, etc. In fragile economies, the reason for such a grid is almost always a bottom-up expansion to ensure better supply (since the traditional utility grid is either not available or unreliable).

Since such mini and microgrid systems are booming in both economies, there needs to be a way to benefit from them. First, such smaller grids can trade among themselves, especially if they are close to one another. Then they need to be compatible with the traditional utility grid, ensuring two-way flow through the interconnection point. Such an approach will ensure that the bottom-up expansion in DERs and other innovative concepts will be contained at the mini or microgrid level. Future top-down expansion will only benefit from such mini/microgrids. It will ensure that it is self-contained and resilient and can support others.

8.5.2 The Aggregator Is the New Actor

Aggregation is the method of grouping consumers or prosumers or any mix thereof to act as a single coordinated entity to engage in the power system operations. The aggregator is a private company that performs aggregation and acts as a VPP. They act as an intermediary between the end-users/DER owners and other actors in the power system (e.g. DSOs, TSOs, market operators, etc.). They create economic value for themselves and their members (consumers/prosumers, etc.) and create value for the power system by providing energy and grid services. Typical grid services provided by the aggregator include: (i) load shifting by providing mechanisms to move the load patterns (hence reducing peak demand) of the end-users based on price signals, (ii) balancing services by providing capacity or ramp-up/down services, (iii) flexibility (if there is a local market) by providing the mechanisms to consider grid parameters while creating market offers and executing market orders, (iv) lowering the marginal cost of energy by providing the means to deploy cheaper DERs over-ramping large, costly centralized generators, and (v) optimizing investments by avoiding the need for the addition of new expensive, large-generation capacities.

A liberalized market is necessary (reformed wholesale market or unleased DMO market) to empower the aggregators. Such markets will allow better participation of the aggregators to provide both energy and other grid services. Another crucial enabling factor is the digitalization of the distribution system and the presence of smart meters. Aggregators need data to function. They need to collect data from DERs to prepare market offers/bids, send this data to the market operator, and if the bid is awarded they need to execute the offer in (near) real time and relay the data to all the parties for billing. A very robust digital infrastructure is required to collect and communicate data. Aggregators need two-way communication with DERs and other parties like the market operator.

Aggregators are the key actors of the current/future power system for enabling DERs. The market for the aggregator was ~$762 million in 2016, and is expected to reach $4597 million by 2023 (based on a compound annual growth rate of 25.9% from 2017 to 2023). It will only grow in the coming years. Bundling of DERs through VPPs is required in future power systems. The aggregators create a sizeable capacity to participate in the current wholesale markets; they provide energy and other grid services. They also ensure the fixed costs and complexity of coordination and dispatch of smaller DERs are spread to multiple parties within the VPP.

8.5.3 Peer-to-peer (P2P) Trading and Localized Energy Markets (LEMs)

As mentioned in Chapter 1, P2P trading mechanisms are emerging globally as a potential solution to address DER issues. It involves buying and selling energy or services between two or more connected parties. For example, I can sell my power to my neighbor, or my neighborhood brewery can sell their solar energy to the nearby grocery store. Such trade can happen between individuals or mini and microgrids or any of them. So why do we need this? Take an example of rooftop solar, a form of DER. In most places, we currently sell excess solar to the grid for a small feed-in tariff. However, in a liberalized society, such trade should have no boundaries, at least in theory. Hence, the concept of P2P was introduced. In P2P trading, individuals or groups buy/sell energy and other services to another set of individuals or groups.

LEMs can be used as a mechanism for buying/selling energy, activating flexibilities from DERs, and preventing grid congestion at a local scale. LEMs can be operated by a distribution operator or by an independent entity. It can be run at a community level. The concept has just started, and the ideas are limitless. For example, in a local community-level market, the VPP and households could submit offers to buy and sell electricity. The distribution operator could also send signals indicating forecasted grid congestions or requests to procure flexibilities. And the LEM clears it. The LEM could then run a multicriteria optimization to maximize local exchanges and harvest the flexibilities.

Exercise 8.14   Which of these newer elements are hot topics in your region? Are there other elements? What are they?

8.6   Innovation and the Power System

Most countries and regions are undergoing an energy transition. The power system actors (see Table 8.2) find the necessity to innovate now. Innovation is the art of creating a new and viable offering. Such innovation is happening, especially to address (i) the stalling growth in electricity demand, (ii) requirements for better management of DERs, (iii) the push for value-based offerings, and (iv) increasing cyberthreats.

We highlight some of the technology, business, and other innovations that are currently happening here:

- New business innovation for protection against loss of revenue from stalling demand. Growth in electricity demand is stalling in most parts of the world. The actors are innovating solutions to manage declining demand and increase in reverse flows from DERs while staying within the constraints of the regulatory structure. Some utilities are exploring newer business models to become operators of mini and microgrids, while others are developing analytical tools to improve forecasting and supporting real-time operations by providing ancillary services. We see significant innovation continuing in this sector where the actors would contact software vendors or universities and think-thanks continually innovating and changing their business models. We also see business innovations happening where traditional oil companies are starting to diversify their investments by looking into newer fuel sources like hydrogen (H_2) and DERs.

- Low-voltage management system (LVMS) for the better management of residential DERs. Before the emergence of DERs, there was no pressing need to have visibility and control of the low-voltage (household level voltage) distribution network. However, DERs have generally emerged in the LV network. DERMS emerged as a potential solution. However, they have not been able to address the LV network adequately. Hence, an LVMS is a vital innovation. An LVMS system is currently in its infancy, and not many vendors provide such a system. However, once matured, it should at least include SCADA, network monitoring, energy accounting, real-time forecasts, Historian and playback, load shedding, system automation, and special modules for DER monitoring and control.
- Product system and experience-based innovations to address the customer push for including value-based offerings. To address the increasing demand for providing value, the actors are focusing more and more on services other than just selling energy. Two of the prominent examples are (i) the grid reinforcements happening to accommodate batteries and newer DER technologies and (ii) creating better transparency by communicating, collecting data, and adequately controlling the customer premise by using. We see more of this happening in the future.
- Awareness and compliance to standards for addressing cyber concerns. With the convergence of IT and OT and the need to have transparency and better control, we have more intelligent technologies connected to the network (Internet or otherwise). The electricity sector is hence one of the targeted sectors for malicious attacks. Because of this, more and more actors in the power system are currently innovating their tools and techniques, introducing new business processes, and adhering to cyber standards. In the future, we will see more stringent cyber standards in place. More focus will be given to cyber issues related to the distribution system, where this topic has historically not been given as much significance as in the transmission world. Cybersecurity in the distribution system is not yet a focus of the regulators.

Exercise 8.15 List some of the product system, process, and business level innovations that are happening in your power system. What is working and what is not? What could have been done better?

Generation, operation, and control as we know them have changed due to economic incentives on flexibility, resiliency, and reliability. Inverter-based resources are the new norm. Several distributed generation sources offer baseload power. This chapter has looked at the "distributed" framing of traditional power generation, operation, and control.

We have looked at a range of operational, policy, and financial interventions worldwide to make the power system more flexible, cleaner (in addition to being reliable, affordable, and secure). We identified the challenges and opportunities in this transformation, and it outlined the policy, regulatory, and market-related pathways that can be implemented to mitigate these challenges. In doing so, we highlighted the importance of mini and microgrids, aggregators, flexibility services, P2P trading, and local energy markets and innovation to achieve this transformation.

8.7 Visioning

We started this chapter by saying that AI was more profound than fire or electricity. We don't know about the fire, but there is no AI without electricity. But we need AI to enhance the future of the electrical grid. We need to focus and act (create laws, strategies, legal frameworks, policies, and frameworks; make decisions; enforce new market rules; and change retail pricing structures, operation protocols, and connection codes) to make the future power system more flexible. We need to leverage the trends like AI, innovation, customer engagement, lower costs of renewables, etc.

A range of operational, policy, and financial interventions already exist to make the power system more flexible and cleaner (in addition to being reliable, affordable, and secure). In this chapter, we looked at a summary of this transformation. You should now be:

- Adequately acquainted with the current global transformations in the power system of this century.
- Adequately acquainted with the challenges and opportunities associated with them.
- Able to use this knowledge to make an informed decision to make or support making policy, regulatory, and financial interventions.

Case 8.1 *General Management and Energy: Supermarket Case* You work for a supermarket and report to the general manager of energy. Your leadership sees value in adding megawatts of solar at your warehouses and retail shopping centers. You are required to pilot this concept at one of your warehouses in your hometown.

As you know by now, installing solar requires the installation of inverters. Installing this solar will lower your electricity bills, demonstrate your organization's commitment to sustainability, and reduce its reliance on the grid. You might also want to install a battery for technical (to smoothen your solar output) and economic reasons (participation in the market).

Please collect the following information and summarize it in fewer than a five-page report for creating a business case:

- Resources provided in the public domain to choose a quality solar + battery system.
- Availability of financial incentives and funding.
- Summary of benefits.
- Potential barriers, including information regarding community attitude toward such technology.

As mentioned elsewhere in this chapter, the price of solar has significantly fallen, and the price of batteries is also steadily declining. However, the financial conditions might not line up, depending on one's circumstances. If it does not line up, how and where should you look for additional sources of revenue.

Hint: You could investigate not only energy consumption but also trading. One could trade both energy and ancillary services into the market (if it exists).

Case 8.2 *Intrapreneur in an Energy-related Software Company* You are an intrapreneur working for a Fortune 500 company. As an intrapreneur you have mandates to innovate product development activities. You are tasked to create a software solution to support the future grid, and the company has identified a low-voltage network (household level voltage) as a priority. Do you agree? Disagree? Investigate if there are LVMSs available on the market. How do the utilities currently gain visibility of their low-voltage network? Conduct interviews with ~5–10 stakeholders/users (utility providers, retailers, DSOs, DNSPs) and summarize the finding in report of no more than 10 pages.

Case 8.3 *Flexibility in the Power System* Traditionally, the power system had large rotating machines that produced the electricity, and there were loads like heaters, lightbulbs, and industrial appliances. Over time, both loads and generators (like solar and battery from Case 8.1) are becoming more inverter/converter based. What are the plans of your government and other stakeholders to address the issues that such systems could cause? Are there any policies, laws, and regulations to promote "flexibility" in the power system? Are there any policies, laws, and regulations to promote grid-edge devices? Why (not)? Create a two-page PowerPoint summary and present it to the class.

Case 8.4 *Renewable Targets* According to the IEA, in Quarter 1 of 2020, nearly 28% of global share of electrical energy came from the renewables [7]. For achieving a better carbon-neutral future, a common goal, we need to increase this percentage share. Several countries worldwide have legislated mandatory renewable energy targets (MRET) to increase this percentage share gradually. Over 60 countries around the globe have such mandates. Does your country/region have one? Are the targets adequate? Are they achievable? What is an achievable target? Assuming you should strive for 100% renewable, how should your country/region plan it? Can it be 100% renewable? Conduct a brainstorming session with a team of five people and summarize the discussion in a meeting minute.

References

1 Clean Energy Ministerial (2015). Power systems of the future. https://www.nrel.gov/docs/fy15osti/62611.pdf (accessed 22 February 2022).

2 IEA (2019). Status of power system transformation 2019. https://www.iea.org/reports/status-of-power-system-transformation-2019 (accessed 22 February 2022).

3 Parkinson, G. (2021). Rooftop solar pushes state's entire local network into negative load for four hours. https://reneweconomy.com.au/rooftop-solar-pushes-states-entire-local-network-into-negative-load-for-four-hours (accessed 22 February 2022).

4 US Department of Energy (2022). The history of solar. https://www1.eere.energy.gov/solar/pdfs/solar_timeline.pdf (accessed 22 February 2022).

5 Corbett, C. (2021). Customer driven innovation. https://thehub.agl.com.au/articles/2020/11/customer-driven-innovation (accessed 22 February 2022).

6 Energy and Policy Institute (2022). Overview of court cases. https://www.energyandpolicy.org/wind-health-impacts-dismissed-in-court/overview-of-court-cases (accessed 22 February 2022).

7 IEA (2020). Report extract: renewables. https://www.iea.org/reports/global-energy-review-2020/renewables (accessed 22 February 2022).

9

Influence of Wholesale Energy Markets in Policy and Pricing Discussions: Part 2

9.1 Introduction

Elsewhere in this book, we discuss the fundamentals of wholesale energy markets:

- Why someone should care about wholesale markets in energy policy discussions.
- Whether energy markets influence policy or is it the other way around.
- Why the decision to join an energy market is not an easy one given the market implementation costs, neither is leaving the market.
- Why it is possible to have multiple markets within a country and even within a state or a region.
- How wholesale energy markets exist in international locations.
- What happens in countries without organized markets for reliability and capacity purposes.

Given that background, we now drill down into energy, capacity, and ancillary services markets (ASMs) specifically. In some parts of the world, capacity is included as an ancillary service in the real-time or day-ahead markets (DAMs) and referred to as "reserves." In such a case, there are two markets – the energy and ASM – during real-time and DAM operations. As we also need capacity for longer-term (years), there is a separate market for this. In this chapter, when we talk about capacity markets in isolation with ancillary services, we generally refer to these long-term capacity markets where capacities are traded.

Additionally, since this chapter mostly provides examples from the United States (US) electrical energy market, here is another upfront note for the reader. The Federal Energy Regulatory Commission (FERC) is the market regulator for all energy markets that cross state boundaries in the US. The North American Electric Reliability Corporation (NERC) is the reliability authority. Hence, there is a clear distinction between the market regulator and the reliability regulator. When the

Modern Electricity Systems: Engineering, Operations, and Policy to address Human and Environmental Needs,
First Edition. Vivek Bhandari, Rao Konidena and William Poppert.
© 2022 John Wiley & Sons Ltd. Published 2022 by John Wiley & Sons Ltd.
Companion website: www.wiley.com/go/bhandari/modernelectricitysystems

text refers to major market policies, it is referring to FERC's role. And when the text discusses major reliability events, it is referencing NERC's mandate.

Wholesale energy markets provide price transparency with transparent market prices like the locational marginal price (LMP) at each market node.[1] There are three components of LMP: (i) marginal energy price, (ii) congestion price, and (iii) loss components. Similar to the LMP, in Australia the wholesale electricity market provides a regional reference price (RRP). US markets are cleared per electrical nodes/buses, and Australian markets are cleared per region (like Victoria, New South Wales).

In either case, energy markets are important to reduce the overall wholesale price of energy because an energy market means competition to serve the demand. The energy market matches both supply and demand, respecting transmission limitations. In addition to competitive energy prices, energy markets have a track record of reducing carbon emissions because competitive markets bring the latest technologies [1].

Readers know energy markets are hardly that simple, and they vary from one geography to another. Generally, there is a real-time market (RTM) for energy purchases and sales in five-minute intervals (not until recently in European markets[2]) in real-time and a DAM, which is for energy purchases and sales in the hour-ahead intervals for the next day [2]. Or there is just a real-time or spot market for energy, and ancillary purchase and sales, e.g. in Australia. Yet another example is Israel, which has a high level of similarity with the US market. The power market there has day-ahead and RTMs.

Where it exists, the intent behind the DAM design is an assurance to the control room operator for resources to match up the forecasted demand. And RTMs exist because things happen, such as demand peaks (due to a cold front), transmission line outages (due to a tree branch falling), or underperforming generation (due to boiler failure).

On an annual average, more than $5 billion was exchanged in the New England ISO energy markets over the decade from 2008 and 2020, as shown in Table 9.1.

The energy markets and their limitations are highlighted by increasing energy storage penetration and distributed energy resources (DERs). For example, an energy storage resource can charge during off-peak energy price at $5 per MWh (hypothetically) and discharge during peak demand when the market clearing price is $500 per MWh. This market opportunity exists and is called price arbitrage.

But the hypothetical example of energy storage bidding in an energy market is not simple. Depending upon which US energy market is discussed, there are

---

1 According to Daniel Moller Snenum, in Europe there is a distinction along those lines: aggregate/large geographical areas' nodes = wholesale market, and high geographical resolution with many nodes = LMP.
2 According to our external reviewer, Daniel Moller Snenum.

Table 9.1 New England Independent System Operator (ISO) market revenue 2008–2020.

Year	New England ISO Energy market revenue ($billion)
2008	12.1
2009	5.9
2010	7.3
2011	6.7
2012	5.2
2013	8.0
2014	9.1
2015	5.9
2016	4.1
2017	4.5
2018	6.0
2019	4.1
2020	3.0
Average	6.3

variations. For the most part, the concepts are the same, but there are distinctions in actual bidding energy storage resources. That is because each independent system operator (ISO) and regional transmission organization (RTO) in the US has the flexibility to implement storage in their market participation models as they see fit.

Readers are directed to the FERC Order 841 ISO/RTO compliance links for details on each US grid operator's implementation plans [3].

9.2 How Do Energy Markets Coordinate Reliability?

By coordinating reliability within the balancing authority (BA) boundary, most energy market operators ensure reliability. An organized market's reliability coordination looks different from a utility's control room.

Energy markets need to ensure reliability. Without reliability, there is no transmission planning or the day-ahead, real-time, or spot market (Figure 9.1).

Some of the jobs found in the Pennsylvania–New Jersey–Maryland Interconnection (PJM) control room are in the following order of workstreams [4, 5]. For example, the master coordinator works with the meteorologist to schedule the next day's power:

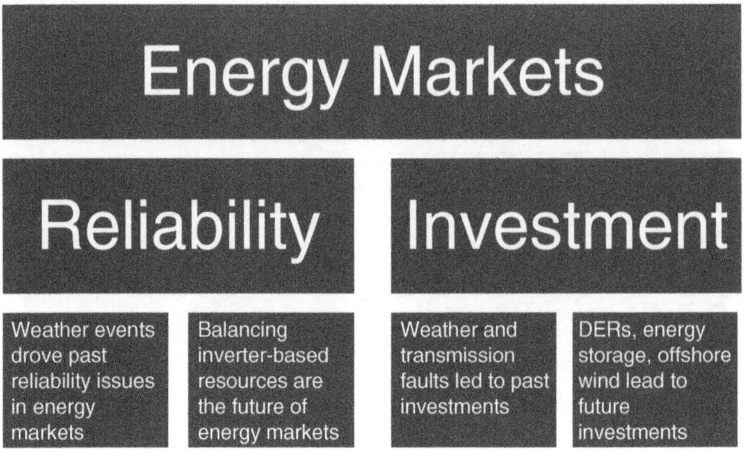

Figure 9.1 Energy markets discussion roadmap.

- Reliability engineer: Responsible for real-time reliability studies in the control room (the model runs 4000 contingencies every minute), including understanding the neighboring system's impacts on PJM operations.
- Master coordinator: Starting role for a PJM dispatcher. A master coordinator is an energy scheduler who takes what is offered in the DAM and schedules energy flows for real-time operations.
- Meteorologist[3]: Works closely with the master coordinator because of the weather impacts on the next day's power demand. The PJM meteorologist takes the weather data from multiple sources and determines what is relevant for the PJM market [6].
- Generation dispatcher: Primarily responsible for balancing supply and demand in real time, including ensuring enough reserves are maintained on the system.
- Master dispatcher: Responsible for power flows over the transmission lines, including preparing for contingencies when a line trips due to weather.
- Shift supervisor: Must work at one or more of the above positions before being placed in a supervisory role. Oversees the entire control room operation and interface with the other utilities and regulators.

Extensive eight weeks of annual training (every six weeks PJM operators train for a week), NERC certification, and PJM certification is needed to be in those control room positions. When COVID-19 hit the US, there was a concern raised about control room operators because of their critical role in electric grid operations. The New York grid operator sequestered critical operations personnel [7].

---

3 According to LinkedIn, there are meteorologists at PJM, California ISO, MISO, and ERCOT.

9.2.1 What Past Reliability Issues from Energy Markets Have Influenced Policy?

Energy markets are dormant for the most part. They hardly attract any attention on a normal day. But when things go wrong like blackouts or energy prices increasing due to natural disasters, or if political leaders are unhappy due to their energy prices, energy markets are suddenly in the news. The following instances illustrate the point about energy markets in recent news:

- Cold weather in Texas in February 2011 led to natural-gas-fired unit challenges in the southwest US for the Electric Reliability Council of Texas (ERCOT). This event is known as the February 2011 Southwest Cold Weather Event. Both FERC and NERC staff conducted a lesson learned report [8]. As a result, FERC initiated electric and natural gas market coordination proceedings.
- Cold weather in ERCOT in February 2021 also led to widespread disruption in daily life in Texas, including the loss of life. FERC and NERC conducted another lesson learned report [9]. It remains to be seen if we see specific NERC standards on winterizing generation equipment any time soon. Please refer to Chapter 3, where we discuss this event.
- Black System South Australia 28 September 2016 Event. On 28 September 2016, tornadoes hit some transmission lines in Australia, tripping them. This event led to a sequence of faults in succession and resulted in voltage dips. The protection settings of several wind farms in mid-north South Australia exhibited a sustained reduction of power ($\sim$456 MW), leading to an increase in the import of power from Victoria and overloading the interconnector, therefore tripping it and islanding South Australia. There was not any fast-acting automatic load shedding scheme in South Australia to balance the generation and load. Therefore, South Australia was blacked out [10]. Following this incident, several rules and policy changes related to system management, system inertia, frequency control services, and system strength have been proposed and implemented in the National Electricity Market (NEM) in Australia.
- When California had blackouts in August 2020, initially renewable energy was blamed. Later, it was found that insufficient imports into the California Independent System Operator (ISO) was the leading cause [11]. As a result, FERC approved California ISO's emergency filing on pricing the imports right after the Aug 2020 California blackout [12].

As these instances indicate, energy markets undergo policy changes when something breaks down for the most part. And the leading cause for something to break down is the weather, which explains the meteorologist's role at most US grid operators. By nature, the electric industry is very resistant to change, but when something bad happens on the grid, everyone's attention is there – policy changes are enacted.

9.2.2 Balancing Inverter-based Resources Is the Future for Operations in Energy Markets

DERs are inverters (or converters) based; so are utility scale wind or solar farms. Washing machines have inverters/converters, so do our fridges. In summary, the power electronic inverter/converter technology is used both in generation and load technologies. Let us understand this problem of balancing by taking examples of DERs. We will also briefly illustrate this problem for grid-scale renewables using virtual synchronous generators (VSCs).

The electric utility industry is in the early stages of DER integration in operations. Hence, there is insufficient data on DER-specific events that impact energy markets, but with forecasted estimates of DERs we can expect big policy changes in the energy markets for DERs.

Some impacts in operations are illustrated by the ones that have large DER participations. New York ISO, when compared to other US FERC-jurisdictional market operators, is farther along in allowing for DERs participation in wholesale markets [13]. California ISO is a close second. To accommodate DERs, New York ISO took a phased approach by creating aggregation concepts, running pilot programs, changing metering policies, enhancing measurement and verification techniques, improving forecasting mechanisms, creating more granular pricing mechanisms, and supporting and changing market rules. We can expect other grid operators to learn how to accommodate DERs in electricity markets from these entities.

Here is another example. We can see that the smart inverter-based standard is a mandatory requirement due to the German experience. In Germany in the year 2011, inverters were tripping due to voltage and frequency fluctuations. This "50.2 Hz problem" as it was called, due to frequency tripping at 50.2 Hz value, resulted in Germany and the rest of Europe changing all the inverters to smart inverters at a significant cost [14].

By smart inverters, we refer to inverters capable of "riding through" minor fluctuations in the voltage and frequency. The German experience is the leading factor behind IEEE 1547:2018 Distributed Generation (DG) interconnection standards in the US. We can find most new solar installations insisting on IEEE 1547 compliant inverters.

Like smart inverters, VSCs are typically used in large renewable farms to make grid-connected inverters/converters work as synchronous machines (e.g. with tunable inertial properties). They represent a promising solution for a future grid where inverters/converters need to mitigate the grid issues that could arise due to further reduction of large rotating generators.

Exercise 9.1 As you know, Jay is an international energy consultant. He volunteered to serve as an expert on a trip to Tegucigalpa, Honduras. This Bureau of Energy Resources at the US Department of State and NARUC led workshop has Costa Rica, El Salvador, Guatemala, Honduras, and Panama regulators. Staff from Honduras's National Electric Energy Commission (Comisión Nacional de Energía Eléctrica, or CNEE) Guatemala's National Electric Energy Commission (CNEE), El Salvador's General Superintendency of Electricity and Telecommunications (SIGET), Costa Rica's Regulatory Authority for Public Services (ARESEP), and Panama's National Public Services Authority (ASEP) may be in attendance.

Prepare two 15-slide presentations with an appendix not exceeding five slides on two topics. Distributed energy resources and demand side management are the topics of interest to the Central American regulators. For each topic, (i) identify common barriers such as policies, technological, economics, and financial; (ii) identify the role of the regulator in the monitoring of corporate power purchase agreement (PPA) in dispute resolution; relevant regulations; and (iii) consider implementing various energy prices to incentivize consumer behavior change and how to involve the utilities in the process.

9.3 How Do Energy Markets Facilitate Grid Investments?

Typically, market operators also serve as ISOs/RTOs. For the smooth operation of these markets, they require more transmission lines to transport the electrical energy reliably. Hence, they promote grid scale investments that are vital for their value proposition.

At the time of writing (2021), we take for granted large utility-scale wind energy markets. That is because of the past decade's renewable policies and incentives for wind integration on the electric grid. Those policies resulted in large-scale transmission build-out. Texas's Competitive Renewable Energy Zones (CREZ) is a prime example of high-voltage transmission for renewables integration.

Another key example of policy influencing transmission planning is the multivalue project (MVP) portfolio implementing renewable portfolio standards (RPS) in each Midcontinent Independent System Operator (MISO) state. This is a MISO MVP portfolio of transmission projects for integrating renewables serves as a blueprint for planning. Key elements of this transmission planning include:

- Planning engineers involving regulators and legislative leaders to understand the RPS and find out where economic development is needed in their state/region.
- Planning engineers understand the capabilities of the current transmission system to find where grid investments are needed.

Transmission planning figures are prominently reflected in any political leaders' statement in the US to achieve clean energy goals [15, 16]. PJM advocates a "transmission first" approach when referring to offshore wind interconnections needed to accommodate a PJM mandate [17]. This planning approach truly takes the saying, "If you build it, they will come" to a whole new level.

9.3.1 What Major Events Have Influenced Transmission Policies?

Large-scale historical events like blackouts do not directly impact transmission planning policy. This is because it is hard to pin down a specific outcome, such as the construction of a transmission line being the direct result of a major blackout. However, compliance standards and the planning assessments can be a direct outcome of a major event.

After researching 16 major events analysis reports available on the NERC website that occurred in North America and compiling transmission planning ("T planning") recommendations in our professional work, we suggest the following take ways [18]:

- Most changes to studies and analyses result after a major event (please note we are not suggesting causality but rather a correlation).
- All the outcomes are not as straightforward as NERC standards for tree trimming in the US, one of the Northeast blackout outcomes in 2003.

The following research explains why US transmission providers (TPs) run underfrequency studies because of the 10 August 1996 blackout in Arizona and other western states that impacted 7.5 million customers, 28 000 MW of load, and lasted up to nine hours. The Western Interconnection separated into four separate islands when underfrequency load-shedding relays tripped [19]. Like the South Australia 28 September 2016 event, the 10 August 1996 blackout in Arizona resulted in protection and control (PRC) NERC standards around underfrequency (Table 9.2).

In summary, 9 out of 16 events were caused by weather (see Figure 9.2), and transmission system faults caused six. Only one event (the August 2003 northeast blackout) had multiple causes, and only two events (January 2014 polar vortex, October 2011 northeast snowstorm) resulted in specific findings of transmission line outage scheduling and trees falling on power lines as a major cause of transmission line outages.

In summary, based on these 16 events collected by NERC in the US, the majority of the events occur due to weather and result in changes to transmission planning studies and assessments.

Table 9.2 Summary of T-planning-related recommendations from 16 major event analysis reports.

S No	Event	Major cause	T-planning impacts
1	May/June 2021 Odessa Disturbance Report	A single line-to-ground fault	No major T-planning impact finding was called out. But the report recommends improvements to the FERC generator interconnection agreements "accompanied by clear requirements for accurate modeling and sufficiently detailed studies during time of interconnection, including electromagnetic transient (EMT) studies where necessary (most cases to ensure appropriate ride-through for BPS fault events)" [20]
2	July 2020 San Fernando Solar PV Reduction Disturbance Report	A single line-to-ground fault	"Interconnection requirements should ensure that the models provided during the interconnection study process are able to account for all forms of tripping by inverter-based resources so that sufficiently accurate studies can be conducted by the Transmission Planner TP and Planning Coordinator PC. In most cases, this will require the collection of accurate, plant-specific electromagnetic transient (EMT) models" [21]
3	January 2019 Eastern Interconnection Forced Oscillation Event Report	Oscillations caused by a faulty input to a combined cycle unit's control system	No major T-planning impact finding was called out
4	January 2018 South Central Cold Weather Event Report	Cold weather caused "183 individual generating units experiencing either an outage, a derate, or a failure"	"The Team recommends seasonal studies that consider more-severe conditions, modeling same-direction simultaneous transfers and other stressed but realistic conditions, and sharing the results with operations staff to aid in planning for more extreme days like January 17" [22]

(continued)

Table 9.2 (Continued)

S No	Event	Major cause	T-planning impacts
5	April and May 2018 Fault Induced Solar Photovoltaic Resource Interruption Disturbances Report	"A 500 kV transmission line fault caused by a failed splice" and "a 500 kV transmission line fault due to insulator flashover caused by a buildup of bird nesting material"	"Transmission entities should review load SCADA points following grid disturbances for those load banks that have relatively high penetration of DERs" [23]
6	September 2017 Hurricane Irma Event Analysis Report	Hurricane Irma caused forced outages of "over 100 high-voltage transmission lines, including one 500 kV line, 48 230 kV lines, and a total of 69 138 and 115 kV lines"	No major T-planning impact finding called out [24]
7	August 2017 Hurricane Harvey Event Analysis Report	Hurricane Harvey impacted "225 transmission assets" in Texas	No major T-planning impact finding called out [25]
8	October 2017 Canyon 2 Fire Disturbance Report	Canyon 2 fire caused phase-to-phase faults on a 220 and a 500 kV T line	"Electromagnetic transient EMT studies should be performed by the affected Generator Operators (GOPs), in coordination with their transmission Owner(s) (TO(s)), to better understand the cause of transient over voltages resulting in inverter tripping" [26]
9	August 2016 1200 MW Fault Induced Solar Photovoltaic Resources Interruption Disturbance Report	Blue Cut fire caused 13 – 500 kV line faults and 2–287 kV line faults [27]	No major T-planning impact finding was called out. But this event provides a helpful reason for planning and operations modelers to work together specifically on stability studies
10	April 2015 Washington DC Area Low-Voltage Disturbance Event	Two independent and separate protection systems failed [28]	No major T planning impact finding was called out.
11	January 2014 Polar Vortex Review	Extreme cold weather caused untended loss of >10 000 MW of generation over three days	"Review generation and transmission outage scheduling processes to limit planned outages during possible peak winter periods" [29]

Table 9.2 (Continued)

S No	Event	Major cause	T-planning impacts
12	October 2012 Hurricane Sandy Event Analysis Report	Hurricane Sandy caused more than 264 transmission assets to trip and made >20 000 MW unavailable [30]	No major T planning impact finding was called out.
13	October 2011 Northeast Snowstorm Event	Snowstorm	"By far, the leading cause of transmission line outages during the October snowstorm was trees or tree branches falling onto power lines from outside and inside utilities' rights-of-way" [31]
14	September 2011 Southwest Blackout Event	System disturbance led to cascading outages	"Not identifying and studying the impact on Bulk-Power System reliability of sub-100 kV facilities in planning and operations." [32]
15	February 2011 Southwest Cold Weather Event	Cold weather caused 210 individual generating units to experience "either an outage, a derate, or a failure to start"	"Planning authorities should augment their winter assessments with sensitivity studies incorporating the 2011 event to ensure there are sufficient generation and reserves in the operational time horizon" [33]
16	August 2003 Northeast Blackout Event	Multiple causes	"System planning and design studies, operations planning, facilities ratings, and modeling data accuracy were ineffective preparations for 8/14 event" [34]

If we shift our focus to a blackout in Italy (28 September 2003 Italy blackout) [35] and a power system disturbance in Europe (4 November 2006 Europe system disturbance) [36], the main causes for both these events are not directly transmission planning related either, but, similar to US experiences, they involve planning studies and assessments (Figure 9.3).

Similar to NERC in the US, the European Network of Transmission System Operators for Electricity (ENTSO-E) in Europe has major blackouts and event analysis reports.

Although most of the events occur due to weather, it does not mean major policies in transmission planning are all related to weather and tree branches. Allocating the costs of major transmission projects and planning for the entire region is a major rule in the US called FERC Order 1000 ("O1000"). O1000 was a

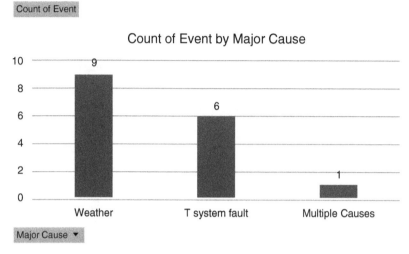

Figure 9.2 Weather is the major cause of major events in the US.

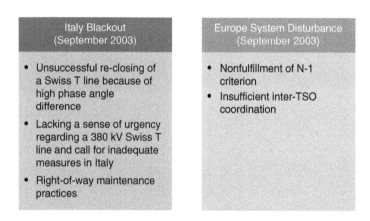

Figure 9.3 Italy blackout and Europe system disturbance root causes.

major event in T planning because FERC opened 345 kV T projects and above for competition at RTOs.

Exercise 9.2 Jay impressed his client at the Bill & Melinda Gates Foundation with his energy markets knowledge. His client wants him to dive into energy markets. Can you help Jay with this latest assignment?

His client wants to fund an energy market for sub-Saharan Africa. Without reliability, there can be no energy market. Hence, can you help Jay think through past events in Africa from an electricity grid perspective and lay out a roadmap for investments to be prioritized. Those investments should prepare the region for an energy market. Prepare 2–3 pages of research brief outlining your findings for Jay.

9.3.2 DERs, Energy Storage, and Offshore Wind Drive Future Grid Investments in Energy Markets

The future of energy markets and planning is incorporating emerging technologies like energy storage and DERs. The energy storage market size in the US is estimated to reach $8.9 billion in 2026, according to a Wood Mackenzie estimate [37].

DERs are increasing on the electric grid at both transmission and distribution levels. For example, the DER market size in the US was estimated to reach $110 billion in 2025, according to a 2020 estimate [38]. As described in other sections, current wholesale market rules are centered around large central generators. Given this scale of investment in DERs, the energy markets need to enact new rules for operations and planning to consider the capabilities and characteristics of DERs actively. Adoption of such rules requires heavy investments in energy markets.

FERC Order 2222 on DER aggregation ("DERA") influences energy policy discussions around opportunities for DERs participation in the wholesale markets. With this order, the line between wholesale and retail electricity markets blurs as consumers become "prosumers" (i.e. producers as well as consumers of energy).[4]

It is early in the process to report on the FERC Order 2222 ISO/RTO compliance plans. Most compliance plans should be available by mid-2022. Each ISO/RTO has started engaging its stakeholders on how to comply with the FERC Order 2222.

DER penetration is not unique to the US. As mentioned elsewhere in this book, DER participation in energy markets in India, Vietnam, Mexico, and other countries is also being discussed.[5] For example, like FERC Order 2222, the concept of energy communities that is emerging in Europe is their method of creating a market mechanism for getting maximum benefits out of DERs. Yet another example is the wholesale demand response (WDR) and VPP program run by the Australian Energy Market Operator (AEMO). WDR allows energy retailers, third-party providers, technology innovators, and large commercial/industrial customers to participate in demand response. The VPP trial program allows aggregation and disaggregation of DERs, storage, and loads to provide grid

---

4 A note from our reviewer indicates that DERs make the market more liquid by increasing the amount of participating actors. That is well aligned with economic theory on making markets more efficient. Source: Daniel Moller Snenum.
5 DERs have been participating actively in markets for almost two decades in Northern Europe. Combined heat and power plants (some sub-10 MW scale) have participated in wholesale and other markets, e.g. referenced in Snenum's thesis [39]. Similarly, wind turbines have offered their production on several different markets (e.g. running half capacity to provide ramp-up/down services) for the last 10 years. Source: Daniel Moller Snenum.

services (energy and ancillary service). These global examples clearly illustrate that DERs drive future investments in energy markets.

Energy storage is another reason for investment in energy markets. For example, FERC Order 841 mandates electric storage resource (ESR) participation in the following manner:

- ISO/RTO must establish an ESR market participation model for eligible ESRs to provide all capacity, energy, and ancillary services they are technically capable of providing.
- ISO/RTO must ensure ESRs can be dispatched and set the wholesale market clearing price as a wholesale seller and wholesale buyer.
- ISO/RTO must account for the physical and operational characteristics of ESRs through bidding parameters or other means.
- ISO/RTO must establish a minimum market participation size requirement that does not exceed 100 kW.
- The sale and purchase of electric energy to ESRs must be at the wholesale LMP.

Looking to Europe, the map in Figure 9.4 shows projects classified as "under consideration" out of the 154 transmission and 26 storage projects. These projects are from the ten-year network development plan (TYNDP) that ENTSO-E publishes every two years, looking ahead 10–20 years [40].

Offshore wind (OSW) deployment on the bulk electric system is increasing because of several factors: technology costs are dropping, higher capacity factors, and the number of jobs generated. According to one estimate, the OSW market size in the US will reach 25 000 MW in 2029 [41]. At a levelized cost of energy (LCOE) of $95/MWh [42], this 25 GW OSW translates to a $2.375 trillion market.

There is an advantage that OSW offers to a transmission planning and operations engineer, which is, by interconnecting all the OSW at onshore transmission substations, the engineer knows where the OSW would inject energy and at what time.[6] This information is valuable for planning grid investments.

Grid operators and market engineers have experience with wind integration in the energy markets. When the wind was deployed initially, there was an outcry about wind curtailments. With more wind incorporated, the market and stakeholders, in general, are comfortable with wind integration (Figure 9.5).

---

6 Comment from Daniel Moller Snenum: "There is much discussion ongoing regarding OSW. Hereunder the energy islands, which will act as large hubs for collecting farms and feeding onshore through single connection. Also, there are discussions ongoing regarding the build-out of grids, where overbuilding transmission capacity for the initial projects will enable savings in subsequent projects."

Figure 9.4 ENTSO-E energy storage and T projects under consideration. Source: ENTSO-E website.

9.3.2.1 Modeling Energy Storage Is Increasingly Relevant in Transmission Planning

Another emerging trend is transmission grid-connected energy storage. Previously only pumped hydrostorage units were connected to provide system stability around nuclear plants. Because nuclear plants cannot cycle, pumped storage plants charge and discharge in 12-hour cycles. Now, battery energy storage systems can provide these services. Batteries are even providing grid stability for renewables.

Before an interconnection is approved, steady-state and dynamic system stability conditions need to be satisfied. In dynamic system stability studies, simulations of oscillations occur with a disturbance on the grid due to losing a major power plant. Damping these oscillations is crucial to reduce the chance of broader power

Figure 9.5 Offshore wind turbine Source: [43]/BOEM.

system impacts like brownouts and blackouts. For example, in the MISO system, energy storage resources with advanced inverter capabilities are potential solutions to traditional dynamic stability constraints.

With a 100% renewables study, MISO identifies steady-state and dynamic stability constraints with 10–20% blocks of new renewable interconnections. Energy storage can add value to transmission-constrained locations on the MISO network, such as the Upper Peninsula of Michigan and the southern region of MISO. Baseline reliability projects address NERC reliability criteria. Generator interconnection projects address reliability issues with new or existing generator interconnection requests.

In summary, a battery energy storage system (BESS) could provide dynamic stability, improve the reliability of the power systems, and smoothen the impacts of renewables. Hence, their modeling is crucial for transmission planning.

Exercise 9.3 Jay wants to impress a colleague who works at Consumer Advocates of the PJM States (CAPS) with his energy markets knowledge. This CAPS person is only interested in PJM energy markets. Can you help Jay with this latest assignment?

The CAPS is a nonprofit organization whose members represent over 65 million consumers in the RTO region. CAPS's main role is keeping consumers' electricity costs lower across the PJM region.

CAPS wants to know if they should hire an outside consultant to help them think through the following three topics: (i) review and evaluate the benefits PJM provides to consumers in its role as a regional transmission planner; (ii) evaluate whether PJM's value statement on transmission planning is increasing, decreasing, or staying the same; and (iii) provide recommendations where consumers can see more benefits. What should be Jay's recommendation for CAPS?

9.4 An Introduction to Capacity Markets

Why do we need capacity markets? Proponents of capacity markets argue that energy markets do not provide long-term (one to three years) market signals for resources. Market revenues from energy markets alone are not enough for asset owners to become market participants (MPs) [44]. The capacity needs are so critical that they could even distort the competition during power system operations or planning in some markets.

For example, during power system operations in the Israeli market, the prices are settled based on the units that provide spinning reserve (we refer to this as ancillary service elsewhere in this chapter), not on the marginal units. From a US market perspective, it is a form of market distortion, hence they are working toward changing it. Spinning reserve is a form of capacity. This market also allows the generators to bid their capacities outside the day-ahead and RTMs.

PJM's capacity market is called the reliability pricing model (RPM), which holds an auction for the next three years' energy supply [45]. PJM ensures enough capacity is available for the next three years by paying the generators (they don't have to be generators: the cleared resources could be demand-side resources) cleared in the capacity auction a capacity payment.

Not all capacity markets are created equal, as we shall see Section 9.5. Some ISO/RTOs do not have capacity markets (e.g. ERCOT, California ISO, SPP). MISO has a voluntary capacity auction. At grid operators where there are no capacity markets, the state regulators are responsible for ensuring enough capacity in the long term (Figure 9.6).

Capacity markets serve a valuable reliability function in RTOs who administer capacity auctions, such as the New York ISO [46]. At New England ISO, the forward capacity market (FCM) is held every year for the next three years' capacity needs on the system [47].

As Table 9.3 indicates, billions of dollars are exchanged in these capacity markets. And the capacity market revenue at New England ISO has increased over the years, indicating capacity issues. New England is heavily reliant on natural gas capacity, and whenever natural gas prices are high, there is a direct impact on energy. Hence the higher average of $6.3 billion over the same 12 years compared to the $1.8 billion capacity market average.

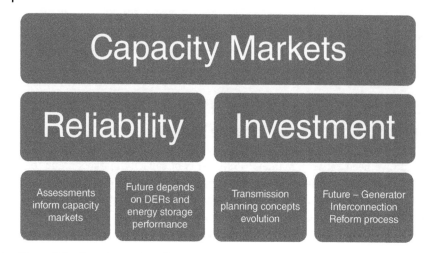

Figure 9.6 Capacity markets discussion roadmap.

9.5 How Do Capacity Markets Ensure Reliability?

Capacity markets ensure reliability by incentivizing both demand side and supply side resources to participate in the market. An example of such an incentive is special case resource (SCR) in the New York ISO installed capacity market (ICAP). SCRs are a type of demand response resources that offer their unforced capacity (UCAP) into the New York ISO capacity market. A resource's UCAP MW amount shows the available capacity after forced outages. "The Unforced Capacity methodology estimates the probability that a Resource will be available to serve Load, taking into account forced outages" [49]. SCRs can be paid as high as $500/MWh for the reliability service they provide in New York ISO's capacity market [50].

Based on how far into the future the capacity market is looking at, reliability assessments are conducted by the grid operators to provide some insight into the market size.

If the function of an energy market is to ensure reliability in real time, i.e. the next five minutes to an hour and day-ahead, a capacity market's function ensures reliability over for the next one- to three-year period. Like an energy market, where the grid operator commits the resources a day in advance and then, economically dispatches them in real-time, in the capacity market, resources are cleared for the year 2 and 3 in the three-year capacity market, but it is the year 1 resources that are "committed" to the market. New England ISO's FCM is run in this manner.

As we learned earlier, New York ISO runs a capacity market called ICAP. However, the difference with New England ISO is that New York ISO's ICAP looks ahead in summer and winter seasons and months ahead, not years [51].

Table 9.3 New England ISO capacity market revenue over a decade (2008–2020).

Year	New England ISO Forward Capacity Market Revenue ($billion) [48]
2008	1.5
2009	1.8
2010	1.6
2011	1.3
2012	1.2
2013	1.0
2014	1.1
2015	1.1
2016	1.2
2017	2.2
2018	3.6
2019	3.4
2020	2.7
Average	1.8

Because of the longer time horizon, the supply and demand side capacities are a major concern in the capacity markets.

What metrics does New York ISO rely upon to understand future available capacity (especially if the future capacity is coming from renewable resources)? There is a need for capacity factor, capacity credit, and effective load carrying capability (ELCC).

The capacity factor is a ratio of the capacity that is available to the maximum nameplate capacity. For example, a wind unit with a nameplate capacity of 100 MW and dispatched at 30 MW, then the wind unit's capacity factor is 30% (30 MW divided by 100 MW).

The capacity credit is the value of how much capacity can be relied upon by the control room operator for a unit during peak demand. For example, if the capacity credit is 15% for the same wind unit, only 15 MW of that 100 MW of nameplate capacity is counted toward capacity in the capacity market.

An ELCC puts a value on the variable resource based on the past performance at the peak time.[7] For example, based on wind performance during the past year's top 8–12 peak demand hours, an ELCC for wind at a regional level could be 15%.

---

7 MISO's ELCC definition, "Effective Load Carrying Capability (ELCC) is defined as the amount of incremental load a resource, such as wind, can dependably and reliably serve, while also considering the probabilistic nature of generation shortfalls and random forced outages as driving factors to load not being served" [52].

Even though the examples cited here are for supply-side resources, the metrics of capacity factor and capacity credit apply to demand-side resources such as demand response (DR). However, there are not a lot of DR that are competing with renewable energy resources. Hence, we focus our examples on renewables, especially wind and solar, for capacity calculations.

Additionally, with increasing renewable penetration, the capacity market's reliability concerns center around having enough "dispatchable" capacity to meet peak demand 10–15 years from 2020. For example, New York ISO is concerned enough to start a "Grid in Transition" process to revamp the capacity market [51]. New England ISO has a similar process underway with two separate tracks, one called "Future Grid Reliability Study" and the other "Pathways to the Future Grid" [53].

Exercise 9.4 Jay wants to contribute to a blog on the ELCC topic. Can you help him write this blog? For an outline, Jay was thinking of starting with current ELCC metrics at US organized markets. He wants to make a case for solar ELCC in Africa and knows some RTOs have wind ELCC, solar ELCC, wind + storage ELCC, solar + storage ELCC, and multiple duration storage ELCC. What other data should he gather to make this blog posting top read?

9.5.1 How Do Reliability Assessments Inform Capacity Markets?

Capacity markets by design procure capacity needs on the system for a year out at the very least. The system's reliability on a day-ahead basis is governed by market rules, such as "must offer." Must offer means the resource must bid its capacity in the DAM.

A reliability assessment is conducted to assess whether there would be enough resources, i.e. capacity in the capacity market in the next season or year, tabulating all the resources available and expected peak demand. This reliability assessment is conducted annually. This assessment shows where there might be a need for capacity in the future. This need for future reliable capacity forms the basis of running capacity markets.

Reliability assessments show upcoming demand and capacity, planning reserve margin, and the risk associated with loss of load under specific scenarios. There are three main types of reliability assessments:

- Summer assessment: The primary objective in a typical summer assessment is to ascertain whether there is enough reserve margin going into the summer. The summer assessment is usually released in June, right before the start of the summer season. NERC, FERC, and all US grid operators release a summer assessment annually. The NERC summer assessment [54] looks solely at each regional entity, whereas the FERC summer assessment [55] includes a market and natural gas discussion, given its role as the federal energy regulator. Each ISO conducts their summer assessment which goes into detail for their region [56].

- Winter assessment: Whether there is enough reserve margin going into the winter months is less of a concern in the winter assessment. The primary concern is how dependent the electric system is on natural gas in regions where heating demand is high and how past events re-occur, whether the region is prepared to handle the contingencies.[8] Like summer assessments, NERC [57], FERC [58], and US grid operators [59] each conduct their winter assessments.
- Long-term reliability assessment (LTRA): Compared to seasonal assessments, an LTRA is released by NERC typically in December [60]. An LTRA looks 10 years into the future and identifies upcoming capacity shortfalls in each NERC region. Since each RTO conducts its transmission planning that looks ahead 10 years, RTOs typically do not release an LTRA.

Examples of reliability issues raised in past reliability assessments are:

- ERCOT raised the possibility of a higher probability of initiating energy emergency alerts given the tight reserve margins and forecasted peak summer demand in 2020 [61].
- "SERC Southeast entities have experienced loop flows from a high regional transfer between MISO North and MISO South. As a result, the impacted utilities along with MISO developed an operating procedure to address potential reliability issues that could result from high MISO regional transfers" [62].
- New England ISO raised the possibility of a loss of generation and firm load-shed if fuel oil supplies were tight in 2020–2021 winter, like New England ISO experienced in the 2017–2018 winter [57, p. 5].

In addition to the summer, winter, and long-term reliability assessment, each ISO is free to conduct its regional surveys. Examples include:

- Organization of MISO States (OMS) MISO survey: Because MISO does not have a true capacity market that looks ahead into capacity needs three years out (like New England ISO, for example), MISO surveys five years' capacity projections with OMS staff. OMS is a body of MISO state regulatory commissioners and their staff. This survey is conducted annually and is usually released in June [63].
- New England ISO PV survey: Like MISO, New England ISO conducts a survey with its state regulatory commission staff on assessing the growth of DG

---

8 With increased electrification of heating, the summer/winter may look increasingly similar. Is there enough capacity to run heat pumps for heating and cooling? Source: Daniel Moller Snenum.

(primarily PV solar) in its region. This New England ISO survey is also conducted annually and released in April [64].

Exercise 9.5 Jay wants to present why summer and winter assessments are important to understand seasonal capacity needs in Uganda. His lecture is part of the Uganda Energy Mix Diversification Strategy for the Uganda Electricity Generation Company, Ltd. (UEGCL).

The United States Energy Association (USEA) implements an activity under the United States Agency for International Development (USAID) Power Africa Initiative. As part of this activity, all the key stakeholders in Uganda are meeting, such as the Ministry of Energy and Mineral Development (MEMD), the Electricity Regulatory Authority (ERA), the Rural Electrification Agency (REA), the Uganda Electricity Transmission Company Ltd. (UETCL), and the Uganda Electricity Distribution Company Ltd. (UEDCL), and Uganda's main electricity distribution company (Umeme). How should Jay convince the Uganda regulators and utility executives about the importance of seasonal assessments? What data should he gather to make this presentation?

9.5.2 The Future Role of Operations in Capacity Markets

As we think about the future with more DERs, it is a well-known fact that generator interconnection queues (GIQs) at most ISOs are congested with many renewable energy project requests. Unless those resources have capacity resource designation, the future capacity needs are not assured. This future capacity is a cause for concern for the regulators and market opportunity for service providers.

Load forecasts are critical in the future capacity markets because of the increasing penetration of DERs and energy storage resources. Grid operators across the US are enhancing their load forecasts to determine the impact of these DERs. This challenge is increasing in complexity because some DERs are not registering with the market operator [65]. After all, it is not economic for them, because grid operators insist on expensive metering and telemetry equipment for transmission grid connections to dispatch each DER.

ERCOT estimates 800 MW of unregistered DERs at the end of third-quarter 2020 [65], 650 MW of those are less than 50 kW solar. This data point illustrates that future reliability assessments and capacity surveys must account for small scale solar; otherwise, the control room operator would not have visibility into DER operations.

Hence, operations' future role in capacity markets depends on how well DERs and other emerging technologies perform in the next 5–10 years. Should the rise in DERs be considered while allocating the capacity needs? The answer depends on regional needs and policies.

9.6 How Do Capacity Markets Facilitate Grid Investments?

As mentioned in Section 9.5, market operators also serve as ISOs/RTOs. They require more assets to carry the planned capacity. More assets mean more grid-related investments. Hence, they promote grid scale investments that are vital for their planning and operations.

Planning the system for a capacity market purpose entails focusing on the transmission system that delivers the load's capacity resources.

9.6.1 Past Transmission Planning Experience for Future Capacity Markets

Before the reader understands how past transmission planning issues have influenced capacity market policy discussions, they need to understand how transmission planning has changed over the years with energy markets. This discussion can be found in Chapter 4.

Most of the planning concepts introduced during the energy markets discussion are still relevant to the capacity market. Some of the complexity, however, has increased. A generator of the future could not only supply active and reactive power but also consume them. This concept of production and consumption is an interesting consideration for capacity market planning, hence some policies are influenced due to such nuances. This knowledge is essential to understand where policy meets the planning needs in capacity markets discussion.

9.6.2 Generator Interconnection Reform for Transmission Planning in Capacity Markets

In some North American energy markets, especially PJM, capacity markets and the role of RTOs in states meeting their carbon-free goals are topics of much serious discussion. With policy incentives and mandates for fossil-fuel-free generation, the US's grid operators must deal with large-volume GIQs.

As of February 2021, PJM alone has 2000+ projects in its queue.

There is no doubt that GIQ reforms are needed to speed up renewable projects' interconnection (because 90% of the projects are renewables) in the organized markets of the US. This queue problem is also a problem in international energy markets. For example, any large generator that wants to connect to an organized market needs to perform a generator performance study (GPS) in Australia. This GPS is a lengthy process that involves sitting in a long queue, conducting the study, and sitting in long negotiations – only then can the construction begin.

Figure 9.7 Generator interconnection queue process flow.

We understand from other chapters that transmission access rights are the reason for any new resource to go through the GIQ. Once a resource obtains transmission access in the market, it becomes a network resource. The resource is now available for the entire network.

The reader should understand by now that wholesale markets influence energy policy, and vice versa. This GIQ reform process is a classic example of how wholesale energy markets have influenced policy because policy initiatives underway speed up the interconnection of renewable projects stuck at most GIQs. And the fact that more than 2000 projects are waiting to be connected at PJM RTO shows that energy policy impacts energy markets.

There are three main topics for the reader to understand the GIQ, as illustrated in Figure 9.7.

9.6.2.1 Multiple Engineering Studies

The first topic, multiple engineering studies, is needed to clear out the projects in GIQ.

When an interconnection customer (IC) initiates the request to connect to the transmission system, it is the grid operator's responsibility as the TP to study that MW injection for any reliability threat to the system. In an ideal scenario, the grid operator's interconnection engineer finds no reliability overload on a T line due to this request and allows the project to construct. However, with so much interest to connect to the T system, there are multiple IC requests, sometimes at the same substation.

Hence, there is a need to study the request. With multiple requests, there is more than one study that is conducted. While the first step in any interconnection study is to determine the feasibility of the project with the feasibility study, it is not until you get to the impact study that the RE developer has a clear picture of what is expected of their project for upgrade costs.

With a feasibility study, an engineer is giving a broad-brush engineering analysis without spending too much time. At this initial stage, the point of interconnection (POI) voltage, the size of the project, the technology, the location, and other engineering details are gathered. It is also important for the RE developer to show that they have "site control," meaning they have access to the RE project site at least for the project's duration.

With the impact study, both the developer and the interconnection engineer are investing more time. And the outcome of the impact study is to understand the impact of this project on other projects in the GIQ and overall electric grid system. These multiple study stages are also essential to provide a "drop-out path" for any RE developer who is not interested in the project (most likely due to the potential upgrade costs).

Finally, in the last step of engineering studies, the facility study is the endpoint. Because the facility study report documents the project and its impacts on the grid plus provides a list of network upgrade costs associated with this request.

This last step is where newer technology options such as dynamic line ratings (DLRs) are coming into the picture because they cost less, take less time to install, and in most cases do not need the interconnecting transmission line to be on planned outage for installation. Please refer to Chapter 7 for additional details on DLRs.

A DLR, which increases the rating of a T line, is an effective mitigation option for network upgrades that cost more than $10 million for generator interconnection projects. If the IC still wants to go through the process, the next step is negotiation.

9.6.2.2 Negotiation
The second topic the reader needs to understand about the GIQ is the back-and-forth negotiation between the IC, transmission owner (TO), and the TP documented in generator interconnection agreements (GIAs). This negotiation takes as long as 150 days at MISO [66]. The negotiation in generator interconnection studies pertains to network upgrade costs, not the reliability, economic, and policy-driven transmission projects discussed in Chapter 4.

After a detailed engineering study process, negotiating an agreement takes time because of the cost implications. Added to the RE developer's cost burden, since engineering studies were run on projects in a group, those studies must be rerun if any of the projects drop out of the group. As a result, negotiating agreements is impacted because there is a new cost estimate.

Projects have the option to drop out at each stage of the engineering study. Thus the list of the interconnecting projects is changing, impacting the cost and time for planned network upgrades.

This agreement takes time because this is a three-party agreement: the TP, the TO, and the IC. Hence all three parties have to agree at the end of the day.

9.6.2.3 Construction
The third and final point is the time it takes to build network upgrades is not trivial. For a project size of less than 20 MW, lesser upgrades are expected. But for larger projects such as 100–500 MW utility-scale renewable projects, one can expect higher transmission project costs and longer construction times because of

the need to upgrade the surrounding electric system to accept that 100 MW injection without causing any reliability issues.

Other factors are engineering staff availability, contractor availability, renewable developers submitting multiple requests at the same POI, model building delays, and interconnection engineers at the grid operators must rerun the studies. But the ones mentioned above – multiple engineering studies, GIA negotiations, and network upgrade costs – are the major driving factors for GIQ delays.

This process is not rigid, and the IC has the flexibility should circumstances change. For example:

- Project size can be downsized (e.g. a 100 MW can change to a 50 MW request) but not upsized.
- Large deposit amounts are paid upfront for the project to be studied (e.g. $5000 per MW).
- Generator technology can be modified during the GIQ stages (e.g. turbine replacement from Vestas 2.0 MW wind turbines to GE 2.82 MW).
- Generator fuel type can be replaced (e.g. replace a coal unit with solar at the same location).

A newer variation of the GIQ reform process includes giving ICs the option to build network upgrades if TOs agree, and ICs can apply for transmission service for any surplus interconnection service (if a project applied for 100 MW interconnection but is now using only 80 MW, the remaining 20 MW is surplus and someone else can apply to use that) (Figure 9.8).

Moving onto international markets, Eastern Africa Power Pool (EAPP) is responsible for generator interconnections in the Common Market for Eastern and Southern Africa (COMESA) region. EAPP, with Power Africa support, has developed an interconnection code compliance program, which is implemented in stages [67].

EAPP plans to integrate 50 000 MW of renewable projects in the region with high-voltage transmission projects [68]. Like PJM, EAPP needs an interconnection code that efficiently interconnects 50 000 MW of renewables (Figure 9.9).

Moving from Africa to Europe, ENTSO-E, the European Union mandate-driven organization, also has the network code for generator interconnections called "requirements for generators" RfG [69]. The European Network Code Requirements for Generators (RfG NC) covers interconnection requirements for the entire EU. The individual countries, i.e. Member States (MS), had to retire their country-specific interconnection requirements. This RfG NC is different in the US because individual states have distribution system interconnection requirements, but each TO must gain acceptance from the federal agency FERC when it comes to transmission interconnections.

Figure 9.8 Eastern Africa Power Pool (EAPP) logo. Source: EAPP.

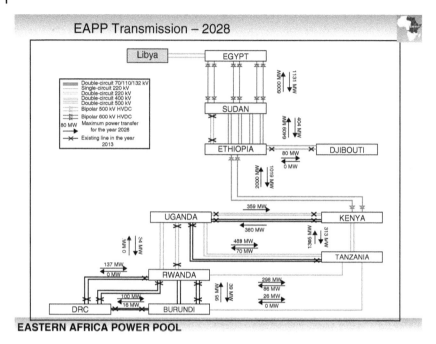

Figure 9.9 Eastern Africa Power Pool (EAPP) transmission plan for 2028. [68]/IRENA – International Renewable Energy Agency.

Exercise 9.6 Jay's work with the Bill & Melinda Gates Foundation on energy markets caught a prominent Texas investor's eye. This new client wants Jay to dive into capacity markets. Can you help Jay with this latest assignment?

His client wants to invest in assets that would give a high rate of return when Texas moves to a capacity market. Until now, Texas has had an energy-only market. Hence, can you help Jay think through other capacity markets from an electricity grid perspective and layout a roadmap for prioritized investments in Texas for his client? Prepare 2–3 pages of research brief outlining your findings for Jay.

9.7 An Introduction to Ancillary Services Markets (ASMs)

A good amount of market revenue could also come from ASMs. More ancillary services products are needed with higher percentages of renewables on the system (Figure 9.10). For example, Australia's NEM has frequency controlled ancillary service (FCAS) as ancillary services further divided into regulation and contingency services.

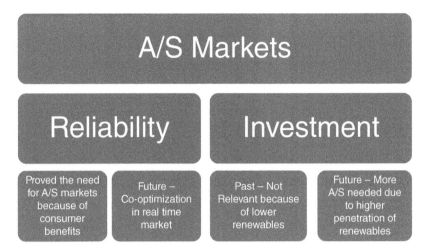

Figure 9.10 Ancillary services markets discussion roadmap.

ASMs are a complex topic because what constitutes ancillary services varies according to countries and regions, and as a result some services are compensated in the market compared to others paid by a contractual arrangement. And yet, there are services like inertia that are not compensated widely.

The definitions of what constitutes ancillary services can vary from country to country. A couple of observations from Table 9.4:

- Both Vietnam and Independent Electricity System Operator (IESO) Canada have black start identified as the ancillary services, but the black start service is not part of the ASM at most ISOs in the US.
- Both Vietnam and MISO have spinning reserves identified as ancillary services, but IESO Canada has spinning reserves in the "operating reserves" market. As we will see later, spinning reserves in one market are called synchronized reserves in another market.
- Since must-run generation is a challenge for Vietnam, they have identified must-run generation services as part of the ancillary services, but neither IESO Canada nor MISO do this. Must run is not an ancillary services called out specifically at most North American markets. Each generating unit is expected to reflect its must-run capacity segment in its day-ahead and real-time offers. Must-run is also called self-commit or self-schedule in some markets.
- All three (Vietnam, IESO Canada, and MISO) have regulation services in the ancillary services category, but with variations. Vietnam has a voltage regulation service, IESO Canada has a regulation service separate from "reactive support and voltage control service," and MISO has regulation-up and regulation-down services (Table 9.4).

Table 9.4 What is defined as ancillary services varies by country and market.

Vietnam	IESO Canada	MISO
Fast start-up service	Certified black start facilities	Regulation Up
Must-run generation services	Regulation service	Regulation Down
Spinning reserve service	Reactive support and voltage control service	Ramp Up
Frequency control service	Reliability must-run	Ramp Down
Voltage regulation service		Spinning Reserves
Black start service		Online Supplemental Reserves
		Offline Supplemental Reserves

Source: Based on [70].

Also, IESO Canada has an operative reserves markets separately defined from ASMs [71].

Another complexity is from different definitions for the same service. Take the example of the "spinning reserves" definition from different countries.

In India, spinning reserves is defined as "the capacity which can be activated on decision of the system operator and which is provided by devices which are synchronized to the network and able to effect the change in active power" [72].

At PJM, there is no spinning reserve definition in the ASM construct. Rather it is called synchronized reserves. Synchronized reserves are defined as "the reserve capability of generation resources that can be converted fully into energy or Demand Resources whose demand can be reduced within 10 minutes from the request of the Office of the Interconnection dispatcher and is provided by equipment that is electrically synchronized to the Transmission System. Synchronized Reserves are supplied from 10-minute synchronized generating resources (i.e. Synchronous Reserves) and 10-minute demand-side response resources" [73].

An added complexity arises if a service is compensated via market price or a contractual arrangement. For example, regulation is a service that is procured via a contract at IESO Canada. In contrast, regulation is part of the market-based ASM at MISO and most ISOs. In contrast, active power/energy is cleared based on who can provide Israel's regulation (spinning reserve). Similarly, voltage control and reactive support is a service procured via transmission settlements at MISO. But at IESO Canada, this voltage control is part of the ASM (Table 9.5).

Another example to illustrate this additional complexity in ancillary services is primary frequency response (PFR) compensation. Frequency response is compensated as a service at ERCOT, Vietnam, Australia, and German markets but not even contractually at other regions. The reason for variation in frequency response compensation is the lack of a current need in some regions.

Table 9.5 Variation exists in ancillary services procured via contract and market.

Services	Procured via contract	Procured via market
Regulation	IESO Canada	MISO ASM
Voltage control and reactive support	MISO schedule	IESO Canada
Frequency response		ERCOT, Vietnam, Germany, Australia

On the other hand, Fingrid, a Finnish grid operator, has multiple frequency response services it procures from the market based on how much time is needed to activate a reserve.[9] An automatic frequency restoration reserve(aFRR) is a centralized automatic frequency activation service with full activation of five minutes [74]. To qualify for fast frequency reserve (FFR), the minimum activation time for a unit is five seconds [75]. Activation time is also called response time.

And finally, Fingrid also has a frequency containment reserve (FCR) both for normal operations, called FCR-N, and disturbance conditions, called FCR-D [76]. The activation time for FCR-N is three minutes and 30 seconds for FCR-D. In addition to the multiple frequency response services, Fingrid posts units designated for providing reserves to the market, as seen in Figure 9.11. These units are not used for commercial electricity production, only for providing reserves [77].

Regarding the importance of ancillary services in the US, there is plenty of research, including highlighting the role of demand response in providing ancillary services [78], how high wind power drives the need for a tighter ASM design [79], and a survey of ASMs in the US [80].

In international settings, regions with high wind penetration first initiated ASM design concerns, such as Spanish and Indian electric market design frameworks [81].

As Figure 9.12 indicates, the need for ancillary services increases under both situations when wind forecasts don't track wind production data. The situation on the left-hand side happened on a Monday, 11 February 2008, when wind production was 3200 MW higher than the forecast during a period of low demand during the early-morning ramp hours on the Spanish grid. The situation happened too fast in the two-hour duration (5:00 to 7:00 hours) for the Spanish operator for reserves to be reduced and thermal plants to shut down, hence wind energy had to be curtailed (from 7:22 to 9:30 hours).

Conversely, on the right-hand side chart, wind production data was 6000 MW lower than the forecast data on Friday, 23 January 2009, when Storm Klaus hit

---

9 Common across Scandinavian countries, see Danish example here: https://en.energinet.dk/ Electricity/Ancillary-Services (accessed 11 December 2021). Source: Daniel Moller Snenum.

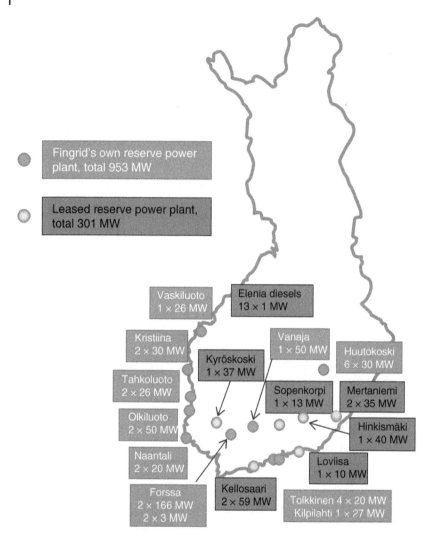

Figure 9.11 Fingrid's map of reserve plants (updated 3 December 2018). Source: [77]/Fingrid Oyj.

the Iberian Peninsula. Even though most wind turbines in Spain shut down due to overspeed production (wind speed at 220 km/h), the Spanish grid could handle the demand on the Friday evening ramp hours [82].

ASMs get complicated quickly with concepts that cross over operations and markets, such as the operating reserve demand curve (ORDC). This ORDC is discussed in Section 9.7.1.

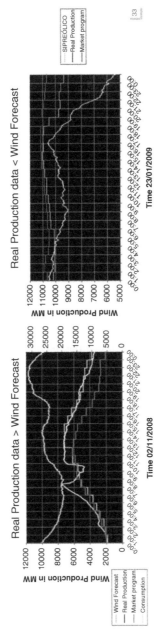

Figure 9.12 If wind forecasts do not track with production, ancillary services cost increases. Source: [82]/Grupo Red Eléctrica.

9.7.1 Operating Reserve Demand Curve (ORDC)

The energy market's fundamental principle is balancing supply and demand. As explained in Chapter 1, balancing electricity supply and demand is a delicate act based on economic principles. The demand curve that meets the supply curve is the market clearing price, also sometimes known as the spot price.

So, the question now becomes: what is the price the market is willing to pay for operating reserves? We also learned in Chapter 4 that operating reserves are for a shorter duration (less than two weeks) than planning reserves.

Different customers have different tolerances for what prices they are willing to pay for backup power and running during an outage. An industrial customer in the middle of an aluminum smelter process is not willing to take any power interruptions during the process but is willing to schedule their next batch process based on the wholesale price of electricity, especially if they can participate in a demand response product.

This customer preference is where wholesale and retail prices meet because the customer decides the interruption costs if the information is provided. Compared to an industrial customer, a residential customer is most likely willing to put up with a service interruption, and hence the cost would be lower. However, a residential customer might complain based on the outage duration – days instead of hours.

The question for the market operator, i.e. the grid operator, is: how much to pay for MPs willing to provide this service? A not-for-profit market operator must recoup this money from the market to pay the ancillary service, i.e. the reserve service provider.

For example, the ERCOT market has value of lost load (VOLL) set at $9000 per MWh, which is $9 per kWh. At MISO, this VOLL is set at $3500 per MWh [83]. The reader should compare the ERCOT $9 VOLL price with $0.12 per kWh on a normal day because the average residential retail price of electricity in Texas was 12 cents per kWh in August 2021 [84]. This $9 per kWh price means that the reserves must be procured up to a "price cap." Above that price cap, the customer is not willing to pay, and the market does not buy the reserve product. And the reserve supply and price curve that shows this price is known as the ORDC (Figure 9.13).

As the ORDC shows, when there is plenty of available reserves MWs in the market, the price of reserves is almost zero or close to zero. But, as reserves become scarce, the price for them starts to climb as we move from higher to lower reserves. And it climbs up to that price cap set at VOLL. Each market might have to decide what that price cap is for them. VOLL is one factor that determines this ORDC. Another factor is the loss of load probability (LOLP).

As we learned in Section 9.5.1, a reliability assessment includes an LOLP analysis. This LOLP analysis determines the current reliability of the system according to the standard of 1 day in 10 and the future reliability of the system under multiple

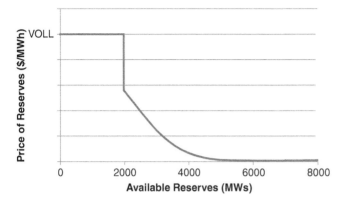

Figure 9.13 The operating reserve demand curve (ORDC) from ERCOT training deck Source: [85]/Electric Reliability Council of Texas.

scenarios of load growth, transmission support, or other sensitivity cases such as high solar penetration.

Between the market-based VOLL and reliability-based LOLP, ASMs meet policy, because an energy policy can set an administrative price based on sound reliability and market analysis like LOLP and VOLL, or an energy policy can be set by fiat.

Exercise 9.7 Jay ran into a colleague in the industry who works at PJM Interconnection. As you know, PJM is one of the RTOs in the US and hence has a lot of credibility with federal regulators. This PJM person talked about changing their ORDC and applying that for both DAMs and RTMs. Jay is intrigued. Can you help Jay with this latest assignment?

Jay wants to research how other US-based ISO/RTOs are approaching ORDC with the amount of knowledge gained in this chapter. If ERCOT has only one ORDC, and PJM moves to two ORDCs – one for a day-ahead and another for an RTM – what are the others doing in this grid operator space? How different are the VOLLs in each of these markets? Like ERCOT, do other markets set a VOLL for the entire market or have individual zones? Prepare 2–3 pages of research brief, outlining your findings, for Jay.

9.8 How Can ASMs Ensure Reliability?

There is a greater reliability need as we move from an inertial system (e.g. large rotating generators like coal) to an inverter/converter-based power system (e.g. renewables). This reliability can be ensured by procuring ancillary services whereby MPs are paid to act if there is a reliability issue.

Table 9.6 . ERCOT ASM average annual day-ahead prices (five-year history).

Ancillary service\year	2016 ($/MWh)	2017 ($/MWh)	2018 ($/MWh)	2019 ($/MWh)	2020 ($/MWh)
Responsive reserve	11.10	9.77	17.64	26.61	11.40
Nonspin reserve	3.91	3.18	9.2	13.44	4.45
Regulation up	8.2	8.76	14.03	23.14	11.32
Regulation down	6.47	7.48	5.19	9.06	8.45

Source: Based on [86].

Let us take an example of a frequency event near a solar farm (e.g. a large industrial load tripped) that is providing ancillary services. If so, the inverter of the solar should quickly detect, react, and rectify the issue. This issue is an example of an ancillary service. Another example is if this solar farm created a system frequency issue, the market operator would take remedial actions and most likely penalize the solar farm. In general, the more the inverter/converter-based systems are introduced to the power system and the traditional generators are made obsolete, the more the need for ancillary services. In ASMs, this directly translates to an increase in the market price for ancillary services.

Take ERCOT as an example, which has a lot of wind and solar in the market. Look at Table 9.6 for ASM prices derived from ERCOT's independent market monitor Potomac Economics "state of the market" reports in 2018, 2019, and 2020 [87]. The average annual price for all four ERCOT ASM products have increased over the past four years. The regulation-up price increased by three times in the past four years.

There is no RTM yet for ancillary services at ERCOT. Hence Table 9.6 indicates day-ahead ASM prices.

Another example of a grid operator with higher penetration of renewables is California ISO. California ISO and ERCOT are different. The former has more distributed-scale solar, and the latter has more utility-scale wind. We expect to see higher annual day-ahead prices on average, but the 2019 data shows California ISO ancillary services prices reduced compared to past years [88]. According to the California ISO market monitor, the decrease in 2019 prices and the increase in 2020 prices is due to operating reserve requirements [88, 89]. To keep the apples-to-apples comparison, the reader must compare the DAM prices of ERCOT in the previous table with the California ISO day-ahead prices in Table 9.7.

Both California ISO and ERCOT are single-state grid operators, and both have ancillary services co-optimized. Hence it makes for a straightforward comparison

Table 9.7 California ISO ASM average annual day-ahead prices (five-year history).

Ancillary Service\Year	2016 ($/MWh)	2017 ($/MWh)	2018 ($/MWh)	2019 ($/MWh)	2020 ($/MWh)
Spin	5.65	10.13	9.16	7.39	9.50
Nonspin reserve	0.39	3.09	3.05	0.75	3.61
Regulation up	10.84	12.13	12.79	13.27	13.10
Regulation down	8.34	7.69	12.05	11.74	10.97

Source: Based on [90].

of their average annual ancillary services prices. ERCOT is moving to an RTM in 2024.

9.8.1 A Single-entity Administrating Ancillary Services Provides Benefits to Consumers

In Chapter 4, when discussing how joining an energy market is not a decision to be taken lightly, PJM's Value Proposition Study was provided as an example to illustrate PJM member benefits. PJM does not call out two specific ASM components in its value proposition (not yet) study, but MISO does.

MISO's 2020 Value Proposition Study includes a range of $128–$142 million, and $60–$67 million for regulation and spinning reserve member benefits [91]. The dollar values are not that important compared to the concept. In addition to being an ISO, MISO is also a NERC BA for the entire region.

A BA is "the responsible entity that integrates resource plans ahead of time, maintains load-interchange-generation balance within a Balancing Authority Area, and supports Interconnection frequency in real time" [92]. The current BA areas in North America are shown in Figure 9.14.

As a result, MISO members who were BAs before no longer must procure and maintain spinning reserves and regulation ancillary services on their own. MISO does that. Hence there is an efficiency from MISO running an ASM.

The MWs are significant in the case of MISO. Before MISO became a BA and ran an ASM, the MISO members who were BAs had a 1559 MW of regulation requirement. Similarly, the average spinning reserve requirement at MISO before an ASM was 1482 MW. But, with ASM, the entire MISO BA has 396 MW and 934 MW for regulation and spinning reserve requirement, respectively. This reduction in ancillary services requirements is a huge benefit for MISO members.

Exercise 9.8 As you know, Jay is an international energy consultant. He volunteered to serve as an expert on a trip to Rwanda, Africa. This USAID-, USEA-, and

Figure 9.14 Balancing authority areas in North America. Source: NERC.

Power Africa-led workshop has representatives from Kenya, Uganda, Tanzania, Rwanda, DR Congo, and Ethiopia in the EAPP. Power Africa is a US government partnership coordinated by the USAID.

Prepare a 15-slide presentation with appendix if needed but not exceeding 10 slides: (i) identifying key issues in ancillary services trading in East African countries, (ii) identifying which of these key issues should be handled within these countries' jurisdictions and which issues should be addressed at the regional level, and (iii) considering whether current policy frameworks in these countries provide enabling environments for trading in ancillary services and deploying energy storage solutions. First two to three slides of your presentation should be for the executives of USEA/USAID/EAPP. Can you do this for Jay?

9.8.2 Real-time Co-optimization Is the Future for Operations in ASMs

What is real-time co-optimization (RTC)? In the RTM, energy offers from a MP reflect their view of what price they are willing to sell for the next MW of energy. And if the load is willing to buy, then the market clears at that price. However, what if the MP wants to bid into the ASM for providing reserves at the same time?

As we discuss in Section 9.7, different kinds of reserve products make up the ASM. So, how does the MP decide which market to bid on: energy or ancillary services? Or both? What if they can make a bid that maximizes their profit? While bidding, they investigate the market prices for energy and ancillary services, history of clearance, market conditions, time of day, holiday events, stock exchange performance, any events, or holidays and create a bid for energy and ancillary service. For example, a power plant can operate normally at 30 MW. Its maximum capacity is 35 MW, and its minimum is 20 MW. The market allows an energy and regulation-up and regulation-down bid to be put. In such a case, it could:

- Put the energy bid for supplying 30 MW and nothing for regulation, if it has a history of not getting cleared for ancillary service.
- Put an energy bid for supplying 30 MW and regulation up for 5 MW and regulation down 10 MW.
- Bid for 33 MW and 2 MW up and 3 MW down, assuming it hopes to make more money from energy bids.
- Curb its output and run at 25 MW, put an energy bid to supply 25 MW with a regulation up bid for 10 MW. Assuming it sees an opportunity to make more money from a regulation -p bid and nothing from a regulation-down bid.

This example was a very simple case. It could quickly be complicated as the number of market products (energy and ancillary service) increases. While the MP makes every effort to maximize its gains, in some cases the market operator also tries to ensure that the MPs don't lose revenue in one market when they bid in another. This process is RTC.

With RTC, MPs don't lose the opportunity to earn revenue in one market when they bid in another. California ISO's and MISO's ASMs are co-optimized. New England ISO's is not. ERCOT is moving to RTC because the entire market benefits when the cost of procuring ancillary services is reduced. Without an RTC, ERCOT must run an energy-only market and a separate market for procuring ancillary services. In this scenario, ERCOT might consider issuing an RFP to procure ancillary services. As the RFP news example from Kenya Electricity Generating Company PLC (KenGen) suggests, the generating company KenGen is providing ancillary services for free to the distribution company Kenya Power ancillary services [93]. But with the increase in RE, this is no longer viable. Hence someone must pay for those ancillary services in the future. But what is the amount of the ancillary services MWs to be procured? An ancillary services requirements study is needed. This study is normal at most grid operators.

Exercise 9.9 Jay ran into a colleague in the industry who used to work for the market monitor for ERCOT. As you know, ERCOT went through a winter storm in February of 2021 that caught the world's attention. This ERCOT market monitor talked about how they suggested the RTC of ancillary services in the ERCOT market in June 2018. Jay is curious. Can you help him with this latest assignment?

Jay wants to research how ERCOT is going to see a reduction in ancillary services costs (estimated $155 million), reduced transmission congestion costs (estimated $257 million), and a reduction in energy costs (estimated $1.6 billion), thereby improving overall ancillary services management. Like ERCOT, do other markets co-optimize ancillary services? Other international market operators are looking into ancillary services co-optimization in addition to MISO (which does) and New England ISO (which doesn't). Prepare 20–30 pages of research brief, outlining your findings, for Jay.

9.9 How Do ASMs Facilitate Grid Investments?

For the smooth operation of these markets, the market operator requires more transmission: lines and newer products that can offer ancillary services, e.g. synchronous condensers, batteries, and wholesale demand response. Transmission allows the market operator to transport the electrical energy reliably. Hence, they promote grid-scale investments that are vital for their value proposition.

One academic paper finds that HVDC connection can help reduce the cost of ancillary services by 70% [94]. This paper is an example of a study that shows HVDC projects can provide ancillary services.

However, a FERC staff report to the US Congress notes the challenges and the opportunities with high voltage transmission [95]. The report notes that the benefit of high voltage transmission "improve frequency response and ancillary services throughout the existing system." Simultaneously, the challenges are, "future transmission development in existing transportation corridors may be restricted by routing limitations, including state and local prohibitions and restrictions, and safety and technical considerations."

The reader's key takeaway is that transmission projects, especially high-voltage transmission, provide better, faster, and cheaper access to ancillary services. But it is not easy to connect countries and regions within a country with high-voltage transmission, because it is tough to convince all the state and federal regulators, including stakeholders in those countries and regions, of the merits of doing so.

This need for ancillary services is also discussed under the operational "flexibility" name in most markets. In the US, FERC staff have issued a paper starting the dialogue on future operational flexibility needs [96].

9.9.1 Past Transmission Planning Experience for Future ASMs

Transmission planning has been carried out for quite a while, even before energy markets were introduced. As we learned in Chapter 4 (Section 4.8.5), some of the T planning terms (like capacity benefit margin and available flowgate capacity) continue to be relevant for the market to nonmarket regional purposes. And new T planning terms were added, such as the Energy Resource Interconnection Service and the Network Resource Interconnection Service. Markets for ancillary services evolved after the energy market was implemented at most US RTOs. Hence it is tough to find T planning instances from ASMs that have influenced energy policy because ancillary services that like supplying inertia are relatively new and the markets are still evolving.

9.9.2 More ASM Products Will Be Needed in the Future

Most transmission grid systems in the US are running with an increasing percentage of renewables compared to two decades ago.

The US federal regulator FERC issued an order 888 in 1996 that opened energy markets, including ancillary services [97]. Regarding ancillary services, FERC differentiated two kinds of ancillary services based on what the TP can provide. The first kind of service that a TP can provide to all their transmission customers was scheduling, system control and dispatch service and reactive supply and voltage control from generation sources service.

The second kind of service that a TP could offer but is not required to do, and FERC anticipated third-party service providers or transmission customers themselves could provide these services include, regulation and frequency response service, energy imbalance service, operating reserve-spinning service, and operating reserve-supplemental service.

Later, in 2007, in Order 890, FERC added another ancillary service called a generator imbalance service. And FERC issued Order 819 for third-party provision of PFR service in 2015 [98].

Hence in US, market compensation for ancillary services and the categories for ancillary services has evolved over the past 20 years. And with more RE on the system in the next 20 years, following the same trajectory, it is reasonable to expect that additional ancillary services and an expansion of current ASM products would be needed.

An example to illustrate this change is ERCOT. Once wind penetration increased in ERCOT market, the grid operator increased the granularity of their reserve services.

Before 2018, ERCOT had 157 to 687 MW of regulation, 2300 to 3200 MW of responsive reserve service (RRS) and 967 to 2361 MW of non-spin reserves defined as ancillary services. Hence the overall ASM was 3807 to 5958 MW.

With the ancillary services changes, ERCOT did not change the regulation MW amount. But divided the RRS into three categories: fast frequency response (FFR), load resources on underfrequency relay (UFR), and PFR. And the total amount of RRS did not change: it remained 2300 to 3200 MW.

The FFR would be triggered at 59.85 Hz and full response is needed in 15 cycles. The reader should note that 60 Hz is 60 cycles per second. So, the FFR would be triggered in one-quarter of a second. Whereas the load resources on UFR would be triggered at 59.7 Hz and full response is needed in 30 cycles, i.e. half a second.

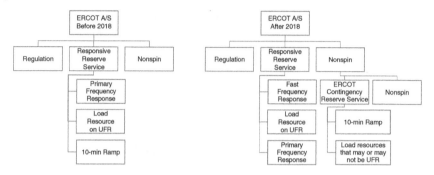

Figure 9.15 ERCOT ASM changes. Source: Modified from [99].

An additional change that ERCOT needed was dividing the nonspin further into the ERCOT contingency reserve service (ECRS) at 508–1644 MW and the nonspin requirement range fell to 0–1180 MW. The ECRS is defined to include the 10-minute ramp service plus any load resources that may or may not be on the UFR (Figure 9.15).

All these changes at ERCOT show what to expect at other market operators running a low inertia grid.

Exercise 9.10 Jay's work with the Texas investor on capacity markets caught the attention of a German company specializing in providing renewable forecasting services. This new client wants Jay to take a deep dive into ASM. Can you help Jay with this latest assignment?

His client wants to understand the market potential for ancillary services in India. Until now, the Indian grid operator was using only homegrown production forecasting tools. Owing to recent US and Europe events, Indian grid operators and renewable developers are convinced of the need for sophisticated renewable forecasting services. Jay's German client wants to know which ancillary service would be the primary driver for increasing renewables forecasting in India. Prepare 2–3 pages of research brief, outlining your findings, for Jay.

9.10 Visioning

It is tough to realize a market that can satisfy all the MPs. But we know that renewables command a significant share of the power grid in the future. Some regions are leading this change; others are following them.

Grid operators should seek lessons from one another while refining their existing rules. For example, grid operators in the US are seeking lessons learned from European grid operators. Similarly, the whole world is learning from the Australian grid operator.

The grid operators who are creating new markets, we believe, can leapfrog. For example, the African market could skip a few steps and directly go to a system with energy and ASM instead of starting an energy market first and adding ancillary markets later.

In both cases, we realize that populist sentiments from citizens and citizen groups influence international markets. After reading this chapter, you should be aware of how wholesale markets influence policy. Therefore, you can now contribute to the transformations in the wholesale market in your region or globally.

Case 9.1 *Market Benefits of an Anchor Tenant Model of Independent Transmission*   Your client is SOO Green, an independent transmission developer. SOO Green is developing an HVDC underground transmission line that takes renewable energy from the Midwest and dumps it into PJM markets. But, your client has run into objections from PJM and PJM's market monitor.

Your task is to find out what those objections are, how you can help your client lay the groundwork for bringing along PJM and PJM's market monitor to your side, and since FERC's approval is necessary for this line, gaining FERC's approval.

Analysis

Some lines of research could include:

- How old is the Anchor Tenant model, and what are the other models of independent transmission development?
- Understand the history of the SOO Green project. How did it start?
- What are the options for SOO Green if FERC rejects this line?
- What are the options for the subscribers to this line's capacity?
- Who else is in SOO Green's boat?

Possible Solutions

Your research may find that the initial FERC application did not address the topics PJM and PJM's market monitor raised. Your analysis may find that PJM has no intention of accepting renewable energy from the Midwest. Ask yourself why. What's in it for PJM and PJM TOs?

Case 9.2 *Your Market Region's Operating Reserve Demand Curve (ORDC)*

Situation

Following to the winter storm in Texas in February 2021, the entire Texas market is being redesigned from scratch, with a new set of Public Utility Commission of

Texas commissioners. Among the market design elements under consideration is ERCOT's ORDC. What should be the value for ORDC in your market region?

Analysis

Some lines of research could include:

- ORDC in other markets with and without capacity markets because Texas does not have a capacity market.
- How would the need for operating reserves in the future impact the ORDC?
- Does what qualifies as operating reserves deserve more attention compared to nonoperating reserves?
- At what price signal would fast-acting demand flexible resources show up versus supply-side resources? Would this lead to fewer or more emissions?

Possible Solutions

Your research may lead you to conclude that the Public Utility Commission of Texas (PUCT) proposal for a lower ORDC is a political decision. A broad discussion of ancillary services in the organized markets is warranted with more renewables on the transmission system.

Case 9.3 *Cluster Study Process for Interconnecting Distributed Energy Resources*

Situation

FERC Order 2222 mandates RTOs accept DERs in their market participation models. But DERs should not enter the FERC GIQ process because they are distributed and connected at the distribution system, hence FERC has no jurisdiction over distribution utilities, but states do. Design a process for DER interconnections at a multistate RTO/TSO.

Analysis

Some lines of research could include

- The current cluster study process or group study at several RTOs for generator interconnection.
- How does the cluster study work for network upgrade cost allocations?
- What changes would the state PUCs have to experience at their end?
- What should be the cost of these interconnection studies?
- How would the study assess the impact of neighboring RTOs' DER interconnection requests?

Possible Solutions

Your research may find that an interconnection portal that takes in the requests and processes of DERs is needed at the state level. This interconnection portal would talk to RTOs and the distribution utility portal.

Case 9.4 *South African Power Pool Capacity Market Construct Proposal*

Situation

Looking at the new renewable projects planned for countries within the SAPP and EAPP, the SAPP wants to put together a capacity market proposal. This proposal needs the approval of the governing members of the SAPP board. What are the capacity market design elements that are "must have" in this proposal, and which design elements would fall under the category of "nice to have"?

Analysis

A detailed research summary of capacity market constructs from Europe and North America is a good starting point. Which design elements from the US capacity markets, such as PJM, would work for SAPP? Does Vietnam's direct power purchase agreement (DPPA) offer SAPP lessons?

Research penalties for nonperformance in the capacity market. Additionally, capacity credit calculations using the ELCC method must be documented for renewable resources planned for the SAPP market.

Possible Solutions

A transition plan that shows how to transition from the current situation to a fully functioning capacity market for the SAPP is a possible solution. This capacity market proposal's benefit-to-cost study would help engage stakeholders and funders alike.

References

1 PJM (2021). The value of markets. https://www.pjm.com/-/media/about-pjm/newsroom/fact-sheets/the-value-of-pjm-markets.ashx?la=en (accessed 11 December 2021).
2 Midcontinent Independent System Operator (2021). Real-time displays. https://www.misoenergy.org/markets-and-operations/real-time--market-data/real-time-displays (accessed 11 December 2021).

3 PJM (2021). Electric storage participation: FERC Order 841. https:// www.pjm.com/committees-and-groups/issue-tracking/issue-tracking-details.aspx?Issue=%7B736CAC88-9404-4421-B178-BD392366098F%7D (accessed 11 December 2021). For SPP, visit: https://spp.org/spp-documents-filings/?id=175274. For California ISO, http://www.caiso.com/Documents/ BusinessRequirementsSpecification-FERC841-RequestedAdjustments.pdf. For MISO, https://www.misoenergy.org/stakeholder-engagement/issue-tracking/ storage-participation--ferc-order-841-compliance. For New York ISO, https:// www.nyiso.com/view-blog/-/asset_publisher/5397qT1ac7HE/content/energy-storage-frequently-asked-questions. (All accessed 11 December 2021.)

4 PJM Interconnection (2021). Who's who in PJM control room. https://www .youtube.com/watch?v=MF0a8JKPA6A&feature=youtu.be (accessed 11 December 2021).

5 PJM Interconnection (2021). A day in the life of dispatch. https://www.youtube .com/watch?v=hDNouhlJWa4 (accessed 11 December 2021).

6 North American Electric Reliability Corporation (2014). Hurricane Sandy event analysis report: benefits of on-site meteorologists section. https://www.nerc .com/pa/rrm/ea/Oct2012HurricanSandyEvntAnlyssRprtDL/Hurricane_Sandy_ EAR_20140312_Final.pdf (accessed 11 December 2021).

7 New York ISO (2021). Press release. https://www.nyiso.com/-/grid-operators-isolated-but-unified-in-keeping-new-york-lights-on (accessed 11 December 2021).

8 Federal Energy Regulatory Commission and North American Electric Reliability Corporation (2011). Outages and curtailments during the southwest cold weather event of February 11, 2011. https://www.ferc.gov/sites/default/files/ 2020-05/ReportontheSouthwestColdWeatherEventfromFebruary2011Report.pdf (accessed 11 December 2021).

9 Federal Energy Regulatory Commission (2021). February 2021 cold weather event in Texas. https://www.ferc.gov/news-events/news/ferc-nerc-staff-review-2021-winter-freeze-recommend-standards-improvements (accessed 11 December 2021).

10 Australian Energy Market Operator (2017). Black system South Australia 28 September 2016. www.aemo.com.au/-/media/Files/Electricity/NEM/Market_ Notices_and_Events/Power_System_Incident_Reports/2017/Integrated-Final-Report-SA-Black-System-28-September-2016.pdf (accessed 11 December 2021).

11 California Independent System Operator (2020). Preliminary root cause analysis. http://www.caiso.com/Documents/Preliminary-Root-Cause-Analysis-Rotating-Outages-August-2020.pdf (accessed 11 December 2021).

12 California Independent System Operator (2019). Initiative: FERC Order 831: import bidding and market parameters. https://stakeholdercenter.caiso.com/

StakeholderInitiatives/FERC-Order-831-Import-bidding-and-market-parameters (accessed 11 December 2021).

13 New York Independent System Operator (2021). FERC order no. 2222: New York ISO responses to FERC data request. https://www.nyiso.com/documents/20142/25957407/FERC%20Order%202222%20-%20NYISO%20Response%20to%20FERC%20Data%20Request.pdf (accessed 11 December 2021).

14 DNV (2022). The German 50.2 Hz problem. https://www.dnv.com/cases/the-german-50-2-hz-problem-80862 (accessed 22 February 2022).

15 New England States Committee on Electricity (2021). New England governor's statement on electricity system reform 2020. http://nescoe.com/resource-center/govstmt-reforms-oct2020 (accessed 11 December 2021).

16 Office of the Governor (2021). Governor Walz appointed Chair of Midwestern Governors Association. https://mn.gov/governor/news/?id=1055-450179 (accessed 11 December 2021).

17 PJM Inside Lines (2020). PJM shares perspectives on offshore wind development. https://insidelines.pjm.com/pjm-shares-perspectives-on-offshore-wind-development (accessed 11 December 2021).

18 North American Electric Reliability Corporation (2021). Major events analysis reports. https://www.nerc.com/pa/rrm/ea/Pages/Major-Event-Reports.aspx (accessed 11 December 2021).

19 North American Electric Reliability Corporation (2021). Examples of major power outages. https://www.nerc.com/pa/rrm/ea/August%2014%202003%20Blackout%20Investigation%20DL/BlackoutTable.pdf (accessed 11 December 2021).

20 North American Electric Reliability Corporation (2021). Odessa disturbance. https://www.nerc.com/pa/rrm/ea/Documents/Odessa_Disturbance_Report.pdf (accessed 11 December 2021).

21 North American Electric Reliability Corporation (2021). San Fernando disturbance, p. viii. https://www.nerc.com/pa/rrm/ea/Documents/San_Fernando_Disturbance_Report.pdf (accessed 11 December 2021).

22 North American Electric Reliability Corporation (2021). The South Central United States cold weather bulk electric system event of January 17, 2018, p. 12. https://www.nerc.com/pa/rrm/ea/Documents/South_Central_Cold_Weather_Event_FERC-NERC-Report_20190718.pdf (accessed 11 December 2021).

23 North American Electric Reliability Corporation (2021). April and May 2018 fault induced solar photovoltaic resource interruption disturbances report, p. viii. https://www.nerc.com/pa/rrm/ea/April_May_2018_Fault_Induced_Solar_PV_Resource_Int/April_May_2018_Solar_PV_Disturbance_Report.pdf (accessed 11 December 2021).

24 North American Electric Reliability Corporation (2021). Hurricane Irma report. https://www.nerc.com/pa/rrm/ea/Hurricane_Irma_EAR_DL/September%202017%20Hurricane%20Irma%20Event%20Analysis%20Report.pdf (accessed 11 December 2021).

25 North American Electric Reliability Corporation (2021). Hurricane Harvey report. https://www.nerc.com/pa/rrm/ea/Hurricane_Harvey_EAR_DL/NERC_Hurricane_Harvey_EAR_20180309.pdf (accessed 11 December 2021).

26 North American Electric Reliability Corporation (2021). 900 MW fault induced solar photovoltaic resource interruption disturbance report, p. 21. https://www.nerc.com/pa/rrm/ea/October%209%202017%20Canyon%202%20Fire%20Disturbance%20Report/900%20MW%20Solar%20Photovoltaic%20Resource%20Interruption%20Disturbance%20Report.pdf (accessed 11 December 2021).

27 North American Electric Reliability Corporation (2021). Blue cut fire report. https://www.nerc.com/pa/rrm/ea/1200_MW_Fault_Induced_Solar_Photovoltaic_Resource_/1200_MW_Fault_Induced_Solar_Photovoltaic_Resource_Interruption_Final.pdf (accessed 11 December 2021).

28 North American Electric Reliability Corporation (2021). Washington DC area report. https://www.nerc.com/pa/rrm/April%202015%20Washington%20DC%20Area%20LowVoltage%20Disturban/Washington_DC_Area_Low-Voltage_Disturbance_Event_of_April_7_2015_final.pdf (accessed 11 December 2021).

29 North American Electric Reliability Corporation (2021). Polar vortex review, p. 20. https://www.nerc.com/pa/rrm/January%202014%20Polar%20Vortex%20Review/Polar_Vortex_Review_29_Sept_2014_Final.pdf (accessed 11 December 2021).

30 North American Electric Reliability Corporation (2021). Hurricane Sandy report. https://www.nerc.com/pa/rrm/ea/Oct2012HurricanSandyEvntAnlyssRprtDL/Hurricane_Sandy_EAR_20140312_Final.pdf (accessed 11 December 2021).

31 North American Electric Reliability Corporation (2021). NE outage report, p. 46. https://www.nerc.com/pa/rrm/ea/October%202011%20Northeast%20Snow%20Storm%20Event/NE_Outage_Report-05-31-12.pdf (accessed 11 December 2021).

32 North American Electric Reliability Corporation (2021). AZ outage report, pp. 5–6. https://www.nerc.com/pa/rrm/ea/September%202011%20Southwest%20Blackout%20Event%20Document%20L/AZOutage_Report_01MAY12.pdf (accessed 11 December 2021).

33 North American Electric Reliability Corporation (2021). FERC/NERC staff report on the 2011 southwest cold weather event, p. 198. https://www.nerc.com/pa/rrm/ea/ColdWeatherTrainingMaterials/FERC%20NERC%20Findings%20and%20Recommendations.pdf (accessed 11 December 2021).

34 Cauley, G. (2021). August 14, 2003 blackout, p. 28. https://www.nerc.com/
pa/rrm/ea/August%2014%202003%20Blackout%20Investigation%20DL/Gerry_
Cauley_Blackout_%20Presentation_to_%20IEEE_Tampa_2-26-04.pdf (accessed
11 December 2021).

35 Union for the Co-ordination of Transmission of Electricity (2021). Final report
of the Investigation Committee on the 28 September 2003 Blackout in Italy.
https://eepublicdownloads.entsoe.eu/clean-documents/pre2015/publications/
ce/otherreports/20040427_UCTE_IC_Final_report.pdf (accessed 11 December
2021).

36 Union for the Co-ordination of Transmission of Electricity (2021). Final report
system disturbance on 4 November 2006. https://eepublicdownloads.entsoe.eu/
clean-documents/pre2015/publications/ce/otherreports/Final-Report-20070130
.pdf (accessed 11 December 2021).

37 Wood Mackenzie (2021). US energy storage monitor. https://www.woodmac
.com/industry/power-and-renewables/us-energy-storage-monitor (accessed 11
December 2021).

38 St. John, J. (2020). 5 major trends driving the $110B US distributed energy
resources market through 2025. https://www.greentechmedia.com/articles/
read/5-takeaways-on-the-future-of-the-u.s-distributed-energy-resources-market
(accessed 11 December 2021).

39 Snenum, D.M. (2020). Flexibility in the interface between district energy and
the electricity system, p. 29. https://backend.orbit.dtu.dk/ws/portalfiles/portal/
216181703/2020_DMS_Flexibility_in_the_interface_between_district_energy_
and_the_electricity_system.pdf (accessed 11 December 2021).

40 ENTSO-E (2021). About the TYNDP. https://tyndp.entsoe.eu/about-the-tyndp
(accessed 11 December 2021).

41 Wood Mackenzie (2020). US offshore wind powers up. https://www.woodmac
.com/press-releases/us_offshore_wind_may_deliver_25_gw-by_2029 (accessed
11 December 2021).

42 US Department of Energy (2020). Offshore wind market report: 2021. https://
www.energy.gov/sites/default/files/2021-08/Offshore%20Wind%20Market
%20Report%202021%20Edition_Final.pdf (accessed 11 December 2021).

43 Bureau of Ocean Energy Management (2021). Renewable energy on the outer
continental shelf. https://www.boem.gov/renewable-energy/renewable-energy-
program-overview (accessed 11 December 2021).

44 Patton, D. (2017). Why do capacity markets exist? https://www
.potomaceconomics.com/capacity/why-do-capacity-markets-exist (accessed
11 December 2021).

45 PJM Interconnection (2014). How PJM's market ensures enough power for
the future. https://www.youtube.com/watch?v=Lk9YqmAhcmM (accessed 11
December 2021).

46 New York Independent System Operator (2020). The capacity market's role in grid reliability: frequently asked questions. https://www.nyiso.com/-/the-capacity-market-s-role-in-grid-reliability-frequently-asked-questions (accessed 11 December 2021).

47 New England ISO (2021). Markets. https://www.iso-ne.com/about/key-stats/markets (accessed 11 December 2021).

48 New England ISO (2021). Markets. https://www.iso-ne.com/about/key-stats/markets (accessed 11 December 2021).

49 New York ISO (2022). Manual 4: installed capacity manual. https://www.nyiso.com/documents/20142/2923301/icap_mnl.pdf/ (accessed 6 March 2022).

50 New York ISO (2020). State of the market report. https://www.nyiso.com/documents/20142/2223763/NYISO-2020-SOM-Report-final-5-18-2021.pdf (accessed 11 December 2021).

51 New York ISO (2019). A grid in transition. https://www.nyiso.com/documents/20142/2224547/Reliability-and-Market-Considerations-for-a-Grid-in-Transition-20191220+Final.pdf/ (accessed 11 December 2021).

52 Midcontinent Independent System Operator (2021). Planning year 2021–2022. https://cdn.misoenergy.org/DRAFT%202021%20Wind%20&%20Solar%20Capacity%20Credit%20Report503411.pdf (accessed 22 February 2022).

53 New England ISO (2021). New England's future grid initiative key project. https://www.iso-ne.com/committees/key-projects/new-englands-future-grid-initiative-key-project/ (accessed 11 December 2021).

54 North American Electric Reliability Corporation (2020). 2020 summer reliability assessment. https://www.nerc.com/pa/RAPA/ra/Reliability%20Assessments%20DL/NERC_SRA_2020.pdf (accessed 11 December 2021).

55 Federal Energy Regulatory Commission (2020). Summer energy market and reliability assessment: 2020. https://www.ferc.gov/sites/default/files/2020-06/2020_Summer_Energy_Market_Reliability_Assessment.pdf (accessed 11 December 2021).

56 Midcontinent Independent System Operator (2020). 2020 summer readiness workshop presentation. https://cdn.misoenergy.org/20200428%20Summer%20Readiness%20Workshop444192.pdf (accessed 11 December 2021).

57 North American Electric Reliability Corporation (2020). 2020–2021 winter reliability assessment. https://www.nerc.com/pa/RAPA/ra/Reliability%20Assessments%20DL/NERC_WRA_2020_2021.pdf (accessed 11 December 2021).

58 Federal Energy Regulatory Commission (2020). 2020/2021 winter energy market and reliability assessment. https://www.ferc.gov/media/report-20202021-winter-energy-market-and-reliability-assessment (accessed 11 December 2021).

59 Midcontinent Independent System Operator (2021). 2020–2021 winter resource assessment. https://cdn.misoenergy.org/2020-21%20Winter%20Resource %20Assessment492510.pdf (accessed 11 December 2021).

60 North American Electric Reliability Corporation (2021). Reliability assessments. https://www.nerc.com/pa/RAPA/ra/Pages/default.aspx (accessed 11 December 2021).

61 North American Electric Reliability Corporation (2020). 2020 summer reliability assessment. https://www.nerc.com/pa/RAPA/ra/Reliability%20Assessments %20DL/NERC_SRA_2020.pdf (accessed 11 December 2021).

62 North American Electric Reliability Corporation (2019). 2019 summer reliability assessment, p. 23. https://www.nerc.com/pa/RAPA/ra/Reliability %20Assessments%20DL/NERC_SRA_2019.pdf (accessed 11 December 2021).

63 Organization of MISO States (2020). 2020 OMS-MISO survey results. https://cdn.misoenergy.org/20200612%20OMS-MISO%202020%20Results %20Webinar451924.pdf (accessed 11 December 2021).

64 New England ISO (2021). Distributed generation forecast. https://www.iso-ne.com/system-planning/system-forecasting/distributed-generation-forecast (accessed 11 December 2021).

65 ERCOT (2021). Resources. http://www.ercot.com/services/rq/re/dgresource (accessed 11 December 2021).

66 Midcontinent Independent System Operator (2021). Generator interconnection queue timeline. https://cdn.misoenergy.org/Definitive%20Planning%20Phase %20Schedule106547.pdf (accessed 11 December 2021).

67 Eastern Africa Power Pool (EAPP) (2021). Project reports and publications. https://eappool.org/category/programs-and-projects-reports (accessed 11 December 2021).

68 Eastern Africa Power Pool (2021). Eastern Africa Power Pool, slide 16. https://www.irena.org/-/media/Files/IRENA/Agency/Events/2013/ Nov/9_1/Afriac-CEC-session-2_EAPP_Gebrehiwot_220613.pdf?la=en& hash=D9D28224FA2BD2E4AF7D9A1640EB914704EA8C21 (accessed 11 December 2021).

69 ENTSO-E (2021). Requirements for generators. https://www.entsoe.eu/ network_codes/rfg (accessed 11 December 2021).

70 IESO (2021). Markets and related programs. http://www.ieso.ca/en/Sector-Participants/Market-Operations/Markets-and-Related-Programs/Ancillary-Services-Market (accessed 11 December 2021).

71 SIESO (2021). Markets and related programs. http://www.ieso.ca/en/Sector-Participants/Market-Operations/Markets-and-Related-Programs/Operating-Reserve-Markets (accessed 11 December 2021).

72 Central Electricity Regulatory Commission (2015). The Committee on Spinning Reserve. https://cercind.gov.in/2015/orders/Annexure-%20SpinningReseves.pdf (accessed 11 December 2021).

73 PJM (2022). PJM manual 12. https://www.pjm.com/~/media/documents/manuals/m12.ashx (accessed 22 February 2022).

74 Fingrid (2021). Automatic frequency restoration reserve. https://www.fingrid.fi/en/electricity-market/reserves_and_balancing/automatic-frequency-restoration-reserve/ (accessed 11 December 2021).

75 Fingrid (2021). Fast frequency reserve. https://www.fingrid.fi/en/electricity-market/reserves_and_balancing/fast-frequency-reserve/ (accessed 11 December 2021).

76 Fingrid (2021). Frequency containment reserves. https://www.fingrid.fi/en/electricity-market/reserves_and_balancing/frequency-containment-reserves/ (accessed 11 December 2021).

77 Fingrid (2021). Reserve power plants. https://www.fingrid.fi/en/electricity-market/reserves_and_balancing/reserve-power-plants/ (accessed 11 December 2021).

78 MacDonald, J.S., Cappers, P., Callaway, D.S., Kiliccote, S. (2012). Demand response providing ancillary services: a comparison of opportunities and challenges in the US wholesale markets. https://gridintegration.lbl.gov/publications/demand-response-providing-ancillary (accessed 11 December 2021).

79 Ela, E., Kirby, B., Navid, N., Smith, J.C. (2021). Effective ancillary services market designs on high wind power penetration systems. https://www.nrel.gov/docs/fy12osti/53514.pdf (accessed 11 December 2021).

80 Zhou, Z., Levin, T., Conzelmann, G. (2016). Survey of US ancillary services markets. https://publications.anl.gov/anlpubs/2016/01/124217.pdf (accessed 11 December 2021).

81 Central Energy Regulatory Commission (2013). Staff paper of introduction of ancillary services to Indian electricity market. http://indianpowersector.com/wp-content/uploads/2014/05/CERC-Staff-paper-on-Introduction-of-Ancilliary-Services-in-Indian-Electricity-Market.pdf (accessed 11 December 2021).

82 RED Eléctrica de España (2021). Ancillary services in Spain: dealing with high penetration of RES, slides 33 and 34. http://www.reshaping-res-policy.eu/downloads/topical%20events/de-la-Fuente_Ancillary-Services-in-Spain1.pdf (accessed 11 December 2021).

83 Midcontinent Independent System Operator (2021). Value of lost load (VOLL) and scarcity pricing, slide 6. https://cdn.misoenergy.org/20200910%20MSC%20Item%2005b%20RAN%20Value%20of%20Lost%20Load%20(IR071)472095.pdf (accessed 11 December 2021).

84 EIA (2021). Table 5.6.A.: average price of electricity to ultimate customers by end-use sector. https://www.eia.gov/electricity/monthly/epm_table_grapher .php?t=epmt_5_6_a (accessed 11 December 2021).

85 ERCOT (2021). Archives. https://www.ercot.com/services/training/archives (accessed 11 December 2021).

86 Potomac Economics (2020). 2020 state of the market report for the ERCOT electricity markets, Figure 28. https://www.potomaceconomics.com/wp-content/uploads/2021/06/2020-ERCOT-State-of-the-Market-Report.pdf (accessed 11 December 2021).

87 ERCOT (2019). State of the market report for 2019. https://www .potomaceconomics.com/wp-content/uploads/2020/06/2019-State-of-the-Market-Report.pdf (accessed 11 December 2021).

88 California Independent System Operator (2019). 2019 market performance report, Figure 6.6. http://www.caiso.com/Documents/ 2019AnnualReportonMarketIssuesandPerformance.pdf (accessed 11 December 2021).

89 California Independent System Operator (2020). 2020 market performance report, Section 5.3 . http://www.caiso.com/Documents/2020-Annual-Report-on-Market-Issues-and-Performance.pdf (accessed 11 December 2021).

90 California Independent System Operator (2020). 2020 Annual report on market issues and performance, Figure 5.6. http://www.caiso.com/Documents/2020-Annual-Report-on-Market-Issues-and-Performance.pdf (accessed 11 December 2021).

91 Midcontinent Independent System Operator (2021). 2020 value proposition study. http://2020 MISO Value Proposition Calculation Details521882.pdf (misoenergy.org) (accessed 11 December 2021).

92 North American Electric Reliability Corporation (2021). Rules of procedures. https://www.nerc.com/FilingsOrders/us/RuleOfProcedureDL/NERC%20ROP %20(Without%20Appendicies).pdf (accessed 11 December 2021).

93 African Energy (2021). Kenya: KenGen seeks consultant for ancillary service study. https://www.africa-energy.com/live-data/article/kenya-kengen-seeks-consultant-ancillary-service-study (accessed 11 December 2021).

94 Tosatto, A., Dijokas, M., Weckesser, T. et al. (2021). Sharing reserves through HVDC: potential cost savings in the Nordic countries. *IET Gener. Transm. Distrib.* 15: 480–494. https://doi.org/10.1049/gtd2.12035 (accessed 11 December 2021).

95 Federal Energy Regulatory Commission (2021). Report on barriers and opportunities for high voltage transmission. https://cleanenergygrid.org/wp-content/uploads/2020/08/Report-to-Congress-on-High-Voltage-Transmission_ 17June2020-002.pdf (accessed 11 December 2021).

96 Federal Energy Regulatory Commission (2021). Energy and ancillary services market reforms to address changing system needs. https://www.ferc.gov/sites/default/files/2021-09/20210907-4002_Energy%20and%20Ancillary%20Services%20Markets_2021_0.pdf (accessed 11 December 2021).

97 Federal Energy Regulatory Commission (2021). Electric: order no. 888 . https://www.ferc.gov/industries-data/electric/industry-activities/open-access-transmission-tariff-oatt-reform/history-oatt-reform/order-no-888 (accessed 11 December 2021).

98 Federal Energy Regulatory Commission (2015). FERC orders address sale: provision of ancillary services. https://www.ferc.gov/news-events/news/ferc-orders-address-sale-provision-ancillary-services (accessed 22 February 2022).

99 ERCOT (2018). Overview of renewables in the ERCOT system. https://integrationworkshops.org/winddublin/wp-content/uploads/sites/18/2018/11/1_3_ERCOT_presentation_Julia_Matevosyan.pdf (accessed 11 December 2021).

10

Energy Policy Should Include Considerations of Energy Poverty

10.1 Introduction

What should graduate students, decision-makers, and members of an informed public expect to learn from this chapter? According to the United Nations, 13% of the world's population lacks any access to modern electricity [1]. That is close to one billion people. An even greater number do not have a reliable or sufficient electricity supply. Numbers vary according to data availability and the definition of the acceptable quantity or reliability of power. One source has estimated the total number of people effected as over three billion [2].

This chapter challenges the reader to think beyond traditional definitions of energy poverty. At the outset, the chapter draws a clear distinction between developed and fragile economies. In a developed energy economy, most of the population has access to electricity, but there are pockets of the population with limited access. The chapter starts with an energy poverty definition and immediately asks the question: "Is energy poverty a binary or a relative concept?" In this chapter, the reader appreciates the tradeoffs and distinguishes between social justice and the concept of paying more for cleaner energy in a developed economy versus no energy access in some regions. Issues on the tradeoffs between energy attributes such as cost, reliability, and carbon are covered in even more depth in Chapters 3 and 7.

Is energy poverty different from energy accessibility? We explore this question to lay the foundation for energy reliability versus energy poverty challenges. There are multiple drivers for energy poverty, hence the importance given to this human right by the United Nations' sustainable development goals. The need for "affordable and clean energy" access now ranks number seven by the UN immediately after "clean water and sanitation." One of the drivers is food and using thermal energy for cooking needs. Not everyone cares about advanced material needs or can afford comfort, but they need access to cooked food. Energy

Modern Electricity Systems: Engineering, Operations, and Policy to address Human and Environmental Needs, First Edition. Vivek Bhandari, Rao Konidena and William Poppert.
© 2022 John Wiley & Sons Ltd. Published 2022 by John Wiley & Sons Ltd.
Companion website: www.wiley.com/go/bhandari/modernelectricitysystems

and energy poverty are also related to clean water, medical services, technology for learning and communication, and many of the other things that are taken for granted in strong economies.

The energy consumed in cooking food is not the only driver in the energy poverty context. Food energy (defined as the energy required to prepare food) and thermal energy are linked. In some developing countries, cooking food using cow dung, wood, or charcoal briquettes results in harmful indoor air quality and external carbon emissions. Additionally, there is no refrigeration in these economies.

Energy poverty is a significant issue now, more than ever, because the global south's population continues to have no or limited choices, whereas the developed world's population can afford to choose to be green or sustainable. The tradeoffs between alleviating energy poverty and mitigating global warming are playing out right in front of our eyes in the form of less expensive diesel generators as backup power versus energy storage.

Lastly, this chapter suggests the regulatory and economic environment needed for reducing energy poverty in fragile economies. Speaking from experience, the authors map out what works and what does not in providing a safe, stable, and guaranteed investment environment.

In conclusion, we look to the future as to what is possible. We must remember that up to a quarter of the world's population does not have a reliable electricity supply. This is real poverty and has a devastating effect on people's lives by hampering the growth of their society.

Through this book and others like it, we hope that a generation of energy and technology leaders address and make positive changes regarding energy poverty, social justice, and climate change. Leaders, engineers, and decision-makers need the fundamentals such as this book as a toolkit showing them how energy is generated, financed, regulated, rationed, and stored. Without understanding these pragmatic and technical tools, the important social issues would only be discussed and advocated for but not solved. It is up to you.

Exercise 10.1   Jay is a global citizen, and by now he has worked in many parts of the world. Currently, he heads an energy department in a global nongovernmental organization (NGO) in your country. He is looking to recruit some graduates. You have applied for the job and anticipate your in-depth knowledge on energy issues will be one of the key areas that Jay focuses his questions on.

In preparation, can you plot per capita energy consumption and per unit average electricity costs in your country for the past decade? Compare these figures against those from Costa Rica, DR Congo, Japan, and the United States (US). What looks similar? What looks different? You can continue to explore other aspects of energy poverty in a few more exercises in this chapter.

10.2 Energy Poverty Definition

How is energy poverty defined?

Energy poverty is a lack of access to sufficient quality energy for consumption. There are two keywords in our definition: "access" and "quality."

According to the United Nations, 13% of the world's population lacks access to modern electricity [1]. This lack of access percentage is even higher (70%) in places like rural sub-Saharan Africa and small island nations in the Asia Pacific [3]. Access to energy sources is critical to reducing energy poverty, and we recognize that not all resources have the same quality.

10.2.1 Energy Accessibility

Is energy poverty more than energy accessibility? Again, it depends on the context. We want the reader to appreciate this context because energy poverty is both a binary and a relative concept. In other words, not all energy access is created equal.

One specific example of definition-driven energy access is provided by the World Bank for use for very poor and refugee populations [4]. This World Bank definition draws the line of energy poverty to those until they receive 200 watt-hours (Wh) per household per day. This definition brings in the concept of rationing of energy consumption as well as peak demand. Quantifiable definitions, such as this, are needed for certain projects and programs but do present difficulties.

To give the reader an idea of how precious electricity is in some parts of the world, we point out that 200 Wh/household/day represents the ability in total for a household over a day to charge a phone for five hours plus use five small light-emitting diode (LED) bulbs for five hours and play a small radio or music amplifier. Energy rationing may be necessary from a cost standpoint but leaves little room for growth. In this type of scenario, refrigerators, sewing machines, electric saws, electric cooking, water heating, air conditioning, and most types of small businesses that use electricity are not yet possible.

A different type of example is the rationing of available electrical energy between commercial, institutional, and residential customers. This is done throughout the global south, as the following example illustrates.

Bukavu is a midsized ex-colonial city in the mountains of the Eastern Democratic Republic of Congo in Africa. It has been served with some electricity since colonial days, but systems have deteriorated. The electrical grid is not large enough to serve the influx of refugees that have come into and around the city in recent decades. The original hydropower dams need repair. Some new dams are being built, but the length of time of construction and the cost make universal electrical access something for the distant future.

This city has a major electrical line that serves a large beverage bottling plant that employs 5000 people, some institutional buildings including an orphanage, a hospital, and government buildings. Blackouts are common, and buildings that can afford them, such as the hospital and orphanage, have diesel generators. It is easy to understand why a hospital, orphanage, and place of employment would receive priority electricity. However, in practice, it is hard to reconcile with the thousands of households, small businesses, and clinics that have no electricity access.

When the funding for energy and technology is scarce, whether by intent or by default, the rationing of resources takes place.

Exercise 10.2 A tradeoff: Jay's cousin Sandra manages a regional city for the national utility in a South American country. A chronic problem faces the power dispatch team. The city is growing, but any growth in the power supply is years away. What options does she recommend when faced with rationing or limiting energy at certain times of the day: people's homes or a factory where many work?

10.2.2 Energy Quality Attributes

Regarding the quality of energy, the following examples show some of the quality of the energy attributes: a diesel generator is of moderate cost and is accessible for some customers in a remote area. But the same diesel generator emits harmful pollutant gases such as hydrocarbons, carbon monoxide, and nitrous oxide, leading to air quality concerns at the energy source (especially if the customer has a health condition). Besides, the burning of fossil fuels releases carbon dioxide, which exacerbates global warming.

Nuclear power is a special case with potentially positive and negative attributes. It is an example of a resource that is not accessible for many but produces high-quality power. Nuclear power plants are not cheap in terms of capital costs, situated away from population centers, usually near a water source for cooling purposes. All this means nuclear power needs a high-voltage transmission network (for long-distance power transmission) for customer access. Some may also regard nuclear waste storage or nuclear safety as representing a lower quality. But the quality of electricity produced is high because in-service nuclear plants generate steady power with no carbon dioxide discharge.

The energy quality attributes and tradeoffs of the various generation, storage, and transmission technologies are covered in detail in Chapters 3 and 7.

Electricity in every part of the world is not equal. Just having access to electricity is not enough. The electricity should be of sufficient quality to perform productive work (e.g. improve education, health, and business).

Let us take another example. Rural parts of Nepal are served by microhydros, and the urban population is served by the national grid. The microhydros might,

Figure 10.1 A leaking micro hydropower in Nepal. This generator can't produce quality electricity.

on paper, be operational and provide electricity to the villagers. Similarly, the national grid, on paper, could be supplying electricity to a large urban population. However, the quality of the electrical connection brought about by power outages, low voltages, brownouts, and frequency fluctuations could still be bad (Figure 10.1). The rural Nepali villager might not be able to run their sawmills. The urban dweller might not be able to conduct video calls due to recurring power outages.

Quality is important for both personal consumption and industrial consumption. Industries need to consume quality electricity, and without industry there is no commerce. Therefore, there would not be a productive use for this electricity.

A similar example can be seen for personal consumption. In a household that is supplied with electricity of an inferior quality, schoolchildren might still have to use their fuel-based lamps to read (leading to health problems), the person preparing food may be unable to cook on time (this may lead to domestic violence in a patriarchal society that wrongly expects women to cook on time), or the useful life of the household equipment could be shortened (leading to unnecessary expenditure). Hence, having electricity is not sufficient. It needs to be of adequate quality for both personal and commercial use.

So, consumption being tied to quality is the main point. And we feel strongly about the word "quality" in our energy poverty definition.

The World Bank defined metrics around the energy quality index for households. For a more in-depth discussion on energy poverty metrics and multitiered framework (MTF), readers are directed to the World Bank source [5].

We included the word "consumption" in our definition because an industrial customer's access to energy is critical for commercial purposes. Yet a residential customer might have access to energy but not quality, e.g. brownouts and voltage reduction issues. Hence our definition for energy poverty includes both the access and the quality of energy.

Exercise 10.3 You are continuing to build your knowledge for the interview for a job with Jay. What percentage of the population of your country has access to electricity? What is the quality of electricity consumption (e.g. are there regulations for ensuring quality, are they followed, are there frequent brownouts, is the generation mix strongly renewable, is the price affordable?) Do you know there are indices called the System Average Interruption Duration Index (SAIDI) and the System Average Interruption Frequency Index (SAIFI)?

SAIDI is a system index of the average duration of interruption in the power supply indicated in minutes per customer, and SAIFI is a system index of the average frequency of interruptions in the power supply. These indices serve as valuable tools for comparing electrical utilities' performance reliability. If your utility publishes these indices, what are the SAIDI and SAIFI for your utility? If your utility does not, select an index for a US utility of your choice. Compare it with other utilities in your region or with other utilities in the US. What information do these indices provide you with? How can they be related to energy poverty?

10.2.3 Multiple Definitions of Energy Poverty

There are multiple definitions of energy poverty.

One World Bank expert defines energy poverty as "the minimum energy consumption needed to sustain lives" [6]. Energy poverty is energy deprivation, according to researchers out of the University of Manchester in the United Kingdom [7]. Center for American Progress defines energy poverty as a "lack of access to modern energy services" [8]. Within the European Union (EU) context, energy poverty is defined as the ability to heat homes [9].

These multiple definitions point to a lack of a consistent definition of energy poverty. But they also point out that our definition of energy poverty is on target – our definition focuses on energy quality attributes as well as access or quantity, which these definitions do not. That shows our definition is unique, and we are adding value to this discussion in the industry. And by tying energy access and consumption together in our definition, we have incorporated some of the conventional thinking in this area, such as the World Bank's definition of minimum energy consumption or the Center for American Progress's definition of lack of access.

Does the energy poverty definition or lack of definition impact policy discussions? It depends on the country and the region based on our experience. In a fragile economy, to provide financial capital for investments to improve access or quality of energy sources, the definition matters to the key stakeholders. On the other hand, in our policy context, for the readers to appreciate the push and pull of access versus quality of energy, a precise definition of energy poverty does not matter.

From an engineering and electric operations perspective, where some aspects are absolutes (e.g. Kirchhoff's circuit laws, standard operating procedures), as we read in Chapter 1, a reasonable question to ask is: "Is energy poverty a binary or a relative concept?"

We believe energy poverty and the resulting social injustice issues can be a binary concept as evident in a country like Nepal and a relative concept in a country like the US with pockets of energy poverty. It is also worthwhile to mention in this chapter that we are differentiating social justice and the difficulty in paying for energy in a developed economy versus no access in some regions of the world. Social justice is not served, and energy poverty is not dealt with wherever basic energy quality attributes such as clean air and affordability are not addressed, even within developed countries.

Exercise 10.4   You need to build your understanding further. Read the paper "Energy Matters" by Lauren C. Culver [10]. What are the several ways to measure energy poverty? How is it tied to what we have described in this chapter? Are you prepared for the job talk with Jay?

10.2.4   Developed and Partially Developed Countries with Energy Poverty and Social Justice Issues

Is it possible for a country to be developed and have regions with energy poverty?

Yes, it is possible in countries with a huge gap between rich and poor, making accessibility to energy unattainable. For example, in Canada, a developed country, in the last 17 years, the real household income (especially in low-income households) has fallen behind the energy costs creating within-the-home energy poverty [11].

We think that inexpensive and plentiful energy has always been a birthright of developed countries. Recall, many farming areas of the US did not receive their first connection to the electrical grid until the middle of the twentieth century. This was thanks in part to the Rural Electrification Act (REA).

Another example of a developed country with an energy access constrained region can still be the US. Because according to the University of Texas at Austin research, "Colonia" residents along the US–Mexico border with an average monthly household income of $1600 lack energy access [12].

Also, energy poverty in pockets of a developed economy point to a rural versus urban dimension to the definition of energy poverty, because there are transitional countries with characteristics of both, i.e. urban vs. rural North Africa. A major city like Kigali in Rwanda has both access and quality of energy sources, but as one travels to rural parts of Rwanda there is limited access (12% in rural areas compared to 72% in urban areas [13]) and a limited quality of energy. As a quality example, in rural Rwanda areas that do have power, brownouts can be frequent for electricity and thermal energy for cooking often takes the form of open fires with wood indoors.

Yet another issue is a lack of education and understanding as well as an underlying poverty. In many fragile economies, theft and vandalism are abundant. In some instances, the powerline, transformers, and other equipment for distributing electricity can be vandalized (potentially somebody was trying to steal the line and sell it in thrifty markets). One example from India, was of 100 power transformers in the Maharashtra state being stolen [14]. So just putting together a power station and building lines is not enough. The end-users should understand that having electricity is more valuable than stealing and vandalizing it. When the local population is educated, is out of general poverty, and understands the importance, this situation can be avoided.

Exercise 10.5 A mechanical engineer steps into your office and asks about an article that he has read in a newspaper headline in the *Himalayan Times* [15]. NEA is the electricity authority in Nepal. He is confused about how electricity can leak. How would Jay explain this to him?

Business

NEA reduces power leakage to 9.78pc in H1

By Himalayan News Service
Published: 12:28 pm Feb 17, 2020

Figure 10.2 The Hierarchy Model of Energy Attributes and Access.

10.3   Hierarchy Model of Energy Attributes and Access

The new Hierarchy Model of Energy Attributes and Access offers an integrated and simplified view of both energy attributes and energy access (Figure 10.2). Note the roles that policy, financial capability, and customer choice play as we move through the hierarchy. It is our wish that this model will simplify the conceptualization of this complex but critical topic.

The following is our new Hierarchy Model of Energy Attributes and Access. We offer it for an integrated and simplified view of both energy Attributes and energy Access. Note the role that policy, financial capability, and customer choice play as we move through the hierarchy. It is our wish that this Model will simplify the conceptualization of this complex but critical topic.

Advanced Access

Ex: Highly developed cities, tech industrial and residential customers, high discretionary income. Green power attributes, high customer choice and "beneficial electrification" important.

Fewer cost impediments.

Traditional Developed City or Country Energy Access

Access to reliable energy.

Ex: Middle class residential and industrial in developed cities worldwide. Energy mix includes renewables

Energy Access with Poverty/Social Justice Issues

Ex: Areas of poverty within more developed economies

Energy Poverty

Access to limited energy and frequent brownouts

Ex: Cities within developing economies

Complete Energy Poverty

No regular access to electrical energy Ex: Rural, Countries with fragile economies

ENERGY QUALITY ATTRIBUTES AND CHOICES RISE WITH ACCESS BUT THIS REQUIRES FINANCIAL CAPABILITY:

HIERARCHY OF HUMAN ENERGY NEEDS ARE GENERALLY BETTER SATISFIED WITH IMPROVED QUALITY ATTRIBUTES AND ACCESS:

A SOCIETY'S EMMISSIONS/CARBON FOOTPRINT WOULD BE EXPECED TO RISE WITH TRADITIONAL ENERGY SOURCES ADDED. THEN FALL AS DECARBONIZATION INVESTMENTS ARE MADE:

SOME EXAMPLES TO FURTHER SHOW THE HIERARCHY OF ENERGY ACCESS AND ATTRIBUTES ACROSS VARIOUS SOCIOECONOMIC GROUPS. **Larger portions of the populations in the global south or fragile economies fall in the energy poverty columns of this model. However, we have not placed countries in any particular category since these groups can and do coexist within the same country or area:**

Complete energy poverty:	Energy poverty with limited electrical energy access:	Energy access with poverty and social justice issues:	Traditional developed energy access:	Advanced energy access:
	Poor reliability	Moderate reliability of energy delivery	High reliability of energy delivery	High reliability of energy delivery
Candles, oil lamps, wood fires inside for cooking	Diesel and hydro common if central grid	Green energy may not be widely available	Some green energy	Choice of green energies
Small SHS (solar home systems) of limited capacity is starting to be introduced	Stand-alone diesel generation, micro- and minigrids		Likely to be traditional grid-based electricity and natural gas for thermal needs	The choice to be off-grid
	Green/renewable sources may be most cost-effective if available			
	Energy for health care, critical industry, education	Basic life safety needs usually met		
Energy selection and efficiency are driven by the first cost	Energy selection and efficiency are driven by operating and first cost	National group, race, or refugee status may be an issue	Energy generation selection and efficiency are driven by ROI and public policies	Energy selection and efficiency are driven by cost plus environmental attributes
Price subsidy needed	Price subsidy needed	Affordability a driving factor Limited energy choice	Price subsidy not universally needed	Price subsidy not normally needed
National group, race, or refugee status may be an issue	Electricity for clean water important	Proximate energy sources may be of a low environmental grade		
	Introduction of clean cooking, domestic water heating		Air conditioning common	Home delivery of food and goods

| Open fire/stove cooking | Space heating Lighting for safety | Automated industrial production, etc. | |
| Needed for life safety | Energy may be limited or "rationed" for residences | Mechanical agriculture and food production | Do resources used here take away from necessary basics in other socioeconomic energy groups? |

Exercise 10.6 Based on the accompanying model, what socioeconomic energy group do you "live"? Where have you "lived" in the past? Where have you traveled to? Note the differences of the level of energy access and positive or negative attributes in these areas. Can you afford to "live" in the energy socioeconomic group that you would like to? Is this more a question of geographic privilege or of economics? What tradeoffs economically or environmentally are needed to provide humanitarian justice through universal energy access?

10.4 Importance of Energy Poverty Mitigation as a Priority in the Eyes of NGOs

We are aware that the world has many challenges that are equally important relative to energy poverty. However, key international NGOs' work plans prescribe the number of resources spent on solving the energy poverty puzzle from a financial priorities' perspective. The reader needs to understand these competing priorities to understand expected outcomes and how they would impact their workstream.

Often humanitarian NGO work focuses on mitigating the immediate crisis and sees political problems generating refugees, pandemics, and natural disasters. These are critical and must be addressed immediately. Unfortunately, energy development is extremely important in the long run but sometimes does not represent a "fire to be put out" on the cash-strapped NGOs' agenda. NGOs also need to address direct humanitarian aid, which normally means dealing with fragile populations: displaced people, residences, healthcare facilities, and schools. Ultimately, a robust society needs to support institutions, businesses, and industries that provide jobs and grow to provide welfare to its population. These facilities need energy also. Therefore, we look to grow societies to a state where energy abundance is the norm for all aspects of the economy. And many NGOs are focused on midterm crises and disasters versus long-term infrastructure building.

The United Nations does, however, recognize the importance of energy and now ranks "affordable and clean energy" as goal number seven out of its 17 sustainable development goals to achieve a "better and more sustainable future for all" in the world by 2030 [16]. This puts energy immediately behind clean water in the ranking of developmental importance.

10.4.1 The World Bank Definition

The World Bank defines two goals to achieve by 2030 [17]. The first goal aims to end extreme poverty, and the second goal aims to raise income to promote shared prosperity. As one would imagine, in fragile economies (our word to describe developing economies), multiple issues touch poverty.

For example, the World Bank lists climate change, education, energy, fragility, poverty and violence, urban development, trade, and health. This list shows that energy poverty is not something to deal with in a vacuum. The readers and students of energy policy should be aware of the impact of these competing forces on resource allocation, including financing.

It is not surprising that climate change is at the top of the World Bank's list of priorities to end extreme poverty. According to the United Nations, the energy sector contributes to 60% of global greenhouse gas emissions [1].

The World Bank list of topics is pre-pandemic in 2020. Post pandemic with COVID-19 impact, an updated list is not available yet. However, as industry observers, we should expect priorities to change. Some of the reprioritization occurs due to the pandemic's uneven impact on the fragile economies.

An exhaustive discussion of energy poverty definitions in the EU can be found here [18]. EU context definition for energy poverty focuses on how much of a household income is spent on energy.

10.4.2 Energy Poverty Progress

So, how to increase or sustain the progress on energy poverty? Energy professionals may be frustrated with the lack of progress on their projects. But if relationship building and focus on energy issues are maintained, then we can address energy poverty. It is not uncommon to expect initial relationship-building could take an initial couple of years with an additional three to four years for good working relationships.

In some economies, the local communities and national institutions test the patience of external NGOs and other energy experts to check if NGOs are interested in their region. One possible solution to continue to keep the momentum is keeping the pace of stakeholder meetings and other touchpoints. This translates into funding needs. The key is to sustain the progress on energy poverty – the reader needs to understand that this journey is a marathon, not a sprint.

An oft-repeated example of well-meaning efforts involves international aid agencies funding programs where various third-party vendors have provided energy access via small self-contained SHS. Often with as little as 50–100 W plus storage. More robust home needs are not addressed. In many cases, these are low voltage and not future grid connectable, unless to potential future local microgrids.

However, by a certain measure, one can consider a household to have a form of energy access from these systems. They offer only a minimum for commercial businesses that can provide employment and economic growth. They also usually do not address public energy needs such as safety lighting. This is representative of the fact that lasting energy access requires significant capital expenditure. While possibly well meaning, this represents a short-term solution to energy access that does not provide a lasting and growing start to empower a local economy.

Exercise 10.7 Jay's sister, Juliet, is working as an economist for the World Bank. She is doing a project for the United Nations.

It is estimated that providing a basic level of electrification to the world's non- or poorly electrified population costs between $5 and $15 trillion. A large investment indeed. If we look at the world's approximately eight billion people in total that would be an investment equivalent of over $1000 for each person currently in the world. She is asked to come up with initial scenarios and some timeframes for potential ways this investment in a vital infrastructure could be structured. What are some initial simple scenarios she should consider?

10.5 Significant Drivers for Energy Poverty

To make an impact on the energy poverty policy, the reader should know the most significant drivers for energy poverty, which prompts the question: "Where do people use the most electricity in fragile economies?"

We know where people are not using electricity. People in fragile economies are not using electricity to dry clothes (because they don't have electric dryers). They are not freezing their meat and food products (because they don't have access to refrigerators and freezers), and they are not using air conditioners (because they use natural cooling from tree shade) and other simple techniques that do not require significant financial spending.

10.5.1 Energy Poverty Links with Basic Needs

So, where do people in fragile economies use the most energy? If it is available, in cooking food, in lighting, pumping clean water, and in running medical equipment if they have critical medical needs. According to the United Nations, three billion people rely on wood, coal, or charcoal for cooking and heating [16].

Depending on where they live, many people may not care for energy-intensive air conditioners. They use ceiling and standing fans for almost everything. Or, if they are in temperate places – they depend on nature to cool them. People living in the mountains don't care about transmission towers and wires carrying electricity. Temperate climates need little space heating. Air conditioning is a learned taste. Is it a right or a privilege? So, what is the cause of energy poverty under such conditions?

10.5.2 No One Driver for Energy Poverty

There is no single or binary cause for energy poverty. In many places, especially in fragile economies, the drivers generally are the sociopolitical and economic conditions of the region. At some places, especially in developed economies, the drivers could be the same as those of the fragile world. They could, however, also be the result of choice. The reader should appreciate that some people in developed countries are even happy with less electricity, especially if that means a lower carbon footprint or no high-voltage wires in their neighborhood. Others may want to live off-grid. Therefore, the drivers could be the regional conditions or personal choices. There is no one driver for low energy access.

Broadly, the drivers for energy poverty can be divided into the following major categories: (i) common factors, (ii) country-specific contingencies, (iii) intracountry spatial and social differences, and (iv) personal choice [19]. The top three examples in each major energy poverty driver are:

- Common factors: Include inadequate energy efficiency (energy efficiency pops up in multiple places when discussing energy poverty), low or insecure household income, and high or rising energy prices. Additionally, include the lack of affordable energy generation, transmission, or household electrical systems due to poverty and lack of investment.
- Country-specific contingencies: Include lack of insulation in homes and schools in New Zealand, *Energiewende* (low-carbon transition) increasing energy bills for some households in Germany, and rapid energy price increases since the liberalization of energy markets in the western Balkans. Other examples for some fragile (and other) economies include short-sighted policies and volatile and corrupt governance that hinder investments and planning for access to energy.
- Intracountry spatial and social differences: Example include, rural versus urban differences, such as in the western Balkans, Poland, the UK, China, and Kenya, and spatial variation in urban areas (reduced energy efficiency in central areas of Athens, Greece), and heating affordability and infrastructure divide along north/south China.
- Personal choice: To live off-grid (e.g. Yogis in India or New Age so-called hippies in the 1970s or some Amish population in the US).

10.5.3 Historical and Current Socioeconomic Drivers for Energy Poverty

We have compiled the following list of historical and socioeconomic drivers that impact energy poverty. Anything as pervasive as worldwide energy poverty has several factors and several theories associated with those factors. We would encourage the reader to continue their perspective as to additional factors:

- Geographical and climatological location of the population. Abundant resources for energy can determine economics, water for hydropower, solar, natural gas, wind, and coal.
- Technological history of the population.
- Colonial history.
- More recent fragile political states. This can result in cumbersome governments and rules governing energy production and distribution.
- Fragile economic states. This can result in low average incomes and insufficient capital for investment in energy infrastructure, which is expensive.
- The inability to raise investment funds. Energy requires significant investment. Countries that are deemed to have an elevated risk of not paying back investment or adding government interference are much less likely to have developers or investors building infrastructure without some form of international subsidization or risk mitigation.
- Technology issues. These can change over time, sometimes to the positive. As an example, first wind and now solar have a substantially lower cost than they did a decade ago. This makes them suitable in some cases for initial electrical energy systems.

10.6 Energy Poverty and Ties to Thermal and Food Energy

When talking about energy poverty, there is always a question of which usage of energy is more important than another. Is electricity more important? Or is heat? Although we focus on electrical energy in this book, it is important to look at thermal, or heat, energy. Often, thermal fuels such as natural gas or coal are in economic demand for electricity as well as cooking or space heating. We define food energy as the energy required to prepare food to differentiate the energy required to prepare food compared to the thermal energy required to stay warm.

There is an example from the US Texas energy crisis of 2021. During an unusual period of cold weather, natural gas was in short supply both as a fuel for electrical generation and space heating. This exacerbated the problem for the people served throughout the state. This Texas example shows that unthoughtful prioritization of one form of usage over other, when the sectors are coupled, can cause havoc, and worsen an energy poverty situation. Please refer to Chapter 3, Section 3.2.2, for further details of the Texas example. To lift people from energy poverty, among other things, is to address the energy security issues among the coupled sectors carefully.

Another example is from fragile economies. People in fragile economies lack access to clean and sustainable energy to prepare their food. The reader should recognize the availability of cooking fuels in some countries – cow dung, wood, charcoal briquettes, or biogas. These fuels lead to harmful indoor air quality and external carbon emissions from cooking. So, one way to lift people from energy poverty related to their cooking needs is to provide them access to clean energy for preparing food. This concept can further be expanded and linked to the electricity-related energy poverty issue. To do so, if fragile economies are electrified using clean energy resources, they could use electricity for cooking their food and heating their homes. This could resolve energy poverty issues related to cooking, thermal needs, and electrical needs. However, doing so would couple the sectors, hence sector coupling will need to be understood.

10.6.1 The Need for Multiple Cooking and Heating Fuels

In some regions of fragile economies, solar cooking is growing. Solar cooking means using solar energy to cook food such as rice, vegetables, and some meats. Solar energy is not present at night (even though there is battery storage) when compared to other fuels such as biogas, cleaner-burning wood/briquette, and tank or piped gas. As an energy poverty mitigation strategy, we suggest multiple cooking methods to customers in fragile economies. Access to energy should not be a single point of failure.

Let us say we have solved the access to energy for cooking food. Then the next question to ask is: "Would people remain in food energy poverty if they had access to sustainable fuels to cook?" Our experience indicates that energy consumption needs change, and as a result the consumer might "demand" more access or more quality energy for their revised demand. We have seen the need then turn to solar refrigeration for produce for freshly caught fish in Sudanese refugee populations, for example.

10.6.2 Energy Poverty and Access to Basic Human and Infrastructure Needs

There is a relationship between energy poverty and access to basic human and infrastructural needs. Let us take an example of food. Electricity and thermal energy are needed in many aspects of agriculture and food preparation. Everything from tractors and irrigation pumps to food processing plants uses energy. For example, the lack of refrigeration hurts access to the food and fishing industry. Food waste due to lack of adequate refrigeration is a big issue in the global south. For example, India produces enough food to meet its population's needs, yet a large number are left unfed. This is primarily due to wastage. One of the ways to control wastage is to refrigerate it, which requires electrical, solar, or other forms of energy. There is also a need for infrastructure energy for water pumping, sewerage treatment, security, safety lighting, Internet, communications, and much more.

10.7 Why Is Energy Poverty a Significant Issue Now, More Than Ever?

In all countries, people are keenly aware of the benefits and rights of technology. Most of our technologies require energy to manufacture and to operate. Hence, energy poverty is linked to access to technology. Access to technology is related, in a multifaceted way, to inequalities like the income gap and the digital divide. Factors like the income gap and digital divide make energy poverty an urgent issue.

This income gap and digital divide are driving a lot of social unrest and social justice considerations. The lack of earning potential for young (age group 18–40) educated people in countries leading to underemployment and unemployment is a big issue [20]. This young people unemployment is an issue worth mentioning in the energy poverty context because most of the fragile and global south economies have young populations in the age group of less than 30 years old. Young individuals need gainful employment to be content. Energy policy advocates must pay attention to the individual needs of different demographic groups.

10.7.1 Specific Experiences in Fragile Economies and the Global South

In a developed economy, the population can afford to choose to be "green and sustainable." In a fragile economy, the population has no or limited choices to do so. To cover perspectives of both economies is the reason why we emphasize energy poverty from both an energy access and an energy quality point of view. And energy poverty is a significant issue now because of the increase in young professionals who may be under- or unemployed.

10.7.1.1 Rationing of Energy Services
Limited resources and a larger customer base lead to the "rationing" of services, including energy. The economic rationing of energy as power is a flashpoint for

Figure 10.3 A hotel owner in Nepal without electricity who uses his television as a candle stand.

all people and especially educated young professionals. With rationed energy in a traditional fragile economy context, households receive power during off-peak times and business institutions during peak hours. The healthcare industry receives power more than commercial businesses (ideally!). Macroeconomics models this rationing to identify policy gaps.

10.7.1.2 Brownouts and blackouts

These are the norm in some fragile economies. For example, while growing up in Nepal, one of our coauthors experienced from two to more than 18 hours of planned outages every day.[1] Each day was planned around the planned outage notices by the electricity department.

From a hotel in Nepal, owing to unreliable electricity, the owner rarely had time to power the TV up. It was mostly used as a candlestand (Figure 10.3).

10.7.1.3 Stakeholders

If all stakeholders in the energy ecosystem have a mutual understanding about the lack of opportunities due to the continued brownouts and occasional blackouts, then the task is half done. Sometimes it is difficult even to gain that understanding because there is a fundamental lack of a forum or platform that provides for a healthy exchange of ideas that enables this shared understanding. NGOs are in their silos with their focus on humanitarian topics, and the commercial institutions are focused on their business needs.

---

1 This is true for Bangladesh as well, according to our reviewer Nashmeen Moslehuddin. She wishes to add that folks who can afford backup power generally bought it.

Therefore, to successfully alleviate energy poverty, institutions and frameworks should bring the NGOs and the commercial institutions together to identify their needs and tie these to the energy poverty needs of the region.

For example, to alleviate energy poverty related to the planned power outage in Nepal, the electricity authority, with backing from the government and support from several NGOs, ran a crackdown operation to minimize electricity theft (e.g. from household-scale petty thefts to secretly supplying megawatts of electricity to selected industries without metering), and solidify the supply/demand needs (instead of exaggerating the forecasts). It took just months, to an outsider, to see the end of the decades' long suffering from the planned blackouts.

10.7.2 Need for Ongoing Data

A healthy exchange of ideas happens when there is trust among the stakeholders. We must point to the lack of reliable nonpolitical institutions to collect transmission and distribution outage statistics relatable to other countries as part of the issue here. Unless industry acknowledges the lack of data and lack of confidence in the quality of that data, the energy poverty puzzle will not be solved.

We acknowledge here that, if data is available, there are still issues with prioritizing investments based on key stakeholders' concerns. Having accurate data alone is not going to solve the energy poverty for a fragile economy.

10.7.3 The Role of Social Media

It is worth noting that social media technology is enabling better, faster, and cheaper communication to vent social issues. But the same technology is not solving energy poverty, because we need to address the underlying gaps in education and the flow of information. And that gap is exactly where this book fits in by holistically addressing engineering, operations, and policy perspectives – to close the gap in education and raise the awareness of data needs that should drive the energy poverty policy debates.

10.7.4 Energy and Safety, Disproportionate Effect on Women, Entrepreneurship, and Energy

According to the United Nations, 6 out of 10 women and girls accounted for the 4.3 million deaths in 2012 caused by indoor air pollution from using combustible fuels for household energy [1].

Many household energy issues fall disproportionately on women and girls. Especially in the global world, they are the managers of the household. Whether it is gathering firewood for meals, cooking dinner, purchasing candles, or charging phones, a substandard energy system creates an enormous amount of work for women and girls. This time can leave them more vulnerable to domestic violence and personal safety hazards. This extra work time takes them away from studying, interacting with technology, and other things that improve their social position and lifestyle.

Let us take a simple example. Rural Nepali households use turmeric (yellow food coloring with medicinal values) and salt in their everyday cooking. They cook in open-pit indoor firewood stoves and use a candle-wax lamp called a *jharro* for lighting. Women are generally the cooks, and they carry end-to-end cooking/feeding responsibility. Therefore, the women need to go to the jungle (several hours a day on average) to fetch firewood. They could have spent this time doing other productive work, like reading.

Cooking in an open fire takes a long time and creates much smoke. Therefore, the cook must spend much time inhaling the exhaust smoke of the cookstove. Hence, all the family is vulnerable to carry a lot of airborne breathing issues. The *jharro* emits a yellow light. So, turmeric and salt both look the same inside hazy smoke and yellow light, meaning there is a chance of adding salt twice, which makes the food taste bad. She is now prone to domestic abuse (physical and mental) from other family members. There are several other harms that she could face just because she did not have access to proper cooking conditions. Hence, energy poverty and women's issues are closely intertwined.

10.7.5 Energy Choice and Growth

Something that needs to be said is that, in a democratic world, energy users need to have both an economic and a controlling stake in the energy they purchase and use. Engineers, administrators, financiers, and developed world people must refrain from overly defining "other needs."

As an example, when working in DR Congo, Jay leads a focus group of rural people who would be potential customers for small energy systems. Jay asks what they would use the most electricity on. Many, of course, see the need for using electrical energy for lighting, education, and safety. However, a popular request is to switch from current clothes irons heated over a wood fire to an electrical iron. This request is understandable. There is great pride in that region in having well-cared-for clothing. Unfortunately, thermal energy is much more electrically intensive than modern LED lighting energy and will not likely be possible in the small "rationed" electrical systems that are potentially affordable.

This actual example leads to the point that, within reason, people need to oversee their destiny. Energy systems must be designed technically to provide basics now and to be able to grow with time to help their users achieve their dreams, goals, and aspirations.

10.8 Can Wholesale Energy Markets Help Solve Energy Poverty?

Yes, to a large extent, wholesale energy markets could help alleviate the energy poverty puzzle.

With our focus on access and quality in the energy poverty definition, we believe an energy market provides both access to and quality of resources to address the gap in energy infrastructure in fragile economies. We know intuitively without an economic degree that, in an adequately managed free market, whenever there is competition the customer benefits.

This is shown by the example of Nepal. Nepal Electricity Authority (NEA) is the monopoly electricity provider in Nepal. Primarily because it is a monopoly, the company employs over 10 000 staff to manage ~1 GW of installed capacity and bears a cumulative loss of billions of dollars and has done little to nothing to alleviate energy poverty. And, as highlighted in Chapter 3, the debundling of NEA, the introduction of other independent generating companies together with other market reforms were able to positively alter the grim picture of Nepal (see Chapter 3 for further details).

Take another example, this time from the Philippines. The Philippines power sector went through a serious supply crisis (like the black- and brownouts of Nepal) in the early 1990s. To resolve this, the Philippines undertook major power sector reforms in the 2000s that allowed, among other things, the formation of wholesale electricity markets. The spot market began operations in 2006. The case of the Philippines power market shows that markets can address supply security risks, and over the years the Philippines market has partially delivered competitive outcomes. Hence, with proper market design and adequate monitoring (especially during the initial phases), wholesale energy markets can provide a platform for competition and benefit customers.

In support of our contention that the energy poverty issue is critical now more than ever, we have referred to the rise in young (age group 18–40) professionals in a fragile economy. We believe young people, with access to social media tools and an interest in advocating for energy policies that reduce energy poverty can play a key role in advancing the concept that energy markets and technology can solve energy poverty

10.8.1 Wholesale Markets Provide Price Transparency and Nondiscriminatory Access to Transmission

What do we mean by wholesale energy markets? In the past (and present in some parts of the world), transmission, distribution, and generation assets were owned by a single entity. This entity is often a national utility, especially in fragile economics. In such places, the real-time dynamic buying and selling of power are rarely done. In other places, where these assets are separately owned, free market price competition exists. Such buying and selling utilizing free market practices at the wholesale level is called the wholesale energy market.

For us, an energy market provides wholesale energy price transparency (see Chapter 1 on for additional details), meaning, for example, that transmission customers would know the energy price in East Africa versus West Africa (and even at several places within them as an example) on a daily price bulletin board. We understand that, right from the start, a day-ahead market (DAM) that closes a business day before the next day is asking too much. But this is not stopping countries like India, Vietnam, and Mexico experimenting with wholesale energy markets, though with varying successes.

Transparent prices provide an economic signal or an incentive for quality resources or resources that are dispatchable by the system operator and available when needed on the transmission system (our energy poverty definition includes the word "quality").

To gain trustworthiness in institutions that provide oversight of wholesale energy prices, it is our experience that those institutions should be not-for-profit, with no stake in the wholesale energy transaction. The grid operator or the market operator should function like a neutral umpire in a cricket match. This neutral umpire should ensure nondiscriminatory access (our energy poverty definition includes the word "access") to the transmission system so that both resources and load are getting the best deal possible.

10.8.2   Market Operators Forecast Future Needs

Price transparency and nondiscriminatory access to the transmission system are not the only attributes of robust wholesale energy markets. There is a forecasting or more prevalent transmission planning function in most organized energy markets, at least in the US. Given an energy market operator's experience with both historical and real-time data, it is natural to expect these operators to have a planning function because grid managers must forecast their resource stack and demand components. It is not a stretch to ask the grid operators to extend the forecasting horizon from days to months and years.

10.8.3   Phased Manner of Market Adoption and Market Startup Costs

But this forecasting or planning function does not have to be tied to the market operator in fragile economies. Because we are concerned about the costs of the market setup, and we don't want a planning function to be a necessary cost barrier for the price transparency and non-discriminatory access benefits, we suggest taking these steps in a well-thought-out phased manner, each with the necessary time built in between to gain critical stakeholder consensus. Rushing into a wholesale energy market with all its bells and whistles to then lose political will is not going to serve the end customer well.

Conversely, we don't think a forecasting function alone without the energy market would work, even though an energy market could work without a fully fledged forecasting function. As an example, consider a future capacity need of 100 MW identified in a region. That knowledge alone or someone sitting on that piece of information is not going to propel action. The marketplace or a regulator should act on that information and conduct at the very least a cost-based auction to fill the 100 MW capacity needs.

In our experience, the cost to operate an energy market is $250 million for a 100 000 MW peak load. These are approximate numbers for the sake of making the argument for a phased-in approach to energy market adoption. This estimate of $2500 per megawatt peak load is for a developed economy, where the hourly staffing rate is upwards of $200 per hour with health benefits. We expect a fragile economy to have half of that cost, $1250 per megawatt for market startup costs. In a fragile economy, the $100 million for a 100 000 MW peak load price tag is a genuine issue. Hence, we advocate those multiple countries combine financial resources to set up a genuinely functional regional energy market.

10.8.4 Power Pool Members in the Global South Are Ideal Candidates

Countries or regions with experience in power pools are ideal candidates for wholesale energy markets. There is precedence in the US to show the way from power pool to energy market transformational journeys. In the US Midwest, Southwest Power Pool, a grid operator, based in Little Rock, Arkansas, has the words "power pool" in its name to signify the roots of the organization. PJM grew out of the Mid-Atlantic Area Council (MAAC) power pool. Midwest ISO (now Midcontinent ISO) during early market startup years (2000–2005) "bought" physical space and engineering staff from Mid-Continent Area Power Pool (MAPP) to set up an office in Saint Paul, Minnesota to be closer to Minnesota members such as Xcel Energy.

In the north-east, New England Power Pool (NEPOOL) is still active, and New England ISO grew from that pool structure. Power pools are ideal to discuss energy markets because the key stakeholders in a power pool already have established bilateral energy contractual relationships and know the energy price to expect in a wholesale market platform.

As discussed elsewhere, Eastern Africa Power Pool (EAPP) in North Africa and Southern African Power Pool (SAPP) are ideal candidates for energy market discussions. Please refer to Chapter 4 for examples of wholesale electricity markets and power pools (existing and potential).

In estimating the total market startup cost of $100 million for 100 000 MW peak load, most of the cost is human resources and capital investment in software and hardware for market platforms. A rough estimate is 60% for hardware and software, and the remaining 40% is for salaries and benefits to hire the people. A forty-million-dollar spend rate on human capital is a lot for a fragile economy. This human resource estimate again proves our point that most countries need to look at a regional set up to pool their resources.

10.8.5 Role of an Independent Board of Directors

For wholesale energy markets to function competitively and thereby contribute to reducing energy poverty in fragile economies, an independent board of directors and a robust legal/regulatory structure are necessary. An independent board does not have any financial stake in the energy market itself. A typical director ensures that the market operator does not lose the trust of both supply and demand resources while ensuring nondiscriminatory access to the transmission system.

Often overlooked is the legal and regulatory structure in a fragile economy. In an energy market consideration, a sound legal framework establishing the mandate for energy markets ensures market participants and stakeholders have something to fall back on when the inevitable disputes occur. A robust regulatory body supports broad market rules by staying technology and resource agnostic. In an ideal world, both legal and regulatory institutions should play an active role in confirming the societal benefits of energy markets in reducing the energy poverty gap.

10.9 Can Retail Tariff Reforms Address Energy Poverty?

The challenge of solving energy poverty in a zone or a region within a developed country is different from one within a fragile economy. There are challenges with generator and load "deliverability," leading to generation or load pockets that have constraints to deliver or receive power, for example when transmission constraints or deliverability issues mean that a zone heavy in generation capacity finds it difficult to transfer that capacity to a zone that needs it or when a zone experiencing high demand cannot receive generation from outside.

One way to provide the incumbent electric utility a guaranteed rate of return for the cost of serving the load is retail reform. Another way is to install smart meters with "revenue protection" capabilities to stop electricity thefts and missing payments. The third way is to adequately analyze the energy subsidies and do a subsidy reform that does not burden the poorer communities. There is no silver bullet on the retail side but let us look at some of the retail tariff reforms.

10.9.1 Industrial and Commercial Customers' Guaranteed Tariff

Readers should note that retail electricity rates vary by customer class. There are three customer classes in general. First are industrial customers.

Typically, industrial customers can select a better relative rate, depending on whether they are connected on the high/low side of a transmission/distribution transformer. In some situations, the industrial rate is straightforward based on the kW class of supply. Industrial customers also pay a demand charge based on their peak demand in 15-minute increments.

To reduce energy costs, some industrial customers enter an interruptible load contract. This contract enables them to reduce or stop their business when the utility signals the need to do so. In return, the utility provides them with a discounted price of electricity. There are instances when industrial customers felt that the utilities have called on them to curtail their industrial production, more than the agreed duration.

Some industrial customers in fragile economies have their generation capacity onsite. This customer generation is called behind-the-meter generation. These units are behind the utility meter.

Commercial customers are the next class that pays high rates. Commercial entities like shopping malls have peak load spread over a certain time during the day or on weekends in most cases. Work and commercial buildings also fall under a commercial rate. Like industrial customers, the commercial customer rate varies according to the transmission or distribution connection at the utility substation. In general, higher loads indicate close to a transmission substation. Lower demands mean distribution level interconnection.

These days, high-rise buildings and commercial real estate companies are engaging with the electric utility to reduce their energy costs. This building's initiative is by nature coming from developed economies and big cities with population densities higher than 1000–5000 persons per square meter.

10.9.2 Residential Customers' Tariff

The next class of customers is the residential type. There are variations in residential class, such as multi-apartment dwellings versus single-family homes. Residential customers can benefit by a statutory policy where they are subsidized and can pay a lower per unit rate relative to what it costs to serve their smaller loads when compared to industrial and commercial entities. The challenge in designing an electricity tariff for residential customers is the sheer number compared to other customer class types.

Conventional wisdom shows there are no residential demand charges. But that is not true in some US states, as evidenced by the 2020 order in Georgia [21]. Arizona Public Service, the utility in Arizona, also tried this in 2016.

The increase in penetration of distributed energy resources such as rooftop solar is challenging the expertise of both the developed and fragile economies' energy regulators.

10.9.3 Natural Gas Could be Bridge Fuel

There is also the question of how big a role natural gas plays in the energy mix. Within the developed economy context, sometimes natural gas is referred to as a bridge fuel: a bridge to transition from coal to renewable energy resources [22]. The International Renewable Energy Agency (IRENA) defines renewable energy sources broadly as bioenergy, geothermal, hydropower, ocean energy, solar, and wind energy.[2]

Globally, the share of our renewable energy consumption has increased from 8% in 1990 to 17.5% in 2016, an annual increase of 4.6% [23], and we expect the trend to continue because wind and solar were 91% of new capacity added in 2020 [24]. To transition from a dominant fuel like coal to a renewables-dominant resource mix, we need a bridge fuel like natural gas to keep electricity reliable and affordable. Since our aim is to eliminate energy poverty, natural gas provides both energy access and quality until such time as renewables and future technologies catch up, especially in a fragile economy with energy poverty issues.

10.9.4 Smart Meters' Role in Reducing Energy Demand and Consumption

Even though we are discussing the retail rate's impact on energy poverty, there is a role for the federal policymakers in this discussion as well. A case in point is the passage of the Energy Independence and Security Act of 2007 in the US. This act defined the characteristics of the smart grid and its role in advancing energy policy. This act and similar policy tools mention the word "smart" within the smart grid context. What is meant by "the smart grid"?

A smart appliance has a technology that allows the customer to reduce their energy consumption based on a signal. That signal can be static (a four-hour delay on a dishwasher) or dynamic (when the energy price falls below a certain threshold, start the dishwasher). Even in advanced countries, there is room for growth in

---

2 For more information, visit www.irena.org.

the smart grid world. For example, in the US, a lot of smart meters and automated metering infrastructure (AMI) were installed as part of an economic recovery act called the American Recovery and Reinvestment Act (ARRA) of 2009.

These advanced meters show the energy consumption in minute intervals and the data is uploaded to the cloud in some cases (please refer to Chapter 1, Section 1.8).[3] The primary reason for smart meter installation is a reduction in energy consumption. Additionally, smart meters avoid having the utility send a meter reader. By one estimate, there are close to 100 million smart meters installed in the US [25]. Even with all these millions of smart meters, most of the surveyed utilities are missing the chance to reduce energy consumption, according to the American Council for an Energy Efficiency Economy (ACEEE) study [26].

The fact that $135.6 billion has been spent on smart meter deployment and yet not all American utilities bother to use the data to reduce energy consumption is a developed economy issue [27]. It certainly puts the challenges in a fragile economy into perspective. Smart meters may not initially be useful in a fragile economy, where there are limited resources.

By illustrating the 100 million+ smart meter deployment and the related $100 billion budget, we want policymakers and funding organizations in fragile economies to do their research first before spending any money. The saying that a good tailor should always measure twice but cut once is relevant here: policymakers in fragile economies should think twice before piloting smart meters.

10.9.5   Energy Subsidy Should Make Way for Distribution System Operators

Energy is subsidized in most parts of the world. This skews the economics. Therefore, many countries are removing subsidies. If this can be done in a way that subsidies are removed from high-income people/industries whilst low-income end-users continue to receive an adequate supply of quality electricity, energy poverty issues can still be addressed.

Looking at the wholesale energy market as a solution and tying retail reform to this could be another way to address energy poverty. Let us take an example of an electricity tariff piece mentioned here as an example of a hybrid approach. In this hybrid approach, there can be competition for both wholesale energy and retail energy requirements. Like a transmission system operator (TSO) on the wholesale side, there needs to be a distribution system operator (DSO) on the retail side. We elaborate on this in Chapter 8.

---

3 Generally, settlements are 30 minutes to an hour long. Smaller timeframes are also possible. For example, from October 2021, AEMO needs to go for five-minute settlement. For settling bills for every five minutes, the meters need to at least provide at five-minute or lower resolution.

The function of a DSO/retailer is mainly to ensure price transparency at the distribution system level, on transmission nodes that intersect with the distribution system feeders. We expect large industrial customers in fragile economies need this distribution system price signal, to sell their excess supply back to the grid.

10.10 Can We Get Rid of Energy Poverty in Our Lifetime?

Yes, we can get rid of energy poverty in our lifetime. With our focus on both energy access and energy quality, if the industry focuses on those two attributes, we believe it is possible to solve the energy poverty puzzle.

If we focus on the first attribute – energy access and quality – we propose a strategy that solves energy poverty to include a multifaceted approach along with the energy ecosystem space of generation, transmission, and distribution.

10.10.1 Energy Access: Focus on Power Generation Need

Power generation has taken on an entirely different meaning with the increase in distributed generation (DG) and prosumer (a consumer who produces and not just consumes energy). To solve energy poverty, we in the industry and policy practitioners must think about opportunities for the prosumers along with the traditional incumbent electric utilities generating power. We need generation from station power and DG. And we need both utility-owned generation and consumer-owned generation. This flexibility leads to optionality to solve energy poverty.

Additionally, an energy policy that focuses on solving to balance the humanitarian and the environmental "need" is much better than a policy that focuses on incentivizing just a specific technology like solar or wind. Because the capacity needs to change according to the generation fleet, some units may retire, some may be at the end of their lifecycle, whereas others might be having a tough time finding fuel supply contracts with changing macroeconomic conditions. One example is a solar-battery-powered streetlight in Nepal, a country that has suffered from power cuts for decades. As shown in Figure 10.4, the "need" in this part of the country was to have a busy section of the street lit during the night. The implementation here was a "need" focus rather than a policy to incentivize a particular technology. In other words, if there were another existing technology that would serve this need, the city would rather install that technology.

One way to ensure there is plenty of opportunity for station power, DG, and utilities, and for prosumer ownership, is to put together an all-source request for proposal (RFP) in the marketplace. This RFP would ensure all sources could

Figure 10.4   Street lights in eastern Nepal.

respond to an identified need on the system. Even if DG does not "win" by participating in this RFP, our experience tells us that DG providers learn from that experience and understand what it takes to win those supply contracts. In other instances, the marketplace learns about the costs of emerging technologies. Xcel Energy Colorado, doing business as the public service company (PSCo) all-source RFP, surprised industry observers with a favorable price for solar and solar plus energy storage. PSCo received 417 bids of all resources such as solar photovoltaic (PV), solar PV + storage, wind, combustion turbine, battery storage, and other fossil resources. Owing to this all-source RFP, PSCo received bids for wind at \$0.0107/kWh, solar PV at \$0.023/kWh, and solar + storage at \$0.03/kWh compared to the average January 2021 residential price of electricity in Colorado at \$0.126/kWh [28].

10.10.2 Energy Access: Focus on the Transmission System

In the multipronged approach to solving energy poverty by working with the energy ecosystem (comprising generation, transmission, and distribution), focusing on the high-voltage transmission system is a next step. Like generation, the transmission system is also changing even though the changes on the transmission side are not newsworthy like the generation side developments.

There are mostly two camps in the industry when it comes to transmission developments. One believes we need more transmission to integrate more and more renewables on the grid. The other that believes we don't need any new transmission, and certainly not a high-voltage transmission line that cuts across natural surroundings and pristine areas. Since our focus here is on solving energy poverty for the greater good of society, we believe there should be room for both models in a fragile economy.

By making room for transmission advocates in the energy ecosystem, consumers benefit from the transmission system delivering a diversity of supply. Recall, in Section 10.8, we suggested that, owing to high wholesale energy market startup costs, a regional market makes more sense to pool the financial resources from different countries. And the high voltage transmission system in a fragile economic region can function as a "backbone" for the regional market.

Along the path of providing optionality to solve energy poverty, we believe energy policymakers should encourage business models for "no" transmission proponents as well. In our view, "no" transmission means no "new" transmission projects but includes emerging technologies that enhance the throughput of an existing transmission line or substation equipment.

The lead time to implement a transmission line can be months, even years. Add to this the decade or so to construct and approve this transmission line. And even though we have sometimes seen high-voltage transmission lines being constructed much faster than this, they can still remain off-line because of political or contractual issues.

The US Department of Energy (DOE) funded synchro phasor deployment across electric interconnections after the 2003 blackout [29]. Like smart meters, we are not convinced synchro phasors are needed to solve the energy poverty. But when placed strategically at locations that have historical power system stability issues and blackouts instances on the transmission system, synchro phasors can add value to the TSO. The approximate cost of an overhead high voltage (230 kV) alternating current (AC) transmission line is $1 million per mile in the US [30], whereas the cost of a phasor measurement unit (PMU) is $43 400 and the cost of a phasor data concentrator (PDC) is $107 000 [31]. Since the cost of the synchro phasors is lower compared to the cost of a new overhead transmission line, to solve energy poverty we propose synchro phasors exploration for locations with stability issues.

If a high-voltage transmission system forms the backbone of a solution, there needs to be a low-voltage transmission system or a distribution system to act as a "collector system." This collector system integrates local resources and provides access to local loads. This collector system is closing the gap between people who have electricity and people who don't. We don't want to create a wedge between people drawing power from the high-voltage transmission system and the people drawing power at the low-voltage distribution system.

10.10.3 Energy Access: Focus on the Distribution System

The distribution system is the final portion of the grid that delivers electricity to most customer locations. Focusing on the distribution system is the final energy access approach toward solving energy poverty. A lot happens in the distribution system compared to the transmission system. Distribution losses are more than transmission losses. Distribution system outages are more extended in duration and higher in frequency relative to transmission system outages. Most people notice any problems with the distribution system because it is right there in front of them, unlike a transmission line that loops around a city's outskirts.

Given the proximity of the distribution system to the electric customer, one sure way to solve energy poverty by providing energy access is to beef up the distribution system. But the reality is the distribution system is not straightforward. Distribution system planning entails distribution feeder level modeling and optimizing around customers served from a segment of the feeder. Single- and three-phase fault, relay coordination, circuit breaker coordination – all factor into the distribution system. This coordination means there are more components to the distribution system.

Pay close attention to the distribution system to solve or to put together an energy poverty policy. We also point to, in a developed economy setting, electric utilities are shifting their capital budget spend from generation and transmission categories to distribution system components. This capital spending is done under the umbrella of "grid modernization" in some US states.

Time, t0 = 0 sec, voltage is stable; Time, t1 = t0 + sec, it becomes 250V; Time, t2 = t1 + sec, it becomes ~180V

Time t3, = t2 + sec, voltage becomes stable again.

Figure 10.5 Example illustrating the quality of supplied electricity in Nepal.

10.10.4 Energy Quality: Focus on the Data Institution Model (DIM) Framework

Shifting our focus to the second attribute – energy quality from energy access, to solve the energy poverty puzzle – we propose a data institution model (DIM) framework. Before that let's look at an energy quality topic.

Figure 10.5 shows a reading from a stabilizer. A stabilizer is a quite common instrument in fragile economies. Since the voltage fluctuates very often, a stabilizer is used to bring the voltage to the rated voltage to avoid equipment damage. The photos in the figure were taken within a couple of seconds of voltage fluctuation, the stabilizer worked and brought the voltage to 220 V. It was supplying a refrigerator. Such a fluctuation in voltage could happen tens of times an hour. Hence, the quality of energy supplied is poor.

Coming back to the DIM model, we focus on each leg of the framework (like the three-pronged approach of generation, transmission, and distribution) to propose a solution for energy poverty.

10.10.4.1 D Is for "Data" in the DIM Framework

There is no doubt engineers like data, and see it in black-and-white terms. Policymakers like words and narration. They like to describe a single data point in 500 words. To provide energy quality, we need both perspectives.

Here's why, as an example, collecting data alone in smart meter deployment (our example in Section 10.9.4) does not meet the original objective function of energy consumption reduction. After the data is collected, we need a robust dialogue of what the data is indicating. Starting questions can include which utility has the most deployments, and what are they doing with that data. We also need to know how the consumers are reacting to smart meters, and what are the lessons learned in one customer class versus the others.

10.10.4.2   I Is for "Institutions" in the DIM Framework

To collect quality data, we need trustworthy institutions, which are discussed in Chapter 6.

10.10.4.3   M Is for "Models" in the DIM Framework

The third aspect in the DIM framework is the model, i.e. network representation. There are four reasons why we need a model: (i) a robust model depicts both the high voltage transmission system and the low voltage distribution system, (ii) a model provides policymakers with a tool to verify the assumptions and the results from utilities, (iii) international NGOs can compare a model against other models, and (iv) a model enables financiers to compare their investments in fragile economies and in developed economies.

To advance policies toward energy quality, and thereby reducing energy poverty, we need models that provide a focal point on the results and context around the narrative. Therefore, we need both engineers and policymakers.

10.10.5   Innovation for Energy Access and Quality: Examples

Innovation is another key factor for addressing energy poverty topics. Human history is full of disruptive innovation, from the introduction of electricity to transportation to bridges and buildings to planes. Innovative ideas should also be promoted to tackle access and quality issues. Innovation should focus on eliminating barriers by accessing technology, promoting behavioral change, overcoming financial barriers, and sharing best practices. Let us take an example that encompasses all these components. In the eastern part of Nepal, overcrowded petrol/diesel buses were a typical means of transportation. From around 2015, a new means of transportation was introduced: the City Safari (Figure 10.6).

It was a disruptive game changer. A new technology, a three-wheeler electric vehicle (EV), was introduced that could be purchased at a fraction of the price at ~$3000 and could even be financed by a financial institution at a reasonable rate. It could ferry 4–6 passengers, could be charged using clean (electricity comes primarily from hydro without dams) and cheap electricity (~@10c/kWh), and was accessible to anyone who wanted to start a business. It was cheap and safe for a rider (~$2/km) and alluring for the owner/operator (~$300–$500/month income;

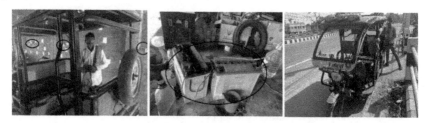

Figure 10.6 City Safari: the three-wheeler EV and its driver at a charging center, charging points are marked (left), batteries taken off for maintenance (center), parked City Safari (right).

Figure 10.7 Penstock pipes from a mini hydropower plant.

alternatively, this person would only earn half this amount). New lines of business that assembled these vehicles provided overnight charging services and maintenance. It created a behavioral impact by engaging the vulnerable population and was a great example of an innovation from the grassroots level.

Other examples of innovation for energy access are micro hydropower (10s of kW to less than a megawatt) and mini hydropower (1–10s of megawatts) in countries where there is abundant hydropower or other microturbine technologies. So, far more than 1000 micro hydropower plants are built in most of the districts of Nepal and most of the remaining electricity is sourced from mini hydropower units (Figure 10.7). When hydropower plants that are close to one other are linked together forming a microgrid, it addresses the energy quality issues. Such an innovation (hydropower or microgrid) is driven by the local community, and it is used to meet their energy access needs, powering their financial growth.

The disruptive innovation of three-wheeler EVs, hydropower, or microgrid technologies, especially that utilizes the lessons learned and evolve over time, is fundamental in solving the puzzle of energy poverty.

10.11   Visioning

In conclusion, our definition of energy poverty is situational. It includes both access to and the quality attributes of energy. We want the reader to appreciate the various definitions of energy poverty but focus on the financial piece of the puzzle to move the needle in the policy debate. Energy poverty can be both a binary concept and a relative concept. This chapter should challenge the reader to explore the rural versus urban dimension.

Our vision in this chapter to advance the energy poverty policy can be summarized in the following manner. We believe:

- Although substantial investment is required, energy poverty is not always a straightforward issue that can be solved if money is thrown at it.
- Up to a quarter of the world's population (approaching two billion people) does not have a reliable, affordable, or clean electricity supply.
- There is no single driver for energy poverty.
- Energy poverty should be defined in both energy access and energy quality contexts.
- The tradeoffs that have or have not been made to supply electricity involve environmental and cost issues, which are discussed throughout this book. As we go forward to address this social justice issue, it will involve calculating the human cost of lost healthcare, industry, education, and life safety. This direct human cost of energy needs to be balanced with global warming and energy source economic tradeoffs.
- International NGOs can reprioritize energy poverty given disasters such as the global pandemic, and this impacts funding for some programs in fragile economies.
- Food energy and thermal energy are linked with energy poverty because of the impact of fuel on preparing, processing, and storing food in a fragile economy.
- Energy poverty is a critical issue now because of the increase in young (age group 18–40) educated professionals in need of power for technology. If we do not recognize the income gap between employed and unemployed people, we run the risk of de-prioritizing energy poverty.
- Wholesale energy markets can help by bringing together fragile economies in a regional market structure. Countries in a power pool are ideal candidates for market startups.
- Retail markets can come to the rescue by focusing on specific customer classes and designing rates appropriately.
- Huge investment and build out from global partners are needed. The regulatory and economic model needed to solve energy poverty must include a guaranteed investment environment.

We can get rid of energy poverty in our lifetime by focusing on both access and quality and this is the challenge to readers and students of this book.

Case 10.1 *Namibia Hydrogen Plans* According to a *Wall Street Journal* article, Namibia plans to generate 300 000 metric tons of green hydrogen using 5 GW of solar and wind by 2030, leveraging the vast desert area and exporting that hydrogen to other African countries [32].

Can you write the feasibility study scope needed for Namibia to evaluate green hydrogen prospects?

Can you also write a blog post advocating for spending Namibian tax dollars on exporting green hydrogen when the access to electricity is 57%, according to the International Energy Agency [33].

Case 10.2 *Uganda's Leader Says Wind and Solar Energy Will Lead to Energy Poverty* The President of Uganda has written an opinion piece that renewable energy alone will not reduce energy poverty in Uganda [34]. They need fossil energy because it is reliable, and only then can Uganda reduce its energy poverty.

Write a brief explaining what is factually correct or incorrect in this argument. Uganda has an installed capacity of 1300 MW. But Uganda's National Development Plan goal calls for 40 000 MW of installed capacity by 2040. An aggressive goal is to aspire to go from 1300 MW installed capacity to 40 000 MW within 20 years. Can Uganda reach that goal without US (or other similar strong economy) support?

Case 10.3 *Does Carbon Pricing Eliminate or Reduce Energy Poverty?* The New York ISO discusses carbon pricing, which it defines as the social cost of carbon [35].

For your country/region, can you prepare a report advocating for carbon pricing in wholesale energy markets?

Case 10.4 *Leapfrogging in a Fragile Economy* Now that you have read this book and are aware of the power systems of strong, emerging, and fragile economies, choose a country in a fragile (or global south) economy. Identify and summarize (no more than five addressable areas) recommendations for its government to leapfrog modern electrical engineering systems to provide reliable, secure, clean, and quality power. Create a five-page policy brief on this.

Hint: It might be easier to select an electrical engineering sector (e.g. generation, transmission, distribution, end-use) and identify and summarize recommendations for one of these sectors.

References

1 United Nations (2021). Sustainable development goals: 7: affordable and clean energy. https://www.un.org/sustainabledevelopment/energy (accessed 12 December 2021).

2 Energy for Growth (2021). 3.5 billion people lack reliable power. https://www.energyforgrowth.org/memo/3-5-billion-people-lack-reliable-power (accessed 12 December 2021).

3 USAID (2021). Power Africa: about us. https://www.usaid.gov/powerafrica/aboutus (accessed 12 December 2021).

4 World Bank Group (2022). Beyond connections: energy access redefined: "tier 2 daily electricity supply". https://openknowledge.worldbank.org/bitstream/handle/10986/24368/Beyond0connect0d000technical0report.pdf?sequence=1&isAllowed=y (accessed 22 February 2022).

5 Bhatia, M. and Angelou, N. (2015). Beyond connections: energy access redefined. *ESMAP Technical Report* 008/15: https://openknowledge.worldbank.org/handle/10986/24368 (accessed 22 February 2022).

6 Khandker, S. (2013). Why energy poverty may differ from income poverty. https://blogs.worldbank.org/developmenttalk/why-energy-poverty-may-differ-income-poverty (accessed 12 December 2021).

7 Bouzarovski, S. and Petrova, S. (2015). A global perspective on domestic energy deprivation: overcoming the energy poverty–fuel poverty binary. *Energy Res. Soc. Sci.* 10: 31–40.

8 Center for American Progress (2009). Energy poverty 101. https://www.americanprogress.org/issues/green/reports/2009/05/14/6142/energy-poverty-101 (accessed 12 December 2021).

9 European Union (2015). Energy poverty and vulnerable consumers in the energy sector across the EU: analysis of policies and measures. https://ec.europa.eu/energy/sites/ener/files/documents/INSIGHT_E_Energy%20Poverty%20-%20Main%20Report_FINAL.pdf (accessed 12 December 2021).

10 Culver, L.C. (2017). Energy poverty: what you measure matters. https://ngi.stanford.edu/sites/g/files/sbiybj14406/f/NGI_Metrics_LitReview%282-17%29.pdf (accessed 22 February 2022).

11 Tardy, F. (2019). Building related energy poverty in developed countries: past, present, and future from a Canadian perspective. *Energy Build.* 194: 46–61.

12 Olmedo, C. (2013). Energy, housing and income: constraints and opportunities for affordable energy solutions in Texas Colonias. http://www.txenergypoverty.org/wp-content/uploads/2015/10/ElectrifyingColonias_FinalReport_10.18.13.pdf (accessed 12 December 2021).

13 USAID (2021). Rwanda: Power Africa fact sheet. https://www.usaid.gov/powerafrica/rwanda (accessed 12 December 2021).

14 Shelke, G., Bhusari, P. (2021). Transformer worth Rs 2 lakh stolen from Bhapkar Mala in Hadapsar; sold as scrap. https://timesofindia.indiatimes.com/city/pune/transformer-worth-rs-2-lakh-stolen-sold-as-scrap-cops/articleshow/87529325.cms (accessed 12 December 2021).

15 Himalayan News Services (2020). NEA reduces power leakage to 9.78pc in H1. https://thehimalayantimes.com/business/nea-reduces-power-leakage-to-9-78pc-in-h1 (accessed 12 December 2021).

16 United Nations (2021). Sustainable development goals. https://www.un.org/sustainabledevelopment/sustainable-development-goals (accessed 12 December 2021).

17 The World Bank (2021). What we do. https://www.worldbank.org/en/about/what-we-do (accessed 12 December 2021).

18 Trinomics (2016). Selecting indicators to measure energy poverty, final report. https://ec.europa.eu/energy/sites/ener/files/documents/Selecting%20Indicators%20to%20Measure%20Energy%20Poverty.pdf (accessed 12 December 2021).

19 Simcock, N., Thomson, H., Petrova, S., and Bouzarovski, S. (2018). Chapter 15. In: *Energy Poverty and Vulnerability*. Routledge https://library.oapen.org/bitstream/id/75b734bc-7bfd-4089-b68d-3c6cf1c0a932/649212.pdf (accessed 12 December 2021).

20 United Nations (2015). Youth population trends and sustainable development. https://www.un.org/esa/socdev/documents/youth/fact-sheets/YouthPOP.pdf (accessed 27 November 2021).

21 Energy and Policy Institute (2020). Georgia power set to automatically enroll new residences in demand charge rates. https://www.energyandpolicy.org/georgia-power-residential-demand-charges (accessed 12 December 2021).

22 Nunez, C. (2021). Can natural gas be a bridge to clean energy? https://www.nationalgeographic.com/environment/energy/great-energy-challenge/big-energy-question/can-natural-gas-be-a-bridge-to-clean-energy (accessed 12 December 2021).

23 United Nations (2022). The energy progress report. https://trackingsdg7.esmap.org/data/files/download-documents/chapter_3_renewable_energy.pdf (accessed 22 February 2022).

24 International Renewable Energy Agency (2021). Renewable capacity statistics 2021. https://www.irena.org/publications/2021/March/Renewable-Capacity-Statistics-2021 (accessed 22 February 2022).

25 Edison Foundation Institute for Electric Innovation. (2019). Smart meters at a glance. https://www.edisonfoundation.net/-/media/Files/IEI/publications/IEI-Smart-Meters-Infographic_Dec2019.ashx (accessed 12 December 2021).

26 ACEEE (2020). Leveraging advanced metering infrastructure to save energy. https://www.aceee.org/research-report/u2001 (accessed 12 December 2021).

27 Walton, R. (2020). Entergy, Duke, ConEd adding millions of smart meters, but overall deployment slowing. https://www.utilitydive.com/news/entergy-duke-coned-adding-millions-of-smart-meters-but-overall-deploymen/569986 (accessed 12 December 2021).

28 Trabish, H.K. (2021). Xcel's record-low-price procurement highlights benefits of all-source competitive solicitations. https://www.utilitydive.com/news/xcels-record-low-price-procurement-highlights-benefits-of-all-source-compe/600240 (accessed 12 December 2021).

29 Office of Electricity (2013). 10 years after the 2003 northeast blackout. https://www.energy.gov/oe/articles/10-years-after-2003-northeast-blackout (accessed 12 December 2021).

30 Xcel Energy (2014). Overhead vs. underground information about burying high-voltage transmission line. https://www.xcelenergy.com/staticfiles/xe/Corporate/Corporate%20PDFs/OverheadVsUnderground_FactSheet.pdf (accessed 12 December 2021).

31 US Department of Energy (2013). American Recovery and Reinvestment Act 2009. https://www.energy.gov/sites/prod/files/2016/10/f33/Synchrophasor_Report_08_09_2013_DOE_2_version_0.pdf (accessed 12 December 2021).

32 Wexler, A. (2021). The world wants green hydrogen: Namibia says it can deliver. https://www.wsj.com/articles/the-worldwantsgreen-hydrogen-namibiasays-it-can-deliver-11639823404?st=1z4d4r2xn7oakh4&reflink=desktopwebshare_permalink (accessed 22 February 2022).

33 IEA (2022). Access to electricity. https://www.iea.org/reports/sdg7-data-and-projections/access-to-electricity (accessed 22 February 2022).

34 Museveni, Y.K. (2021). Solar and wind force poverty on Africa. https://www.wsj.com/articles/solar-wind-force-poverty-on-africa-climate-change-uganda-11635092219?st=qv677pbtrrd8bac&reflink=desktopwebshare_permalink (accessed 22 February 2022).

35 New York ISO (2020). Carbon pricing in wholesale energy markets: frequently asked questions. http://www.nyiso.com/-/carbon-pricing-in-wholesale-energy-markets-frequently-asked-questions (accessed 22 February 2022).

Index

Modern Electricity Systems: Engineering, Operations, and Policy to address Human and Environmental Needs,
First Edition. Vivek Bhandari, Rao Konidena and William Poppert.
© 2022 John Wiley & Sons Ltd. Published 2022 by John Wiley & Sons Ltd.
Companion website: www.wiley.com/go/bhandari/modernelectricitysystems